Proof and other Dilemmas

Mathematics and Philosophy

© 2008 by
The Mathematical Association of America, Inc.

ISBN 978-0-88385-567-6

Library of Congress number 2008922718

Printed in the United States of America

Current printing (last digit):
10 9 8 7 6 5 4 3 2 1

Proof and other Dilemmas

Mathematics and Philosophy

edited by

Bonnie Gold   and   Roger A. Simons
Monmouth University   *Rhode Island College*

Published and Distributed by
The Mathematical Association of America

SPECTRUM SERIES

The Spectrum Series of the Mathematical Association of America was so named to reflect its purpose: to publish a broad range of books including biographies, accessible expositions of old or new mathematical ideas, reprints and revisions of excellent out-of-print books, popular works, and other monographs of high interest that will appeal to a broad range of readers, including students and teachers of mathematics, mathematical amateurs, and researchers.

777 Mathematical Conversation Starters, by John de Pillis

99 Points of Intersection: Examples—Pictures—Proofs, by Hans Walser. Translated from the original German by Peter Hilton and Jean Pedersen.

All the Math That's Fit to Print, by Keith Devlin

Calculus Gems: Brief Lives and Memorable Mathematics, by George F. Simmons

Carl Friedrich Gauss: Titan of Science, by G. Waldo Dunnington, with additional material by Jeremy Gray and Fritz-Egbert Dohse

The Changing Space of Geometry, edited by Chris Pritchard

Circles: A Mathematical View, by Dan Pedoe

Complex Numbers and Geometry, by Liang-shin Hahn

Cryptology, by Albrecht Beutelspacher

The Early Mathematics of Leonhard Euler, by C. Edward Sandifer

The Edge of the Universe: Celebrating 10 Years of Math Horizons, edited by Deanna Haunsperger and Stephen Kennedy

Euler and Modern Science, edited by N. N. Bogolyubov, G. K. Mikhailov, and A. P. Yushkevich. Translated from Russian by Robert Burns.

Euler at 300: An Appreciation, edited by Robert E. Bradley, Lawrence A. D'Antonio, and C. Edward Sandifer

Five Hundred Mathematical Challenges, Edward J. Barbeau, Murray S. Klamkin, and William O. J. Moser

The Genius of Euler: Reflections on his Life and Work, edited by William Dunham

The Golden Section, by Hans Walser. Translated from the original German by Peter Hilton, with the assistance of Jean Pedersen.

The Harmony of the World: 75 Years of Mathematics Magazine, edited by Gerald L. Alexanderson with the assistance of Peter Ross

How Euler Did It, by C. Edward Sandifer

Is Mathematics Inevitable? A Miscellany, edited by Underwood Dudley

I Want to Be a Mathematician, by Paul R. Halmos

Journey into Geometries, by Marta Sved

JULIA: a life in mathematics, by Constance Reid

The Lighter Side of Mathematics: Proceedings of the Eugène Strens Memorial Conference on Recreational Mathematics & Its History, edited by Richard K. Guy and Robert E. Woodrow

Lure of the Integers, by Joe Roberts

Magic Numbers of the Professor, by Owen O'Shea and Underwood Dudley

Magic Tricks, Card Shuffling, and Dynamic Computer Memories: The Mathematics of the Perfect Shuffle, by S. Brent Morris

Martin Gardner's Mathematical Games: The entire collection of his Scientific American columns

The Math Chat Book, by Frank Morgan

Mathematical Adventures for Students and Amateurs, edited by David Hayes and Tatiana Shubin. With the assistance of Gerald L. Alexanderson and Peter Ross.

Mathematical Apocrypha, by Steven G. Krantz

Mathematical Apocrypha Redux, by Steven G. Krantz

Mathematical Carnival, by Martin Gardner

Mathematical Circles Vol I: In Mathematical Circles Quadrants I, II, III, IV, by Howard W. Eves

Mathematical Circles Vol II: Mathematical Circles Revisited and Mathematical Circles Squared, by Howard W. Eves

Mathematical Circles Vol III: Mathematical Circles Adieu and Return to Mathematical Circles, by Howard W. Eves

Mathematical Circus, by Martin Gardner

Mathematical Cranks, by Underwood Dudley

Mathematical Evolutions, edited by Abe Shenitzer and John Stillwell

Mathematical Fallacies, Flaws, and Flimflam, by Edward J. Barbeau

Mathematical Magic Show, by Martin Gardner

Mathematical Reminiscences, by Howard Eves

Mathematical Treks: From Surreal Numbers to Magic Circles, by Ivars Peterson

Mathematics: Queen and Servant of Science, by E.T. Bell

Memorabilia Mathematica, by Robert Edouard Moritz

Musings of the Masters: An Anthology of Mathematical Reflections, edited by Raymond G. Ayoub

New Mathematical Diversions, by Martin Gardner

Non-Euclidean Geometry, by H. S. M. Coxeter

Numerical Methods That Work, by Forman Acton

Numerology or What Pythagoras Wrought, by Underwood Dudley

Out of the Mouths of Mathematicians, by Rosemary Schmalz

Penrose Tiles to Trapdoor Ciphers . . . and the Return of Dr. Matrix, by Martin Gardner

Polyominoes, by George Martin

Power Play, by Edward J. Barbeau

Proof and Other Dilemmas: Mathematics and Philosophy, edited by Bonnie Gold and Roger Simons

The Random Walks of George Pólya, by Gerald L. Alexanderson

Remarkable Mathematicians, from Euler to von Neumann, Ioan James

The Search for E.T. Bell, also known as John Taine, by Constance Reid

Shaping Space, edited by Marjorie Senechal and George Fleck

Sherlock Holmes in Babylon and Other Tales of Mathematical History, edited by Marlow Anderson, Victor Katz, and Robin Wilson

Student Research Projects in Calculus, by Marcus Cohen, Arthur Knoebel, Edward D. Gaughan, Douglas S. Kurtz, and David Pengelley

Symmetry, by Hans Walser. Translated from the original German by Peter Hilton, with the assistance of Jean Pedersen.

The Trisectors, by Underwood Dudley

Twenty Years Before the Blackboard, by Michael Stueben with Diane Sandford

The Words of Mathematics, by Steven Schwartzman

MAA Service Center
P.O. Box 91112
Washington, DC 20090-1112
800-331-1622   FAX: 301-206-9789

Contents

Acknowledgements

We would like to express our appreciation to the Spectrum committee for their patience as we developed this book. We are especially grateful to J.D. Phillips and Sanford Segal, the two members of the committee who volunteered to read each chapter in its penultimate version and suggest changes that they felt would make each more accessible or historically accurate. We also appreciate Gerald Alexanderson's careful reading for misprints in the final version.

We are also grateful to all our authors for their patience as we pressed them multiple times to revise their chapters to make them accessible to a wider community. In addition, we especially would like to thank Professor Charles Chihara for reading and suggesting improvements in the first draft of the historical parts of the introduction.

We would like to thank Jonathan Borwein for the graphic found on the front cover of this volume.

More personally, we would like to thank our spouses, David Payne and Patricia Simons, for their support and patience through this project. What began as a sabbatical project of Bonnie Gold turned into a three-year project for both of us. We both spent many evenings and weekends on this project, reading and commenting on manuscripts, rather than with our respective spouses.

Finally, we appreciate the encouragement we have received during this project from numerous others and from a few of the authors.

Introduction

Bonnie Gold
Monmouth University

Section 1 of this introduction explains the rationale for this book. Section 2 discusses what we chose *not* to include, and why. Sections 3 and 4 contain a brief summary of historical background leading to contemporary perspectives in the philosophy of mathematics. Section 3 traces the history of the philosophy of mathematics through Kant, and Section 4 consists of an overview of the foundational schools. Section 5 is an annotated bibliography of sources for interesting recent work by some influential scholars who did not write chapters for this book. And finally, section 6 consists of very brief overviews of the chapters in this book.

1 The Purpose of This Book

This book provides a sampler of current topics in the philosophy of mathematics. It contains original articles by leading mathematicians, mathematics educators, and philosophers of mathematics written with a mathematical audience in mind. The chapters by philosophers have been edited carefully to minimize philosophical jargon, and summarize many years of work on these topics. They should thus provide a much gentler introduction to what philosophers have been discussing over the last 30 years than will be found in a typical book written by them for other philosophers. We have also included a glossary of the more common philosophical terms (such as epistemology, ontology, etc.). The chapters by mathematicians and mathematics educators raise and discuss questions not currently being considered by philosophers.

The philosophy of mathematics, starting about 1975, has been undergoing something of a renaissance among philosophers. Interest in foundational issues began receding and philosophers returned to more traditional philosophical problems. Meanwhile, some developments in mathematics, many related to the use of computers, have reawakened an interest in philosophical issues among mathematicians. Yet there is no book on these issues suitable for use in a course in the philosophy of mathematics for upper-level mathematics majors or mathematics graduate students, or for mathematicians interested in an introduction to this work. (Hersh's recent collection [Hersh 2005] contains many interesting articles related to the philosophy and sociology of mathematics, and is accessible to a similar audience, but it does not attempt, as we do, to cover the range of current discussion in the philosophy of mathematics.)

Our principal aim with this volume is to increase the level of interest among mathematicians in the philosophy of mathematics. Mathematicians who have been thinking about the philosophy of mathematics are likely to enjoy the variety of views in these papers presented in such an accessible form. Mathematicians who have never thought about philosophical issues but wonder what the major issues are should find several chapters to whet their interest. Those teaching courses in the philosophy of mathematics for upper-level mathematics undergraduates (or others with a similar mathematical background) should find it a useful collection of readings to supplement books on the foundational issues. Moreover, we hope to encourage more dialogue between two communities: mathematicians who are interested in the philosophy of mathematics, and philosophers who work in this field. We expect that most readers will not read every chapter in this book, but will find at least half to be interesting and worth reading.

2 *What is not Included in This Book*

A few words about our selection of topics for inclusion in this book are in order. We have not tried to include every topic that has ever been discussed in the philosophy of mathematics, or even everything currently being worked on. In part because we do not have adequate expertise to edit such articles, we have not included anything on the philosophy of statistics, which is currently a quite active field (although we do have a chapter on the philosophy of probability). More importantly, we have chosen *not* to include articles on the three foundational schools that developed in the late 19th and early 20th centuries: logicism, intuitionism, and formalism. They are described briefly later in this introduction, and much more thorough accounts of them appear in many books, including Stephan Körner's *The Philosophy of Mathematics*, Alexander George and Daniel Velleman's *Philosophies of Mathematics*, Marcus Giaquinto's *The Search for Certainty*, and Dennis Hesseling's *Gnomes in the Fog*. While there is still active work continuing in these fields, in our view the century from approximately 1865 to 1965 was an anomalous one for the philosophy of mathematics. What had seemed, prior to this period, to be the most certain form of human knowledge, mathematics, suddenly appeared to rest on shaky foundations. Thus essentially all work in the philosophy of mathematics during this period focused on trying to determine what basis we have for believing mathematical results. Gradually, problems were found with each of the foundationalist schools. Meanwhile new paradoxes did not appear despite an enormous growth in mathematics itself. As a result, the concern about mathematical coherence decreased, and philosophical attention began to return to more traditional philosophical questions. This book, then, concentrates on this new work, and complements the four books, just mentioned, that quite adequately discuss this foundational work.

Today there are many philosophers actively working in the philosophy of mathematics. A number of the better-known among them were invited to contribute to this book. Some of them declined due to prior writing commitments. However, several very well respected philosophers of mathematics *have* written chapters for this volume, and other viewpoints are well represented by some younger philosophers who were recommended by their mentors. Thus, most current viewpoints in philosophy are represented here. However, a single volume cannot hope to do this in full detail.

3 *A Brief History of The Philosophy of Mathematics to About 1850*

Although this book is concerned with recent developments in the philosophy of mathematics, it is important to set this work in the context of previous work. Thus I have written this historical section despite little expertise in the subject. Much of this material comes from, and is discussed in more detail in, chapter 1 of [Körner 1968]. Moreover, I am grateful to Charles Chihara for his many suggestions on how to improve my first version of this section. Any errors that remain here are my responsibility, not his.

The way a culture approaches mathematics and its use directly influences its philosophy of mathematics. Mathematics has been of interest to philosophers at least since ancient Greece. It has been used primarily as a touchstone to explore and test theories of knowledge. Traditionally, knowledge comes from two sources: sense perception and human reasoning. Mathematical knowledge has generally been taken as the archetypical example of the latter.

Plato is particularly important to any understanding of the history of the philosophy of mathematics, for two reasons. First, he is the earliest known philosopher who saw mathematics (which, for him, was synonymous with geometry) as central to his philosophical discussions. Ancient texts assert he viewed mathematics as so important that above the door of his Academy, Plato inscribed "Let no one who is not a geometer [or, "who cannot think geometrically"] enter." Plato used mathematical examples throughout his dialogues for various purposes. For example, in *Meno*, there is a famous sub-dialogue between Socrates and a slave boy. In it, Socrates leads the slave boy to discover that if you want to double the area of a square, you must take a square whose side is the diagonal of the original square. This discussion is used to explore an idea Plato wants to propose, of knowledge as memory from a previous life. There are several excellent books on Plato's philosophy of mathematics and the mathematics of Plato's time: for example, [Brumbaugh 1954] and [Fowler 1999].

Second, some of Plato's general philosophical views have resulted in his name being given to what is still seen, by philosophers today, as the default philosophy of mathematics, "platonism." That is, "platonism" is the view that (1) there are mathematical objects, (2) these are abstract objects, existing somewhere outside of space and time, (3) mathematical objects have always existed and are entirely independent of people, (4) mathematical objects do not interact with the physical world in any "causal" way—we cannot change them, nor can they change us—and yet, (5) we somehow are able to gain knowledge of them. These properties come from Plato's theory of "Forms," which appears in his later dialogues, primarily the *Republic* and *Parmenides*. Plato was struggling with our everyday world of appearance, trying to discern what is permanent and dependably true. This led him to the idea of the form of an object (say, a table) as a sort of ideal limit toward which objects of the physical world are striving but are imperfect copies. In this realm of forms live the assorted mathematical objects we work with: numbers, geometric objects, and so on. Objects in the realm of the forms are apprehended by reason, rather than by the senses. The appeal of viewing geometric objects, so central to Greek mathematics, this way is apparent. We see imperfect lines and points in the physical world and can easily imagine a perfect point and line. Mathematical statements are necessarily true, because they describe objects in this unchangeable realm. Objects in the physical world "participate in" the forms that describe them, and, because they are only imperfect likenesses, are only approximately described by mathematical theorems.

Aristotle objected to abstracting properties of objects into an independent existence. Rather, you can discuss these abstracted *properties*, but they reside in the objects they're abstracted from. Mathematical statements are then idealizations of statements about objects in the physical world. To the extent that these idealizations are accurate representations of the physical objects they're abstracted from, mathematical theorems can be approximately applied to physical objects. Two other contributions Aristotle made to the philosophy of mathematics were a discussion of infinity, and the beginnings of logic. Aristotle's distinction between potential infinities (basically, what happens when we take the limit as x → ∞) and actual infinities (such as the set of integers, real numbers, etc.) was important historically in mathematicians' hesitation to accept many developments involving actual infinities.

Gottfried Wilhelm Leibniz was, of course, one of the founders of calculus, but he also made a substantial contribution to logic and to philosophy. He believed that by developing a systematic calculational logic (a "calculus ratiocinator"), one could represent much human reasoning and resolve many differences of opinion. He began to develop such a system, and introduced many of the modern logical concepts: conjunction, disjunction, negation, etc. (None of this, however, was published during his lifetime.) For Leibniz, mathematical facts are truths of reason, "necessary" truths whose denial is impossible (as opposed to truths of fact, that are "contingent," that just happen to be true in this world, and whose denial is possible). Mathematical facts are true in "all possible worlds" (a terminology he introduced).

John Stuart Mill was a complete empiricist about mathematics as about everything else. He believed that mathematical concepts are derived from experience and that mathematical truths are really inductive generalizations from experience. There are no necessary truths. Thus every mathematical theorem can, in principle, be found to be false and in need of revision. Mathematical truths are about ordinary physical objects. Geometrical propositions are inductively derived from our experience with space, and are taken to mean that, the more closely physical objects approach these idealized geometrical objects, the more accurately the theorems can be applied to them. Statements such as "$2 + 3 = 5$" is a generalization about how many objects you get when you put together a pile of two and a pile of three physical objects. However, for people, such inductive generalizations are a psychological necessity, because they come from very deep and invariant experiences. These experiences create an appearance of mathematical facts being necessarily true.

For **Immanuel Kant**, mathematics provided central examples for his classification of knowledge. Knowledge of propositions was classified into *a priori* or *a posteriori*. Meanwhile, propositions were classified as synthetic or analytic. A proposition is known *a priori* ("from the former"—before experience) if it is known without any particular experiences, simply by thinking about it. A proposition is known *a posteriori* if knowledge of it is gained from experience or via the senses. An "analytic" proposition is one whose predicate is contained in its subject. For example, "all squares are rectangles" is analytic because the definition of square (as "a rectangle with congruent sides") contains the requirement that it be a rectangle. The canonical example of an analytic proposition is "all bachelors are unmarried." Many mathematical and logical truths are analytic and are known *a priori*, as with "all squares are rectangles." A proposition that is not analytic is "synthetic." According to Kant, most truths about the world—"Mount Everest is the highest mountain in the world," for example—are synthetic, and are known *a posteriori*. It is generally believed that no analytic propositions can be known *a posteriori* (although a modern philosopher, Saul Kripke, has disputed this).

This leaves the category of synthetic propositions that are known *a priori*. According to Kant, our intuitions of time and space, which give us facts about the real numbers ($\mathbb{R}^3$ in particular) and the integers (such as $2 + 5 = 7$), are synthetic, yet are known *a priori*. We do not get them simply by analyzing their definitions, but rather by thinking about space and time. (Frege disagreed with this, at least in the case of arithmetic facts; he viewed them as analytic, and this was part of the point of his *Foundations of Arithmetic*.) Nonetheless, actual experience with space or time is not required to get this knowledge. In Kant's case, by space, he meant Euclidean space; that is,

Euclidean geometry gives us our intuition of space. Thus Euclidean geometry is the inevitable necessity of thought, rather than being of empirical origin. The integers come from our intuition of time in the form of one moment, then the next moment, and so on. (This idea first appeared much earlier, in Plato's *Timaeus* 39b-c.) Another distinction Kant makes is between concepts we can both perceive and construct—such as the concept of two objects—those we can construct but not perceive—such as $10^{10^{10^{10}}}$ —and those we can neither construct nor perceive, but are simply "ideas of reason" because they are consistent—such as actual infinities.

4 The Foundational Problems and the Three Foundational Schools

In the nineteenth century, three events occurred that caused both mathematicians and philosophers to reassess their views of issues such as what mathematics is about, how we acquire mathematical knowledge, and how mathematics can be applied to the physical world.

First came the inconsistencies in the use of limits in calculus. Soon after the introduction of calculus, there were concerns about foundational issues. Derivatives were found by taking a ratio of two infinitesimal quantities and then treating the denominator as if it were zero (even though if the denominator *is* zero, one could not even have formed the quotient). Bishop George Berkeley, in *The Analyst*, 1734, inveighed against "ghosts of departed quantities." Maclaurin responded by showing how one can derive calculus results via contradiction in the "manner of the ancients," the method of exhaustion—thus, calculus is simply a short-cut to legitimate results. Lagrange (1797) responded to the problem by trying to use power series to get rid of limits. This introduced its own problems, in the absence of some way of seriously considering issues such as divergence. For example, Grandi, in 1703 (see [Burton 2003], p. 567) set $x = 1$ in

$$\frac{1}{1+x} = 1 - x + x^2 - x^3 + \cdots \quad \text{to get} \quad \frac{1}{2} = 0.$$

These difficulties were of increasing concern to mathematicians in the 1800s, and the need for developing textbooks for university students eventually led Cauchy, in the 1820s, to give careful definitions of continuity, differentiability, and the integral. But Cauchy did not notice the need for a distinction between pointwise and uniform convergence of functions. As a result, he stated a false theorem about convergence of sums of continuous functions. This led to the development of careful definitions of the limits by Weierstrass and of the real numbers by Dedekind. Fourier series introduced a new set of complications: when did they converge to a function, and what exactly was a function anyway? More broadly, mathematicians began to be concerned about the foundations of mathematical beliefs—how can we be sure that what we develop is free of contradictions? This concern led to logicism, the attempt to reduce all of mathematics to logic. This work began with Gottlob Frege's *Foundations of Arithmetic* in 1884. In his 1879 *Begriffsschrift*, he developed the first fully formalized axiomatic development of the sentential calculus; he also introduced quantifiers and expanded his calculus to a predicate calculus. This was the basis for modern predicate logic, a major advance over the Aristotelian logic that had dominated for centuries. It provided him the language to formalize arithmetic. It was hoped that if there were any contradictions in mathematics, they would inevitably be found before they could cause any damage once everything was reduced to logic. One thing this formalization of arithmetic accomplished was to make statements such as $2 + 5 = 7$ a consequence of the definitions of the numbers involved, and thus turn them into analytic propositions.

The second event was the development of non-Euclidean geometries: Lobachevsky (lectured on in 1826, published about 1835), Bolyai (published in 1831 as an appendix to a book by his father), Gauss (who apparently discovered it earlier but did not publish it), and later Riemann (1854). Apparently because of the obscurity of the journals in which the work of Lobachevsky and Bolyai published, it was not until Riemann's work that the world-view of Kant was finally rejected. (Kant's world-view (see section 3, above) was that Euclidean space is the inevitable necessity of thought, rather than being of empirical origin.) Euclidean geometry was no longer the science of space—it is still far from clear which geometry is best to describe actual physical space. This revolutionary development threw mathematics out of the physical world (though, of course, not out of its usefulness in describing that world). It also led to the use of axiomatics as a way of discovering new mathematics.

The third event causing a revolution in the philosophy of mathematics was the discovery of contradictions (paradoxes) in naïve set theory. These contradictions were discovered not only for the work of Cantor (which many mathematicians were already suspicious of, as it dealt with "completed infinities") but right there in the careful work of Frege. Frege's *Grundgesetze der Arithmetik* (Fundamental Laws of Arithmetic) is a work of logicism reducing the truths of arithmetic to theorems of logic. The second volume was at the publisher in 1903 when Russell wrote to Frege, informing him of the inconsistency of his system via the Russell paradox. (The set consisting of all sets that are not members of themselves both *must be* and *must not be* a member of itself. That is, let $A = \{B : B$ is a set and $B \notin B\}$. Then both $A \in A$ and $A \notin A$ lead to contradictions.) Thus, the elementary step of forming the set of all objects having a given property can lead to a contradiction. Since mathematicians frequently form sets this way, this discovery shook a larger portion of the mathematical world than the others. Instead of the occasional misuses of limits, which were viewed as the result of bad mathematical taste, the view now was that there was a *crisis in the foundations of mathematics*. Was all of mathematics a house built on shifting sands? In response to this crisis, two additional foundational schools, intuitionism and formalism, were developed. Logicism was also further developed, by Russell and Whitehead, Zermelo, and others, trying to mend the problems in Frege's account.

Logicism is the thesis that mathematics is a sub-branch of logic, that all theorems of mathematics can be reduced to theorems of logic. Logic had experienced significant development in the nineteenth century in the work of Boole, De Morgan, C. S. Peirce, and Venn, among others, as well as, of course, Frege. This work made logic a very systematic study of correct rules for reasoning. Therefore, it seemed plausible that if all of mathematics could be deduced from logic, mathematics would be free of contradictions and its foundations firm. In addition, the work of Peano giving axioms for the natural numbers, of Dedekind building the real numbers, and of Weierstrass defining limits, gave logicists much material needed to reduce mathematics to logic. Frege began this work with his *Grundlagen der Arithmetik* (Foundations of Arithmetic) and continued it with his two-volume *Grundgesetze der Arithmetik* (Fundamental Laws of Arithmetic). However, the Russell paradox meant that a different approach needed to be taken. Russell and Whitehead developed one such approach in *Principia Mathematica*, an enormous work comprising three volumes and over 2000 pages. Their hope was to show in this work that all of mathematics (or at least, number theory) could be reduced to logic. Russell had analyzed the Russell paradox and other paradoxes of set theory, and determined that all of them involved defining a set by using a larger set of which the set being defined was a member, which he called a "vicious circle." He believed that, as long as one avoided using vicious-circle definitions (also called "impredicative definitions"), one could avoid paradoxes. To do this, the set theory developed in *Principia Mathematica* builds sets in a hierarchy, a type-theory, with sets of the lowest type being individuals. On the next level are sets composed of these individuals. At each level, sets are built up of members that are sets from previous levels. The known large sets that lead to contradictions cannot be constructed in this system. (For brevity, the description here is significantly simplified. Their actual approach used propositional functions, rather than sets, as the basic objects on which everything else was built, and a "ramified" theory of types. See *Ontology and the Vicious-Circle Principle* [Chihara 1973], chapter 1, for a good description.)

Logicism in this revised form had three significant problems, which largely led mathematicians to lose interest in it. First, with a rigid type-theory, many important mathematical theorems not only cannot be proven, they cannot even be stated. For example, the least upper bound of a bounded set of real numbers is defined in terms of the set of real numbers. Therefore it must be of a higher type than the real numbers. Thus, this least upper bound cannot be, in this type-theory, a real number. To overcome this problem, an additional axiom was added to their system, called the axiom of reducibility. This axiom essentially says that a set that is defined at a higher level using only sets at some lower level is equivalent to some set that appears at the first level above all those involved in its definition. The problem with this axiom is that there is no justification for it within logic (and there are some concerns that it might allow the paradoxes to reappear). Hence, the program of reducing mathematics to logic fails: either you cannot get well-known theorems, or you must add a principle that is not purely logical.

There is a similar problem with the axiom of infinity. For much of mathematics, we need infinite sets. Yet their existence simply does not follow from other axioms. Russell and Whitehead introduced it as an axiom, but cannot justify it based purely on logic. Later logicians have attempted to overcome these two problems, most notably Quine, but no one has managed to build up all of mathematics purely from logical principles.

Third, Gödel's incompleteness theorem dealt a very significant blow to even the *possibility* of deriving all of mathematics from logic. At least for consistent first-order, recursive axiomatizations of number theory, this theorem says that if they are sufficiently strong to prove normal arithmetic properties, then there are theorems that are true but not provable in such systems. Hence, one simply cannot get all of mathematics from (at least first-order) logic.

Intuitionism is the thesis that mathematical knowledge comes from constructing mathematical objects within human intuition. Intuitionism's ancestors were Kant, Kronecker, and Poincaré. Kant contributed an intuition of the integers from our *a priori* intuition of time. Kronecker was famous for his statement "God made the natural numbers; all else is the work of man." He objected to any mathematical object that could not be constructed in a finite way. In particular, he fought Cantor's transfinite numbers. Poincaré viewed logic as sterile, and set theory as a disease. On the other hand, he viewed mathematical induction as a pure intuition of mathematical reasoning.

L.E.J. Brouwer, the founder of intuitionism, believed that the contradictions of set theory came from inappropriate dependence on formal properties, including logic. In particular, the use of the law of the excluded middle (that either a statement P or its negation $\sim$P must be true) with completed infinities or with proofs of existence is illegitimate and dangerous to the coherence of mathematics. Brouwer started with Kant's idea that our intuition of time is the basis for the natural numbers. Mathematical objects are mental constructions, which Brouwer described as "intuited non-perceptual objects and constructions which are introspectively self-evident." ([Körner 1968], p. 120) Completed infinities cannot be inspected or introspected, and so are not part of mathematics. A mathematical statement is true "only when a certain self-evident construction had been effected in a finite number of steps." ([Burton 2003], p. 661) To prove a proposition of the form "P or Q," one needs to prove P or to prove Q. To prove that $(\exists x)P(x)$, one needs to give a construction of an object and show that it satisfies P.

The rejection of completed infinities causes problems in the construction of real numbers. To define a real number, the intuitionist must, for example, give an algorithm that produces a sequence of rational numbers and give a proof that that sequence converges.

Many standard theorems are not intuitionistically true. For example, the standard proof of the Intermediate Value Theorem involves repeatedly bisecting an interval on which the function changes from being below the desired value C to being above it (or vice versa), maintaining that property in the sub-interval chosen. However, intuitionistically, one cannot always *determine* whether a given real number is greater than, equal to, or less than C. For example, let

$$a_n = \begin{cases} 1 & \text{if } 2n \text{ is the first even integer that is not the sum of two primes, } n > 1, n \text{ even} \\ -1 & \text{if } 2n \text{ is the first even integer that is not the sum of two primes, } n > 1, n \text{ odd} \\ 0 & \text{otherwise} \end{cases}$$

Define the real number $r = \sum_{n=2}^{\infty} a_n 10^{-n}$. Both intuitionistically and classically, r is a well-defined real number: to calculate its nth digit, just check if all even integers from 4 to $2n$ can be written as the sum of two primes. If the Goldbach conjecture is true, $r = 0$. If it is false and first fails at a multiple of 4 (i.e., n as used above is even), $r > 0$. If it first fails at an integer congruent to 2 modulo 4, $r < 0$. You can calculate r to whatever degree of accuracy you wish, simply by trying to decompose the appropriate values of n into sums of two primes. But while, classically, r must be either positive, negative, or zero, intuitionistically it is none of these until we decide the Goldbach conjecture. One can easily use r to give a function that shows that the Intermediate Value Theorem is not true intuitionistically, not just that there is a problem with the usual proof.

Intuitionism was developed in the same period that many abstract areas of modern mathematics—topology, functional analysis, etc.—were being developed. Most mathematicians were more interested in exploring these new developments than in retreating inside the shell of intuitionism.

There are many philosophical problems with intuitionism as well. If mathematical objects are mental constructions, there is no good reason to believe that two people will construct the same objects or have the same theorems. It is also not clear why mathematics is so useful in the world. Furthermore, much of modern physics uses mathematical objects (from functional analysis, for example) that intuitionists do not accept.

Intuitionism initially received enthusiastic support from Hermann Weyl (although he fell away from it later). Arend Heyting extended Brouwer's work in intuitionism and made Brouwer's often mystical and obscure writing much more accessible. However, because so many theorems of standard mathematics cannot be proven intuitionistically, very few mathematicians were inclined to adopt intuitionism. It required giving up too much mathematics just to avoid a few contradictions with extremely huge "sets." In the 1960s, Errett Bishop developed a variation on intuitionism, which he called constructivism (see [Bishop 1967]). He developed many theorems that are, using classical logic, equivalent to standard theorems but are constructively true. Thus, at least in analysis, one needs to give up less mathematics to be a constructivist than to be an intuitionist. This led to some renewed interest in the subject, but still has not led very many mathematicians to abandon classical mathematics.

Formalism is less well-defined. It is not clear that many serious mathematicians ever asserted the most extreme version of what is called formalism, that mathematics is just a formal game.[1] This view of mathematics is extremely unhelpful philosophically: it does not explain why we choose the axioms we choose, why mathematics is applicable to the world, why anyone would bother studying mathematics at all.

This extreme characterization of formalism appears to come from combining two parts of Hilbert's work. In his Foundations of Geometry (*Grundlagen der Geometrie*), he fixes some incompletenesses in Euclidean geometry, adopting an axiom system based on three undefined objects—points, lines, and planes—and three undefined relations—incidence (a point lying on a line), order (betweenness), and congruence. He makes it clear that, while the intuition behind the axioms comes from what we call points, lines, and planes, they could just as well stand for any objects—say, tables, chairs, beer mugs—as long as those objects satisfy the axioms. This work of Hilbert is one of the early works of modern mathematics, where, instead of working entirely within one mathematical structure, one sets up definitions and axioms and then proves theorems about the whole class of objects that satisfy the definitions and axioms.

Hilbert's proof of the consistency of his axioms for geometry reduces the question of the consistency of those axioms to the consistency of arithmetic. This brings us to the second part of Hilbert's work that is relevant for formalism. This is the "Hilbert program," aimed at restoring confidence in mathematics after the contradictions, described above, that came from work in the foundations of analysis and from naïve set theory. In part, his program was a reaction to what he considered the pernicious affect that intuitionism was having on mathematicians. He was determined to put mathematics on a sound footing without giving up large parts of mathematics in the process. The program is to first set up each field of mathematics as a formal theory, consisting of undefined terms and axioms. A proof in such a theory is a finite sequence of formulas, each of which is either an axiom or follows from earlier formulas by finitary logical rules of inference. One then investigates several metamathematical questions about the systems thus developed.

First, is the theory consistent? This can be investigated in one of two ways. One is to give a model of the theory. Usually this involves picking an already known mathematical structure (such as the integers). Then one interprets each of the undefined terms of the theory as objects within that structure in such a way that all of the axioms can be shown to be true theorems about the structure. When the structure involved is infinite, this then reduces the consistency of the original theory to the question of whether the axioms for the structure used to interpret the theory are consistent. Thus it is called a "relative consistency proof." Hilbert (in his *Grundlagen der Geometrie*) had given such proofs for Euclidean and non-Euclidean geometry by interpreting them within the real algebraic numbers. Thus, as long as the arithmetic of the real numbers is consistent, so is both Euclidean and non-Euclidean geometry. But this kind of consistency proof

---

[1] An exception is apparently von Neumann, who allegedly said "We must regard classical mathematics as a combinatorial game played with the primitive symbols . . . " [von Neumann 1966, pp. 50–51]. There is a quotation floating around, attributed to Hilbert: "Mathematics is a game played according to certain simple rules with meaningless marks on paper." This quotation appears for the first time in E.T. Bell, without citation—it may well have been made up by Bell. In fact, Hilbert, in [Hilbert 1919, p. 19], said "Mathematics is **not** like a game in which the problems are determined by arbitrarily invented rules. Rather, it is a conceptual system of inner necessity that can only be what it is and not otherwise." (translated by Michael Detlefsen, emphasis mine).

does not rule out the possibility that all of the theories involved are inconsistent. In addition, it is not using strictly finitary reasoning, and thus does not provide the foundation that is needed.

The second way consistency can be investigated is to show, in a finitary way, that it is not possible to derive a contradiction (for example, the statement $0 = 1$) from the axioms. This would then be an *absolute* consistency proof. It would not depend on another system (except, of course, the logic involved, which is finitary and might be acceptable to intuitionists). Of course, if one could give this kind of consistency proof for arithmetic, it would provide an absolute proof of the consistency of geometry, since a relative consistency proof had reduced the consistency of geometry to that of arithmetic.

Second, is the theory complete? This has a syntactic and a semantic meaning. Semantically, can all truths about the structure involved be proven from the axioms? If an axiomatization is not complete, then it has not captured all the relevant features of the mathematical structure it is axiomatizing, and there is a need to find further axioms so as to fully represent the structure. Syntactically, if an axiomatization is complete, every sentence or its negation is derivable from the axioms (since every sentence is either true or its negation is true in the structure).

Third, are the axioms independent of each other, or can some be eliminated? One usually shows independence by giving a structure in which all but one of the axioms are true, and the remaining one fails. This is the least important question, more an aesthetic issue than one central to the adequacy of the theory. But as mathematicians tend to like clean results, it is preferable to find axioms that are independent. Hilbert, in his *Grundlagen der Geometrie*, showed that many of his axioms were independent, though, given the tediousness of going through all combinations, he did not show that all were.

The foundational school called formalism contains as its core the view that to set mathematics on firm foundations, one should investigate these questions for the various structures and theories that make up mathematics. This led to the development of the field called proof theory, which investigates these metamathematical questions.

Unfortunately for Hilbert's program, two results of Gödel showed that the program could not work. His first incompleteness theorem showed that any consistent first-order axiomatization for the natural numbers that can be described recursively (basically, in a finitist way), and that is sufficiently strong to prove most of the standard theorems of number theory, is incomplete. That is, there are truths about arithmetic that cannot be proven within that axiomatization. (Actually, the result Gödel proved required a little more, called ω-consistency; the result was improved by Barclay Rosser to simply require standard consistency.) Thus, one cannot capture all truths about the integers within a finitistic system. His second incompleteness theorem was even more devastating. Given any consistent, recursive system of (first-order) axioms that is sufficiently strong to do a significant amount of mathematics[2], it is impossible to prove the consistency of the system within that system. Thus, there is no point in looking for a finitary proof of consistency. There has been continuing work in proof theory investigating properties of axiom systems, but there does not appear to be any hope of reviving Hilbert's original program. Gödel proved a third important theorem relevant to the Hilbert program, the completeness theorem for first-order

---

[2] Here, "sufficiently strong" represents a technical requirement involving being able to represent the primitive recursive functions within it and derive some standard number theoretical results; for details, see any standard textbook on mathematical logic.

logic. This says that every consistent set of first-order statements has a model. That is, our system of first-order logic is complete: in it, every first-order statement which is true in every model can be proved. Thus, semantic consistency (having a model), for first-order theories, is equivalent to syntactic consistency (not being able to derive a contradiction). Second-order theories, however, may be consistent without having any models.

Of these foundational schools, only logicism can really be called a philosophy of mathematics, as the other two do not really provide answers to all of the traditional philosophical questions: "what is the nature of mathematical objects," "what is the nature of mathematical knowledge," and "how can mathematical results help us understand physical world?" Intuitionism does not answer the last; formalism does not answer the first or the third (and, because of Gödel's results, does not answer the second either). Logicism's answer to all of these questions reduces to the similar questions about logic. However, since there are serious problems in reducing mathematics to logic, logicism does not settle these questions either. But for the first three-quarters of the twentieth century, work on foundations replaced almost all other discussion about the philosophy of mathematics.

More detailed discussions of the three foundational schools can be found in [Burton 2003]; [Körner 1968] and [George/Velleman 2002] are books, aimed at the same audience as this book, devoted to a thorough discussion of these views. Also, [Giaquinto 2002] is an accessible book that gives a good discussion of what work has been done in each of these schools.

4.1 *Other Philosophers in This Period*

There are two philosophers who wrote a substantial amount about mathematics during this period, but were not part of any of these foundational schools. One was **Edmund Husserl**, who developed phenomenology. He had a Ph.D. in mathematics, and his *habilitation* dissertation was *On the Concept of Number* (1887), which was later expanded to *Philosophy of Arithmetic*, published in 1891. This book attempted a psychological foundation of arithmetic, and preceded his phenomenological work, which was first published in 1900 in *Logical Investigations*. Husserl also has a very fine (and influential) essay, called "The Origin of Geometry," that usually appears as an appendix to his *The Crisis of European Sciences and Transcendental Phenomenology*. Derrida's Ph.D. thesis is a response to it. Husserl is quite difficult to read. Richard Tieszen has worked on making Husserl accessible, as well as answering philosophical objections to Husserl's work; see his *Phenomenology, Logic, and the Philosophy of Mathematics* [Tieszen 2005].

Ludwig Wittgenstein is another influential philosopher of this period who is also not easy to read. His work focuses on "language games," or the relation between language, as we use it, and reality. His initial work on this topic in the *Tractatus Logico-Philosophicus* (1922) set the stage for his work on the philosophy of mathematics in *Philosophical Remarks* (1929–30), *Philosophical Grammar* (1931–33), and later in *Remarks on the Foundations of Mathematics* (1937–44). According to the Stanford Encyclopedia of Philosophy, Wittgenstein maintains that mathematical propositions differ from real propositions. Mathematical statements do not refer to anything real, but their content comes from their syntax. "On Wittgenstein's view, we invent mathematical calculi and we expand mathematics by calculation and proof, and though we learn from a proof that a theorem *can* be derived from axioms by means of certain rules in a particular way, it is *not* the case that this proof-path pre-exists our construction of it." (http//plato.stanford.edu/entries/wittgenstein-mathematics/) He views mathematics as a human invention, and no mathematics exists until we discover it. Wittgenstein is thus a precursor of some social-constructivist views of mathematics.

5 More Recent Work That is Worth Reading but is Not Represented Here

As I mentioned in the first section of this introduction, this book consists of original articles by philosophers, mathematicians, and mathematics educators, most summarizing work over a period of years. To put this book together, I invited people whose work I had read and admired to write a chapter for this volume. I got a relatively good response, and thus this volume covers a fairly wide range of contemporary issues. However, in part because I was often asking very senior people in the field, there were a number of excellent writers on the philosophy of mathematics who declined to participate in this project. You'll certainly find suggestions for continued reading on any of the topics in this book in the bibliographies of the individual chapters. However, I want to take some space here to recommend some other very good places to learn more about the philosophy of mathematics. Full bibliographic references for these books and articles are in the Bibliography at the end of this introduction.

5.1 Logicians with a Philosophical Bent

Two logicians who have done a significant amount of very thoughtful and careful work in the philosophy of mathematics have recently collected that work in books: **Solomon Feferman**'s (math.stanford.edu/~feferman/) *In the Light of Logic* [1998] and **William Tait**'s (home.uchicago.edu/~wwtx/) *The Provenance of Pure Reason: Essays in the Philosophy of Mathematics and Its History* [2005]. I recommend both books highly.

5.2 Philosophers

There are many philosophers working in the philosophy of mathematics, almost all of them working on questions of the nature of mathematical objects and of mathematical knowledge: the debate, represented and summarized in this volume by the chapters by Balaguer, Chihara, Linnebo, and Shapiro, of platonism versus nominalism. All of the philosophers listed below have written a lot more than is mentioned here, of course; but I'm pointing to those I think are likely to be interesting to mathematicians.

   Paul Benacerraf, at Princeton (philosophy.princeton.edu/components/com_faculty/ documents/paulbena_cv.pdf) wrote two seminal papers that initiated the move in the 1970s, by philosophers of mathematics, back to traditional philosophical questions and away from foundations: "What Numbers Could Not Be" [1965] and "Mathematical Truth" [1973]. They are still well worth reading. He also, with Hilary Putnam, edited a book of readings in the philosophy of mathematics. It has gone through two editions, with a quite different selection of papers in the second edition. Both editions are worth looking at. His two articles mentioned above are reprinted in the second edition.

   John Burgess (www.princeton.edu/~jburgess/), also at Princeton, works in logic (philosophical and mathematical) and the philosophy of mathematics. What he says about mathematics is very careful and correct. However, since he works in the philosophy department, his interests have been turning more and more toward technical philosophical issues. His "Why I Am Not a Nominalist" [1983] is quite accessible. I have not read his two recent books: *A Subject with No Object* [Burgess/Rosen 1997] and *Fixing Frege* [2005]. I looked at the former and decided that it was far more technical than I could handle without devoting months to it. I do hope to look at the latter once this book is finished.

Imre Lakatos combined the approaches of philosophers of science Thomas Kuhn and Karl Popper and applied it to mathematics. His best-known work is *Proofs and Refutations* [1976], a lively dialogue about Euler's theorem that $v - e + f = 2$ (where v represents the number of vertices, e the number of edges, and f the number of faces) for a polyhedron. It shows how cases that are counterexamples motivate revisions of the hypotheses of the theorem and the definition of polyhedron. This provides, according to him, an example of how mathematics develops.

Penelope Maddy (www.lps.uci.edu/home/fac-staff/faculty/maddy/), at the University of California at Irvine, started her career as a student of Burgess and a platonist. Her first book, *Realism in Mathematics* [1990], described an unusual form of platonism in which mathematical objects are located in the physical world. This view was broadly attacked by other philosophers. Her current direction, a naturalist approach to the philosophy of mathematics (*Naturalism in Mathematics* [1997]), is one that takes science as the standard by which all knowledge is to be judged. Knowledge of anything, including mathematics, must be justifiable through our best scientific theories, in particular, empirical psychology, linguistics, etc.

Charles Parsons (www.fas.harvard.edu/~phildept/parsons.html), at Harvard, works in the philosophy of mathematics as well as in logic and in other fields of philosophy. His article "Mathematical Intuition" [1979–80] is a fairly interesting discussion of how one can have intuitions of mathematical objects such as numbers and sets. However, it will be disappointing if you are expecting something like what Poincaré wrote on the topic. Many of his papers in the philosophy of mathematics are collected in [Parsons 1983].

Hilary Putnam (www.fas.harvard.edu/~phildept/putnam.html), also at Harvard, works on philosophy of mathematics, philosophy of science, and other fields of philosophy. His article, "Mathematics without Foundations" [1967], is one of the early articles moving philosophy of mathematics back from foundations to more traditional philosophical problems. Some of his work in the philosophy of mathematics is collected in [Putnam 1985].

Michael Resnik (http://philosophy.unc.edu/resnik.htm), at the University of North Carolina, Chapel Hill, is a structuralist (of a slightly different sort than Stewart Shapiro, who has an article in this volume). His book, *Mathematics as a Science of Patterns* [1997], is quite readable once you are used to philosophical terminology. Given that many mathematicians assert that mathematics is the science of patterns, the book is worth reading to see how philosophers establish such an assertion.

5.3 People Working in the History and Philosophy of Mathematics

Several people work on the boundary between philosophy of mathematics and history of mathematics. One is **Kenneth Manders** (www.pitt.edu/~philosop/people/manders.html), at the University of Pittsburgh. He is primarily a philosopher (and logician) with a strong mathematical background, but his arguments are very carefully historically based. Unfortunately, he rarely publishes. One published article is "Domain extension and the philosophy of mathematics"

[1989]. I have a very interesting preprint, "Why Apply Math?" from 1999, and another, "Euclid or Descartes: Representation and Responsiveness." Both are very carefully and thoughtfully written, but the only way to get them is to write to him.

Another is **Leo Corry** (http://www.tau.ac.il/~corry/), at Tel Aviv University. He is more of a historian of mathematics, but he asks philosophical questions. For example, his *Modern Algebra and the Rise of Mathematical Structures* [2004] investigates how the notions of what was meant by "algebra" and "mathematical structure" developed over the last two centuries.

A third is **Paolo Mancosu** (http://philosophy.berkeley.edu/mancosu/); see his "On Mathematical Explanation" [2000]. In addition to being interesting in itself, it mentions several other articles on this topic.

A fourth is **Howard Stein** (https://philosophy-data.uchicago.edu/index-faculty.cfm#Stein), at the University of Chicago, who works on the history and philosophy of mathematics and physics. Three articles worth looking at are "Yes, but . . . : Some Skeptical Reflections on Realism and Anti-realism" [1989], "Eudoxos and Dedekind: On the Ancient Greek Theory of Ratios and its Relation to Modern Mathematics" [1990] and (do not be put off by the title) "Logos, Logic, and Logistiké: Some Philosophical Remarks on 19th Century transformation of Mathematics" [1988].

5.4 People Working in Mathematics Education

By the very act of teaching mathematics, one takes a position on how people acquire mathematical knowledge, which has both psychological and philosophical aspects. Therefore many people whose research is in mathematics education have interesting philosophies of mathematics. I contacted several of them, and, as it turns out, the person whose work I find most interesting *has* made a contribution to this book, but several others whose work I also respect were either unwilling or unable to do so.

Ed Dubinsky (http://www.math.kent.edu/~edd/) has worked applying an interpretation of Piaget's work to higher-level mathematics education. He is very active (though now retired), and has gathered a large community of mathematics educators who work with him. He started the Research in Undergraduate Mathematics Education Community, which later branched into the SIGMAA on RUME. He started out as a functional analyst. In his early work in mathematics education, he used a computer-based language, ISETL, to help students understand abstract mathematical objects. For example, see his "Teaching mathematical induction, I/II" ([1986], [1989]). The basic theory that he developed, APOS theory (standing for Action, Process, Object, Schema), describes how students gradually develop more sophisticated concepts through a process of reflective abstraction. A description can be found in "A theory and practice of learning college mathematics" [1994]. It is applied to student understanding of functions in "Development of the process conception of function" [1992], and to abstract algebra in "Development of students' understanding of cosets, normality and quotient groups" [1997]. An overall framework is given in "A framework for research and curriculum development in undergraduate mathematics education" [1996]. Dubinsky views his philosophy as inseparable from his educational theory and practice, and declined to write an article for this volume because he felt that his work already expresses his philosophical position adequately.

Paul Ernest (http://www.people.ex.ac.uk/PErnest/) received a Ph.D. in philosophy of mathematics, but spent much of his career working in mathematics education. He is the editor of the *Philosophy of Mathematics Education Journal*. He has written an article, "The Impact of Beliefs on the Teaching of Mathematics" ([Ernest 1994]; originally written in 1989), suggesting that to make significant changes in mathematics education requires changing beliefs about the nature of mathematics as well as about how it is taught and learned. This then led to a book, *The Philosophy of Mathematics Education* [Ernest 1991]. More recently, he wrote a book setting forth his philosophical views of mathematics itself, *Social Constructivism as a Philosophy of Mathematics* [Ernest 1998]. As I make clear in my review of that book [Gold 1999], I do not view it as a viable version of social constructivism, but not all mathematicians agree with me, and I encourage readers to decide for themselves.

Annie and **John Selden** are the editors of the Research Sampler on MAA Online (http://www.maa.org/t_and_l/sampler/research_sampler.html), which brings selected research in mathematics education to the attention of collegiate mathematics educators. After long careers at various universities in the U.S. and abroad, they are now Adjunct Professors of Mathematics, New Mexico State University. In their own research, they have examined students' ability to solve novel calculus problems ([1989] and [2000]), students' grasp of the logical structure of informal mathematical statements, student difficulties with proofs [2003], and are currently investigating (college) teachers' beliefs about mathematics, teaching, and learning. See also their Research Questions page (http://www.maa.org/t_and_l/sampler/rs_questions.html) on MAA Online.

David Tall (http://www.warwick.ac.uk/staff/David.Tall/) also connects philosophical views of mathematics with his educational work in significant ways. See his "Existence Statements and Constructions in Mathematics and Some Consequences to Mathematics Teaching" [Tall/Vinner 1982], and a book he edited, *Advanced Mathematical Thinking* [1991]. More recently he has looked at the mathematical world as really three different realms [2004], the first coming from our perceptions of the world and thinking about them, the second the world of symbols we use in mathematics, the third the formal axiomatic world.

5.5 Mathematicians

I had a better success rate getting mathematicians who are interested in the philosophy of mathematics to contribute to this book. One who did not was **Saunders Mac Lane**; anyone interested in the philosophy of mathematics will find his *Mathematics: Form and Function* [1986] worth reading.

Chandler Davis has a very interesting view of mathematics, coming from a materialist perspective; see his "Materialist Mathematics" [1974] and "Criticisms of the Usual Rationale for Validity in Mathematics" [1990].

Lynn Steen has written a number of articles (and books) popularizing mathematics that include a philosophical bent. See particularly "The Science of Patterns" [1988].

Ian Stewart has also written a number of popular books about mathematics that have substantial philosophical implications. One of the best in that direction is *Nature's Numbers: The Unreal Reality of Mathematics* [1995].

6 A Brief Overview of This Book

This book is not designed for a straight read from beginning to end, although some readers might choose to do that. It is meant to be dipped into as the topic and writing style appeals to you. Each chapter is self-contained and most are liberally sprinkled with references for those wanting to delve more deeply into a particular topic. We have tried to organize it somewhat by topic, but within each topic the style and point of view of the chapters can be quite different. Thus, we've tried to provide you here with a guide to the chapters. Also each chapter is preceded by a short description of it and a brief biographical sketch of the author.

For mathematicians who have some curiosity about philosophical questions regarding mathematics, but who have not read any contemporary philosophy, a good place to start might be **Philip Davis**'s chapter. He asks a question that we have all wrestled with at some point, after we have done some mathematics and are thinking of writing it up: when is a problem solved? When can we say, OK, let's wrap that up now?

6.1 Views on Mathematical Objects

Barry Mazur's chapter should also be very accessible to mathematicians without much philosophical background. He asks a question that is close to a traditional philosophical question—how can one tell when one mathematical object is really the same as another. However, he looks in a very different direction than philosophers generally do as he traces some category theory from fundamentals to propose an answer this question, with some interesting comments along the way.

The others writing about mathematical objects are all philosophers.

Stewart Shapiro gives an overview of philosophical discussions concerning mathematical objects. This culminates with his view that mathematics is the science of structures (as suggested originally by Bourbaki), or, as it is sometimes called, of patterns.

For mathematicians interested in reading about current work in the philosophy of mathematics, **Charles Chihara**'s chapter is a relatively gentle introduction to the kind of discussions philosophers have. He discusses concerns about the existence of mathematical objects. Finally he turns to how one can develop a structural account of mathematics without being committed to the actual existence of structures in either the world or some ideal platonic realm.

Mark Balaguer gives an overview of the major variations philosophers have discussed over the last thirty years on whether there are mathematical objects; if there are, what their nature is; and how we can gain mathematical knowledge. As a summary of thirty years of philosophical discussion, this chapter is quite dense. However, it very effectively and systematically summarizes the discussion from a wealth of philosophical views.

Øystein Linnebo develops a new view of mathematical objects that allows them to exist in some sense while avoiding some of the traditional objections to a platonist account of mathematical objects.

6.2 Views on Proof

Proof and its relation to mathematical knowledge is an issue that has become an active concern again thanks to computer-assisted proofs and mathematical investigations involving computers.

Michael Detlefsen discusses both the role of empirical reasoning—primarily due to the use of computers—and of formalization in mathematical proofs.

Joseph Auslander discusses the various roles proof plays in mathematics, and how standards of proof vary over time.

Jon Borwein focuses less on proof than on the development of mathematics, and the roles computers may play in that development. Necessarily this includes the role they play in developing proofs.

6.3 What is Mathematics?

Robert Thomas's chapter suggests, as a definition of mathematics, a variation on "mathematics is the science of patterns." He takes mathematics as one extreme end in the spectrum of the sciences, and suggests (read the chapter for what he means) that "mathematics is the science of relations as such."

Guershon Harel approaches the problem of "what is mathematics?" from the viewpoint of a researcher in mathematics education. He proposes an answer that includes not only the theorems, but also the tactics and conceptualizations we use.

6.4 Social Constructivism

Reuben Hersh, one of the few mathematicians to attempt to describe social constructivism in some detail, discusses mathematics and its development (or, as he phrases it, mathematics as "a living organism") as the subject of scientific investigation. This lively chapter includes a beautiful attempt to describe the feeling when an idea for solving a problem suddenly flashes into one's mind.

Julian Cole just finished his Ph.D. thesis in philosophy, working on how one can make social constructivism coherent from a philosopher's standpoint. His chapter summarizes the main points of interest to mathematicians of his work.

6.4.1 The Boundaries Between Mathematics and the Other Sciences (Physical and Social), and the Applicability of Mathematics

Mark Steiner looks at a particular aspect of this question (primarily from the standpoint of a philosopher interested in the application of mathematics to physics) related to generalizations of addition.

Keith Devlin looks at the question of what we currently call mathematics versus what we currently relegate to applied mathematics. He describes how he believes this will change.

6.5 Philosophy of Probability

Alan Hájek discusses some of the fundamental philosophical issues about the nature of probabilities. It is a very accessible chapter.

Bibliography

[Benacerraf/Putnam 1964/1983] Benacerraf, Paul, and Hilary Putnam, eds., *Philosophy of Mathematics*, Cambridge: Cambridge University Press, 1st ed. 1964, 2nd ed. 1983.

[Benacerraf 1965] Benacerraf, Paul, "What Numbers Could Not Be," *The Philosophical Review*, 74 (1965), pp. 47–73.

[Benacerraf 1973] ——, "Mathematical Truth," *The Journal of Philosophy*, 70 (1973), pp. 661–680.

[Bishop 1967] Bishop, Errett, *Foundations of Constructive Analysis*, New York: McGraw-Hill, 1967.

[Brumbaugh 1954] Brumbaugh, Robert S., *Plato's Mathematical Imagination*, Bloomington IN: Indiana University Press, 1954.

[Burgess 1983] Burgess, John P., "Why I Am Not a Nominalist", *Notre Dame Journal of Formal Logic **24*** (1983), pp. 93–105.

[Burgess/Rosen 1997] Burgess, John P., and Gideon Rosen, *A Subject with No Object: Strategies for Nominalistic Reconstrual of Mathematics*, Oxford: Oxford University Press 1997.

[Burgess 2005] Burgess, John P., *Fixing Frege*, Princeton: Princeton University Press, 2005.

[Burton 2003] Burton, David M, *The History of Mathematics: An Introduction*, 5th edition, New York: McGraw Hill, 2003.

[Chihara 1973] Chihara, Charles, *Ontology and the Vicious-Circle Principle*, Ithaca: Cornell University Press, 1973.

[Corry 2004] Corry, Leo, *Modern Algebra and the Rise of Mathematical Structures*, Basel: Birkäuser, 2004.

[Davis 1974] Davis, Chandler, "Materialist Mathematics," pp. 37–66 in Cohen, Stachel and Wartofsky, eds., *For Dirk Struik, scientific, historical, and political essays in honor of Dirk J. Struik*, Boston: D. Reidel, 1974.

[Davis 1990] ——, "Criticisms of the Usual Rationale for Validity in Mathematics," pp. 343–356 in A.D. Irvine, ed., *Physicalism in Mathematics*, Dordrecht: Kluwer Academic Publishers 1990.

[Dubinsky 1986] Dubinsky, Ed, "Teaching mathematical induction I," *Journal of Mathematical Behavior* 5 (1986), pp. 305–317.

[Dubinsky 1989] ——, "Teaching mathematical induction II," *Journal of Mathematical Behavior* 8 (1989), pp. 285–304.

[Dubinsky et al. 1992] Breidenbach, D., E. Dubinsky, J. Hawks and D. Nichols, "Development of the process conception of function," *Educational Studies in Mathematics* 23 (1992), pp. 247–285.

[Dubinsky 1994] Dubinsky, Ed, "A theory and practice of learning college mathematics" pp. 221–243 in *Mathematical Thinking and Problem Solving*, A. Schoenfeld ed., Hillsdale: Erlbaum, 1994.

[Dubinsky et al. 1996] Asiala, M., A. Brown, E. Dubinsky, D. DeVries, D. Mathews and K. Thomas, "A framework for research and curriculum development in undergraduate mathematics education," pp. 1–32 in *Research in Collegiate Mathematics Education II, CBMS Issues in Mathematics Education*, American Mathematical Society, 1996.

[Dubinsky et al. 1997] Asiala, M., E. Dubinsky, D. Mathews, S. Morics, and A. Oktac, "Development of students' understanding of cosets, normality and quotient groups," *Journal of Mathematical Behavior* 16, 3 (1997), pp. 241–309.

[Ernest 1994] Ernest, Paul, "The Impact of Beliefs on the Teaching of Mathematics,", in Bloomfield, A. and T. Harries, eds., *Teaching and Learning Mathematics*, Derby: Association of Teachers of Mathematics, 1994.

[Ernest 1991] ——, *The Philosophy of Mathematics Education*, London: Falmer Press, 1991.

[Ernest 1998] ——, *Social Constructivism as a Philosophy of Mathematics*, State University of New York Press, Albany, NY, 1998.

[Feferman 1998] Feferman, Solomon, *In the Light of Logic*, New York: Oxford University Press, 1998.

[Fowler 1999] Fowler, David, *The Mathematics of Plato's Academy*, 2nd edition, New York: Oxford University Press, 1999.

[George/Velleman 2002] George, Alexander, and Daniel Velleman, *Philosophies of Mathematics*, Malden MA: Blackwell Publishers, Inc., 2002.

[Giaquinto 2002] Giaquinto, Marcus, *The Search for Certainty: A Philosophical Account of Foundations of Mathematics*, Oxford: Oxford University Press 2002.

[Gold 1999] Gold, Bonnie, review of *Social Constructivism as a Philosophy of Mathematics* and of *What is Mathematics, Really?*, *American Mathematical Monthly* 106, 4 (1999), pp. 373–380.

[Hersh 2005] Hersh, Reuben, *18 Unconventional Essays on the Nature of Mathematics*, New York: Springer-Verlag, 2005.

[Hesseling 2003] Hesseling, Dennis, *Gnomes in the Fog: The Reception of Brouwer's Intuitionism in the 1920s*, Birkhäuser Verlag, Basel, 2003.

[Hilbert 1919] Hilbert, David, "Natur u. mathematisches Erkennen", Lectures Fall semester, 1919.

[Körner 1968] Körner, Stephan, *The Philosophy of Mathematics: An Introductory Essay*, London: Hutchinson & Co., 1968.

[Lakatos 1976] Lakatos, Imre, *Proofs and Refutations*. Cambridge: Cambridge University Press, 1976.

[Mac Lane 1986] Mac Lane, Saunders, *Mathematics: Form and Function*, New York: Springer-Verlag 1986.

[Maddy 1990] Maddy, Penelope, *Realism in Mathematics*, Oxford: Oxford University Press 1990.

[Maddy 1997] ——, *Naturalism in Mathematics* Oxford: Oxford University Press 1997.

[Mancosu 2000] Mancosu, Paolo, "On Mathematical Explanation," pp. 103–119 in Emily Grosholz and Herbert Breger, eds., *The Growth of Mathematical Knowledge*, Dordrecht: Kluwer Academic Publishers, 2000.

[Manders 1989] Manders, Kenneth, "Domain extension and the philosophy of mathematics," *Journal of Philosophy* 86 (1989), pp. 553–62.

[Manders unpub.a] ——, "Why Apply Math?", unpublished manuscript.

[Manders unpub.b] ——, "Euclid or Descartes: Representation and Responsiveness", unpublished manuscript.

[Parsons 1979–80] Parsons, Charles, "Mathematical Intuition", *Proceedings of the Aristotelian Society* 80 (1979–80), pp. 145–168.

[Parsons 1983] ——, *Mathematics in Philosophy*, Ithaca: Cornell University Press, 1983.

[Putnam 1967] Putnam, Hilary, "Mathematics without Foundations", *Journal of Philosophy* 64 (1967), pp. 5–22.

[Putnam 1985] ——, *Mathematics, Matter and Method. Philosophical Papers*, vol. 1. Cambridge: Cambridge University Press, 1975. 2nd. ed., 1985.

[Resnik 1997] Resnik, Michael, *Mathematics as a Science of Patterns*, New York: Oxford University Press, 1997.

[Selden et al. 1989] Selden, A. and J., and A. Mason, "Can Average Calculus Students Solve Nonroutine Problems?" *Journal of Mathematical Behavior* 8 (1989), pp. 45–50.

[Selden et al. 2000] Selden, A., J. Selden, S. Hauk, and A. Mason, "Why Can't Calculus Students Access their Knowledge to Solve Non-Routine Problems?", *CMBS Issues in Mathematics Education*, 2000.

[Selden 2003] Selden, A. and J., "Validations of Proofs Considered as Texts: Can Undergraduates Tell Whether an Argument Proves a Theorem?" *Journal for Research in Mathematics Education* 34(1) (2003), pp. 4–36.

[Steen 1988] Steen, Lynn, "The Science of Patterns," *Science* 240 (29 April 1988) pp. 611–616.

[Stein 1988] Stein, Howard, "Logos, Logic, and Logistiké: Some Philosophical Remarks on 19th Century transformation of Mathematics," pp. 238–259 in William Aspray and Philip Kitcher, eds., *History and Philosophy of Modern Math*, Minneapolis: University of Minnesota Press, 1988.

[Stein 1989] ——, "Yes, but . . . : Some Skeptical Reflections on Realism and Anti-realism," *Dialectica* 43 (1989), pp. 47–65.

[Stein 1990] ——, "Eudoxos and Dedekind: On the Ancient Greek Theory of Ratios and its Relation to Modern Mathematics," *Synthese* 84 (1990), pp. 163– 211.

[Stewart 1995] Stewart, Ian, *Nature's Numbers: The Unreal Reality of Mathematics*, Perseus Publishing, 1995.

[Tait 2005] Tait, William, *The Provenance of Pure Reason: Essays in the Philosophy of Mathematics and Its History*, New York: Oxford University Press, 2005.

[Tall/Vinner 1982] Tall, David, and Shlomo Vinner, "Existence Statements and Constructions in Mathematics and Some Consequences to Mathematics Teaching," *American Mathematical Monthly* 89, 10 (1982), pp. 752– 756.

[Tall 1991] Tall, David, ed., *Advanced Mathematical Thinking*, Kluwer: Holland 1991.

[Tall 2004] Tall, David, "Introducing Three Worlds of Mathematics," *For the Learning of Mathematics* (2004).

[Tieszen 2005] Tieszen, Richard, *Phenomenology, Logic, and the Philosophy of Mathematics*, Cambridge: Cambridge University Press, 2005.

[von Neumann 1966] von Neumann, John, *Theory of Self-Reproducing Automata*, A.W. Burks, ed., Urbana: University of Illinois Press, 1966.

I

Proof and How it is Changing

Proof has been an essential part of mathematics since the time of the ancient Greeks. Its centrality has engendered much controversy. What is the role of proof in mathematics? What makes for an adequate proof?

The recent use of computers in developing mathematical conjectures, and in checking cases when there are too many for humans to check in a reasonable amount of time, has led to questions about the role and importance of proof in mathematics, as well as what qualifies as a proof. The chapters in this section give three different views of these and other issues regarding relationships among proof, mathematics, and computers.

1

Proof: Its Nature and Significance

Michael Detlefsen
Professor of Philosophy
University of Notre Dame

From the Editors

In our first chapter, Michael Detlefsen carefully examines the historical tension between inductive and deductive methods in mathematics, and relates it to the current discussion of the roles of each in the development of mathematics. He then turns to the question of whether, in fact, formalization of proofs actually increases either understanding or reliability of proofs. He also summarizes recent work on diagrammatic reasoning in mathematics, and the possible roles of visual experience in proofs.

We have chosen this as the first chapter in the book because we believe it is a fine, careful examination of these questions that virtually every reader of this volume will benefit from reading. For those of us who teach mathematics, an awareness of the fluctuations in the role of proof, and what is considered a proof, can be of use in the classroom. Such awareness can give us both a context in which to set our students' attempts at proof and a historical background we can impart to our students. An awareness of the importance of inductive methods in the development of mathematics is also worth transmitting to our students. In particular, making students aware of the current discussion in the mathematical community about the role of computers in mathematics can help them realize that mathematics is still a growing subject, even if most of the mathematics they study at the undergraduate level is centuries old.

Michael Detlefsen is a Professor of Philosophy at the University of Notre Dame (philosophy.nd.edu/people/all/profiles/detlefsen-michael/). His interests include logic, the philosophy of mathematics, and more specifically the role of proof in mathematics. He has written one book, Hilbert's Program *(1986), and edited two others,* Proof, Logic, Formalization *(1991),* Proof and Knowledge in Mathematics *(1991). Among his articles that are likely to be of particular interest to readers of this volume are "The Four-Color Theorem and Mathematical Proof" in* The Journal of Philosophy *(1980), "Poincare vs. Russell on the Role of Logic in Mathematics,"*

Philosophia Mathematica *(1993), "Mind in the Shadows: Essay Review of Roger Penrose's The Emperor's New Mind (OUP, 1989), Shadows of the Mind (OUP, 1994) and The Large, the Small and the Human Mind (CUP, 1997),"* Studies in the History and Philosophy of Modern Physics *(1998), "What does Gödel's Incompleteness Theorem Say?"* Philosophia Mathematica *(2001), and "Formalism," in* The Oxford Handbook of the Philosophy of Mathematics and Logic *(2005). He's currently working on "The Role of the Imaginary in Mathematics" and has a forthcoming article, "Purity as an Ideal of Proof," to appear in* The Philosophy of Mathematical Practice.

1 Introduction

Recent philosophical work on the topic of mathematical proof has focused on epistemological concerns. Prominent among these are the questions whether

(i) there is a special type of knowledge that proof and proof alone supports, or for which it provides special support,

whether

(ii) the knowledge supported by proof warrants a regimentation of mathematical practice that makes proof the sole legitimate or at least the preferred form of justification in mathematics

and, relatedly, whether

(iii) there is a place for broadly empirical reasoning in the development of mathematical knowledge.

These concerns are not new, of course, but have been of perennial interest to mathematicians and philosophers. Traditionally, responses to (i) have generally been affirmative. Views on (ii) and (iii) have been more mixed, with some arguing that empirical reasoning has little if any place in the development of mathematical knowledge and others (in roughly equal numbers) maintaining that it plays a vital role.

For most of the past three centuries, philosophical work on mathematics has mainly admitted the usefulness of empirical methods while also insisting that they do not provide the same quality of knowledge as classical demonstration or proof.

Recent work has, for the most part, sustained this compromise. It has, in particular, supported the use of various types of empirical reasoning to help solve mathematical problems. Yet while earlier thinkers generally based their support of such methods on considerations of usefulness, convenience or perhaps practical necessity, recent writers have sometimes strengthened this to something verging on psychological or perhaps even physical necessity.

This is nowhere more evident than in the recent controversy concerning the status of the computer-assisted resolution of the four-color problem in 1977. Here some have argued that the proof of the four-color theorem (or any other extremely long and/or complex argument) represents a fundamental departure from traditional standards and methods of justification in mathematics. Others have argued the contrary. My sympathies are primarily with the latter.

An examination of the historical record reveals, I think, that there have long been arguments recommending the use of inductive arguments in mathematics on the grounds of their usefulness and practical necessity. I survey some of this history in section 2 and relate it to more recent work in section 3.

In section 4, I consider a different challenge to the currently prevailing view of proof—one which focuses on rigor and the conditions necessary for its attainment. Of particular interest there is a recent series of papers by the artificial intelligence researcher J. A. Robinson, who argues that proof has two essential aims. One is to convince, the other to explain. He observes that though formalization may assist the convictive aim of proof in certain ways, it can also obstruct its explanatory aim. This being so, it may compromise the greater ends of proof. It may even interfere with rigor since, as he maintains, explanatory coherence is sometimes the best protection we have against serious gaps in our reasoning.

Robinson's overall goal is to explain a notable phenomenon—namely, the apparent gap that exists between the standards of proof that seem to guide mathematical practice, and the more austere standards of formalization that prevail in mathematicians' descriptions of their ideals. If formalized proof is indeed the ideal of mathematical justification, why should ordinary practice remain so far from it? The common response would be that the two are not so remote from each other, that formalization of ordinary proof is largely a routine affair. Robinson challenges this view both with arguments and examples. More positively, he proposes a conception of proof which emphasizes its affinity with performance rather than pure text. The upshot is a view that promises to be at once subtler and empirically more realistic than the currently prevailing views of proof and rigor.

In section 5, I turn to the large, wide-ranging recent literature on diagrammatic reasoning and its place in mathematics. This intersects with the topics treated earlier in that it stresses the role of visual experience in proof and considers how mathematical thinking might make use of such experience while still remaining properly rigorous. Of particular interest in this connection is the radical view of Jon Barwise and his co-author John Etchemendy (and their students), who claim that diagrammatic reasoning can play not only a heuristic but a genuinely justificative role in proof. At the heart of their view is the belief that visual and linguistic representations of the same information can and often do have significantly different properties. In particular, they commonly differ with respect to certain types of efficiency—the diagrammatic variants being generally more efficient in these ways. Describing and accounting for such differences has been a major preoccupation of both their work and other recent work on diagrammatic reasoning.

So too has been the question of how visual and/or diagrammatic reasoning fits with the explanatory goals of proof and the quest for rigor. Of particular interest in this connection is a body of work by the philosopher Marcus Giaquinto, who, more than anyone else, has taken pains to clarify both the senses in and the extent to which diagrammatic reasoning figures in justificatively significant ways in mathematical reasoning. He refines the description of the types of justificative contributions diagrammatic reasoning can make and carefully investigates its justificative limits, particularly in analysis. Finally, he considers the difficult question of how diagrammatic reasoning fits with the explanatory aims of proof.

In section 6, I summarize and conclude. I find that much recent work continues the dominant view of the last three centuries in its view of the place of empirical reasoning in mathematics. The chief novelties concern refinements in our understanding of the nature and role(s) of diagrammatic reasoning and of the proper place of formalization in proof.

2 Empirical Reasoning in Mathematics: Historical Background

The use of empirical evidence and broadly inductive reasoning in mathematics are by no means new phenomena in the history of mathematics. They have, in fact, been a major source of concern for more than two millenia. Archimedes gave an important early defense of the usefulness of empirical methods (particularly, mechanical methods) in solving geometrical problems. He did not see them as altogether supplanting classical methods, but he did see them as useful means of discovery (both of truths and of proper demonstrations).

> ...I have thought fit to...explain...a method, with which...you[1] will be able to make a beginning (*aphormē*) in the investigation (*thēorein*) by mechanics...in mathematics....[I]nvestigation by this method does not amount to actual proof (*apodeixeōs*); but it is...easier to provide the proof when some knowledge of the things sought has been acquired by this method rather than to seek it with no prior knowledge.
>
> [Archimedes], pp. 221, 223[2]

Archimedes thus acknowledged a role in mathematics for reasoning other than proof. Though proof might yield knowledge in its highest form(s), other forms of reasoning might nonetheless yield lesser knowledge, and also aid in the development of genuine proofs.

It was thus common for non-demonstrative methods of reasoning to be classified as methods of invention or discovery, as distinct from methods of justification proper. The terminology is potentially misleading, though, in that it suggests that non-demonstrative reasoning was not viewed as justificative. This is not true. Discovery of a proposition was discovery of its truth. Discovermental methods were thus generally taken to have justificative value, but not so great as that typical of demonstration. This reflected the Aristotelian "causal" ideal of knowledge.

> We...possess unqualified scientific knowledge of a thing...when we know the cause on which the fact depends as the cause of the fact and...that the fact could not be other than it is.
>
> [Aristotle 1908], 71b8–b11

The broad division of reasoning into justificative and discovermental varieties is thus of ancient origin. Philosophers and mathematicians generally marked it and so too did other disciplines in which reasoning featured prominently. As Cicero remarked: 'every careful method of arguing has different divisions—one of discovering, one of deciding' ([Cicero 1894], vol. 4, *Topics*, pp. 459–460)).[3] Methods of the former type were termed *arts of discovery* (*artis inveniendi*), methods of the latter type *arts of justification* (*artis iudicandi*).

16th and 17th century algebraists embraced the distinction. Viète, Descartes and Wallis all stressed the different purposes served by discovery-oriented and demonstrative reasoning and the different standards to which they are rightly held. They also saw in it a reflection of the ancient

---

[1] The person addressed was Eratosthenes.

[2] The reader should bear in mind that the text for Archimedes' *Method* was only rediscovered at the turn of the 20th century and cannot generally be assumed to have been available to earlier thinkers.

[3] Cicero attributed the distinction to Aristotle, who urged a similar division in the *Topics*. It was also marked in Roman law, which distinguished evidence appropriate to the detention, questioning and/or charging of a suspect (*investigatio*), from evidence appropriate to her conviction (*demonstratio*).

distinction between analysis and synthesis. The classical statement of this distinction was given by Pappus of Alexandria in his *Treasury of Analysis*.

> ... in analysis we suppose that which is sought to be already done, and inquire what it is from which this comes about ... until, by retracing our steps, we light upon something already known or ranking as a first principle ...

> ... in synthesis ... we suppose to be already done that which was last reached in analysis, and arranging in their natural order as consequents what were formerly antecedents and linking them one with another, we finally arrive at the construction of what was sought ... [Pappus 2000], pp. 597, 599[4]

They believed, moreover, that symbolic algebra (what they commonly referred to as the *analytic art*) was analysis *par excellence*. It was, in the first instance, a method of discovery widely believed to be efficient and reliable. Wallis described it as 'plain, obvious and easy' ([Wallis 1685], p. 298) and as yielding results that were readily verifiable by classical means (cf. [Wallis 1685], p. 305). In a similar spirit, Leibniz praised it as "a great aid in shortening thought and also in discovery" ([Leibniz 1707], p. 436) and claimed that it could not "lead us into error" (*loc. cit*).

Others (e.g. MacLaurin, cf. [MacLaurin 1742], pp. 47, 49) conceded its usefulness but also emphasized that it was not the justificative equal of classical (synthetic) method.

> In general, it must be owned, that if the late discoveries [of Wallis', in his *Arithmetica Infinitorum* (1656)] were deduced at length, in the very same method in which the ancients demonstrated their theorems, the life of man could hardly be sufficient for considering them all ... [MacLaurin 1742], p. 49, brackets mine[5]

Still, though

> [m]athematicians [may] indeed abridge their computations by the supposition of infinites, ... [they] cannot be too scrupulous in admitting of infinites, of which our ideas are so imperfect. [MacLaurin 1742], pp. 46–47, brackets mine

Still others less qualifiedly opposed the use of algebraic methods. These included Hobbes ([Hobbes 1839–45], vol. 1, pp. 311–312)[6] and, at times, Newton.

> Equations are Expressions of Arithmetical Computation, and properly have no Place in Geometry ... Multiplications, Divisions, and such sort of Computations, are newly received into Geometry, and that unwarily, and contrary to the first Design of this Science. [Newton 1720], p. 229

---

[4] Aristotle made a similar distinction earlier [Aristotle 2000] (cf. III.3, 1112b 15–27). There is also a statement of uncertain origin in the manuscript sources for Book XIII of Euclid's *Elements*. See Heath's historical note on Book XIII ([Euclid 1956], vol. 3, pp. 438–439 and his commentary on Propositions I–V ([Euclid 1956], vol. 3, pp. 441–442) for more on this.

[5] For similar statements during the same period, see that by Christian Wolff ([Wolff 1739], preface, v–vi). A representative statement a half century earlier was that by Johann Christoph Sturm ([Sturm 1700], preface, articles XIV, XX), a half century later that by Charles Hutton ([Hutton 1795], vol. 1, p. 107).

[6] Hobbes and Wallis carried on a well-known dispute concerning the legitimacy of algebraic methods. An interesting account of this dispute is given by Jesseph in [Jesseph 1999].

To counter such charges, algebraists of the 16th and 17th centuries argued that classical geometry would not have been the success it had been had not ancient geometers made regular use of algebraic methods in arriving at their discoveries—a use they then tried to conceal (cf. Descartes [Descartes 1620–28], Rule IV, [Wallis 1685], ch. II, 3, p. 290 and Viète [Viète 1591], p. 27). Wallis thus wrote of Apollonius that

> ... we may well give him the name of *Magnus Geometra*, and look upon him as a man of a prodigious reach of Phansy, if we can think it possible that he could discover all those Propositions, and perplex demonstrations, in the same order they are there delivered, without some such Art of Invention, as what we now call Algebra.
>
> [Wallis 1685], p. 290

This is the traditional view of analytic or algebraic method and its place in classical geometry. But why classify it as empirical in character? Generally speaking, the reason is that it often relied on inductive forms of reasoning. Archimedes and various other ancient and medieval mathematicians appealed to analogies between mechanics and geometry, and algebraists of the early modern era (i.e. late 15th–17th centuries) often used a form of inductive reasoning they associated with the *Principle of Exhaustion*—the idea that if two quantities can be made to differ from each other by less than any assignable amount, they can then be treated as equal.[7]

They often reasoned by analogy as well, extending laws proven for finite magnitudes and collections to infinite generalizations of them. One example of this is Wallis' extension of the law for sums of (finite) arithmetic progressions to sums of infinite arithmetic sequences ([Wallis 1656], p. 155; [Wallis 1685], pp. 285–287, 297, 305–306). Overall the method was inductive and Wallis described it as such.

> The simplest method of investigation, in... various problems... is to exhibit the thing to a certain extent, and to observe the ratios produced and to compare them to each other; so that at length a general proposition may become known by induction.
>
> [Wallis 1656], p. 13

Following this procedure, Wallis "retrieved" the classical law for the area of the triangle (viz. $A = \frac{bh}{2}$)[8] by applying Cavalieri's Method of Indivisibles. He resolved the triangle into a progression of uniformly thin rectangles, reasoned inductively that, compositely, they would come ever closer to matching the interior of the triangle as the individual rectangles became ever thinner, and reasoned analogically to their sum by extending the formula for arithmetic progressions to the "infinite" case.[9] He thus arrived at $\frac{0 \times \frac{h}{\infty} + b \times \frac{h}{\infty}}{2} \times \infty$, thence $\frac{\frac{bh}{\infty} \times \infty}{2}$, thence $\frac{bh}{2}$, the classical law for the area of a triangle.[10]

Wallis repeated the same general form of reasoning—finding the appropriate type of progression and analogically inferring its sum—to obtain solutions to a variety of other quadrature

---

[7] The Principle of Exhaustion was well-known and widely used in antiquity (cf. Def. IV, Bk. V of *The Elements*, which is used to prove another variant in Prop. I, Bk. X). Democritus is generally thought to have been the first to formulate it, though there is evidence that Hippocrates formulated and used it too. Archimedes attributed the first "proof" of it to Eudoxus.

[8] Cf. Wallis [Wallis 1656], Proposition 3; [Wallis 1685], pp. 285–287.

[9] Cf. Wallis [Wallis 1656], pp. 13–15; [Wallis 1685], pp. 280–290.

[10] Cf. Wallis [Wallis 1656], pp. 14–15, [Wallis 1685], pp. 285–287.

and cubature problems as well.[11] There was immediate, sharp criticism from both philosophers and mathematicians, most notably Hobbes, Huygens and Fermat. All three criticized Wallis for his use of inductive reasoning, characterizing it variously as unclear, uncertain, unnecessary and insufficient.[12] Wallis' chief response, stated in a reply to Fermat, was to reaffirm the ancient *two-methods* (discovery vs. demonstration) distinction. His aim, he said, was not primarily one of "Demonstrating things already known" ([Wallis 1685], p. 305), but "to shew a way of . . . finding out . . . things yet unknown" (*ibid.*).

Thus Wallis, and algebraists of 16th and 17th centuries generally, embraced the ancient distinction between an *ars inveniendi* and an *ars iudicandi*. In the next section we'll consider a recent proposal by Arthur Jaffe and Frank Quinn to divide mathematical labor in a roughly similar way between discovermental and more rigorously demonstrational components. We'll consider as well a widely-discussed argument (concerning the computer-assisted proof of the four-color theorem) that empirical reasoning may sometimes be the only humanly feasible means of justification.

3 Empirical Reasoning in Mathematics: Recent Proposals

3.1 Empirical Reasoning and Epistemic Productivity

Arthur Jaffe and Frank Quinn recently offered a new incarnation of the division of methods theme. They acknowledge the benefits of rigor as a constraint on mathematical reasoning and so affirm the virtues of strict proof. It has 'brought to mathematics a clarity and reliability unmatched by any other science' ([Jaffe/Quinn 1993], p. 1). This notwithstanding, it has also at times made progress in mathematics 'slow and difficult' (*loc. cit.*).

Too strong an emphasis on proof may thus be more of an impediment than an aid to the development of new mathematical knowledge. To become more efficient, they suggest, mathematics should follow the lead of physics and permit freer use of intuitive methods of thinking. And this despite the fact that by means of such more liberal reasoning, mathematicians may occasionally go beyond the bounds of what can be strictly established (*op. cit.*, p. 2).

Freer, more efficient 'theoretical' methods[13] should be used to generate initial hypotheses and to outline justifications. These hypotheses and justifications should then be converted into rigorous reasoning by mathematicians particularly skilled in such work.

In Jaffe and Quinn's view, the role of rigorous proof in mathematics is 'functionally analogous to the role of experiment in the natural sciences' (*loc. cit.*). They thus foresee two types of mathematical research—a more intuitive and speculative 'theoretical' type aimed at efficient discovery, and a more rigorous, conventional type aimed essentially at confirmation. The latter is intended (i) to 'ensure the reliability of mathematical claims' (*loc. cit.*), and (ii) to yield, at least occasionally, 'new insights and unexpected new data' (*loc. cit.*).

---

[11] Cf. Wallis [Wallis 1656], *passim*; [Wallis 1685], pp. 285–290, 290–298.

[12] There were admirers too, of course. In addition to those mentioned above, these included Newton (at times) and, nearly two centuries later, Charles Babbage. Such influential admirers notwithstanding, Wallis' inductive and analogical methods did not change the norms of mathematical practice. Succeeding generations of mathematicians for the most part viewed them as falling short of ideal norms of rigor, certainty and precision.

[13] So called because they resemble thinking in *theoretical* physics.

Dividing mathematics in this way, they suggest, may bring about the same rapid advancement in it as it did in physics. This, at any rate, is their hope. Their proposal goes farther, however, in proposing that the division of methods be incorporated into the institutions of professional mathematics, specifically, into its methods of training and its system of rewards.

3.1.1 *The Division of Mathematical Labor*

Fundamentally, Jaffe and Quinn's proposal is one of divided labor and, as such, it embodies the same strategic ideas that schemes of divided labor generally embody. Specifically, it proposes to increase productivity through increased specialization.[14]

Adam Smith's classical statement of the benefits of divided labor maintained that it 'occasions, in every art, a proportionable increase of the productive powers of labor' ([Smith 1776], Bk. I, ch. 1, para. 4). By dividing production into small tasks and 'dedicating' each individual worker to the repeated execution of a single task (or a small number of such), productivity is increased. This is so because the tasks are 'smaller' and, so, more fully within the range of the worker's competence, and because the familiarity that comes from repetition increases the worker's proficiency in performing them. The result is a better product more efficiently produced. Or so the thinking goes.

At bottom, this is what Jaffe and Quinn place their faith in. Mathematics will be divided into specialists in speculation or conjecture (practitioners of 'theoretical' methods in mathematics) and specialists in confirmation (those who convert 'theoretical' reasonings into proofs). With this increased specialization will come increased proficiency, and with increased proficiency, increased productivity. Mathematical knowledge will both improve in quality and grow faster.

There are grounds for caution, however. One is the general lack of evidence for the claims and assumptions that Jaffe and Quinn make. Perhaps the most basic of these is the assumption that the recent growth of knowledge in physics is greater than that in mathematics. Even granting this assumption, though, questions remain. For example, do we know that it's the division of physics researchers into theoretical and experimental that's responsible for its superior rate of growth? And, supposing that it is, is what makes that division effective its separation of the speculative (roughly discovermental) and confirmatory (roughly justificative) tasks? Or might it instead be the increase in financial support for physics (and applied mathematics) research fueled by the race to develop atomic weapons and energy, or the race to put humans on the moon? Is the task of developing a proof for a conjecture in mathematics relevantly similar to that of confirming a physical conjecture? Is it generally as *easy* to devise a proof for a true mathematical conjecture as it is to design and conduct a confirmatory experiment (or body thereof) for a true physical conjecture? Such questions are not easily answered.

In addition to these uncertainties, there is another that may be of even greater significance. It has to do with the costs of dividing labor. Even Smith, the champion of divided labor, acknowledged these costs and that they are considerable.

> In the progress of the division of labour, the employment of the far greater part of those who live by labour . . . comes to be confined to a few very simple operations,

---

[14] It may also be that Jaffe and Quinn believe something like what Archimedes expressed when he claimed that it's 'easier to provide the proof when some knowledge of the things sought has been acquired' ([Archimedes], p. 223). The informal sketches of justifications produced by 'theoretical' mathematics may give the rigorist something to directly build on.

frequently to one or two.... The man whose whole life is spent in performing a few simple operations, of which the effects are perhaps always the same, or very nearly the same, has no occasion to exert his understanding or to exercise his invention in finding out expedients for removing difficulties which never occur. He naturally loses, therefore, the habit of such exertion, and generally becomes as stupid and ignorant as it is possible for a human creature to become. [Smith 1776], Bk. V, ch. 1, para. 178

A more disheartening view of the effects of divided labor would be hard to imagine. Consistent division of labor, on this view, impedes the worker's ability to find fulfillment in her work. In short, it alienates her from her work.

In addition to concerns regarding the accuracy of their comparison of mathematical and physical research, then, their proposal also raises larger moral and social concerns that have not been adequately addressed.

3.2 Empirical Reasoning vs. Proof

We saw in section 2 how broadly inductive reasoning and reasoning from analogy have been used as methods of discovery in mathematics since ancient times. In all cases, however, discovery and justification (at least ultimate or ideal justification) were conceived as distinct tasks. Discovermental arguments were therefore to be only temporary substitutes for proper demonstrations.

In recent times a different role for empirical reasoning in mathematics has been suggested. It is no longer seen as a mere propædeutic to proof, but an alternative to it—in some cases, a necessary alternative.

Views of this type have been inspired by the appearance of extremely long and/or complex proofs. A well-known example is the widely discussed computer-assisted solution of the four-color problem developed by Kenneth Appel and Wolfgang Haken in 1977.

This proof is so large as to seemingly prohibit the type of step-by-step surveyal commonly assumed for mathematical proof. It has thus given rise to a variety of philosophical questions concerning whether proof is indeed the appropriate standard to adopt for mathematical justification.

Thomas Tymoczko wrote a widely read discussion of these issues in his 1980 paper 'The Four-Color problem and its mathematical significance'. He argued that the solution of the four-color problem offered by Appel and Haken (hereinafter, the AH argument or proof) forced a reconsideration of traditional conceptions of mathematical proof and knowledge. His main claim was that

... if we accept the 4CT as a theorem, we are committed to changing the sense of 'theorem', or, more to the point, ... the sense of the underlying concept of "proof".

... use of computers in mathematics, as in the 4CT, introduces empirical experiments into mathematics. Whether or not we choose to regard the 4CT as proved, we must admit that the current proof is no traditional proof, no a priori deduction of a statement from premises. It is a traditional proof with a ... gap, which is filled by the results of a well-thought-out experiment. This makes the 4CT the first mathematical proposition to be known a posteriori and raises again for philosophy the problem of distinguishing mathematics from the natural sciences. [Tymoczko 1979], p. 58

Tymoczko's argument had three main components. The first was an analysis of the traditional conception of proof which identified three key characteristics—convincingness, surveyability and formalizability. The second was an argument to the effect that, of the three ingredients just mentioned, surveyability was the most basic. The third was the claim that the AH proof is not surveyable.

I'll now consider the argument more carefully, focusing, as Tymozcko did, on the two latter components—the claims that surveyability is central to the standard conception of proof and that the AH proof is not surveyable.

In calling a proof convincing, Tymoczko meant that it had the capacity to move a rational prover[15] to belief in its conclusion. This in turn required that the proof be surveyable—that is, that it be capable of being 'looked over, reviewed, verified' (*op. cit.*, p. 59) by human provers, specifically, by members of the human mathematical community (*op. cit.*, p. 60; [Tymoczko 1980], p. 132).

Such a conception of proof was, in Tymoczko's view, seriously at odds with the computer-assisted proof of the 4CT. This proof established the existence of a formalized proof of the 4CT, a formalized proof so large, however, as to debar human survey (cf. [Tymoczko 1979], p. 58). The AH proof of the 4CT thus substituted, at certain point(s), the *results* of unsurveyably long computer runs for the formal computations to which they correspond. So, at any rate, Tymoczko claimed.

To evaluate this reasoning, we need to keep two distinct items separate. One is the actual argument given—that is, written down—by Appel and Haken. This argument is surveyable and was indeed surveyed by a number of mathematicians before being published. I'll call this the *compressed argument*, since it replaces the details of certain computations with (descriptions of) their results.

The unsurveyable argument, on the other hand, is the argument that would result from setting out the suppressed details of the compressed argument in full. Call this the *decompressed argument*.

Judged by Tymoczko's standards, neither the compressed nor the decompressed argument is a proof in the traditional sense. The decompressed argument is not a proof because it is not surveyable. The compressed argument is not a proof because it lacks the explicitness—the full disclosure of premises and inferences—traditionally required of proof. It is

> ... like a mathematical proof where a key lemma is justified by an appeal to the results of certain computer runs or, as we might say "by computer." This appeal to computer, whether we count it as strictly a part of a proof or as a part of some explicitly non-proof-theoretic component of mathematical knowledge, is ultimately a report on a successful experiment. It helps establish the 4CT (actually, the existence of a formal proof of the 4CT) on grounds that are in part empirical.   [Tymoczko 1979], p. 63, brackets added

In Tymoczko's view, then, to accept the compressed argument (i.e. the AH proof) as adequate mathematical justification for the 4CT requires changing the traditional conceptions of proof and

---

[15] By a 'prover', I mean not only one who discovers a proof, but also one who grasps it.

mathematical knowledge (cf. [Tymoczko 1979], p. 58). In particular, it requires relinquishing or modifying each of the following (cf. *op. cit*, p. 63):

(i) all mathematical theorems are known *a priori*
(ii) mathematics has no empirical content
(iii) mathematics relies exclusively on proof and makes no use of experiment, and
(iv) mathematical theorems are certain to a degree that no theorem of natural science can match.

Tymoczko thus took the acceptance of the AH argument for the 4CT to be both a novel and a philosophically significant development.

3.3 The Philosophical Significance of the AH Argument

Items (i)–(iii) center on the question of the compatibility of the commonly supposed *a priori* status of mathematical judgments with the use of broadly empirical considerations in mathematical reasoning. To get a better idea of the plausibility of Tymoczko's claim that (i)–(iii) must be relinquished or modified, we must first get clearer on what he means (or might or should mean) by 'mathematical theorem', 'known *a priori*', 'empirical content' and 'empirical justification'.

Tymoczko assumes that the traditional view of theorems is essentially this:

(i-aux): A proposition is rightly classified as a theorem if and only if there is a known proof of it.

From (i-aux) and the supplementary claim that

(i-supp): If there is a known proof for a proposition, then it (the proposition) is known *a priori*,

(i) follows.

In truth, though, the claim that (i-aux) represents the traditional view is doubtful, at least if it is taken to imply that a known proof is a proof that has been surveyed in all its details. There is nothing in the traditional conception of proof to prohibit joint enterprises where a resolutive task is broken up into parts, and the various parts given over to different persons or groups in such a way that, in the end, no one participant will have surveyed the entire joint proof. The product of such an undertaking could still count as a proof so long as it was known that each part was correctly executed and that, taken together, the several parts solve the original problem.[16]

If this is right, a more reasonable condition than (i-aux) would be

(i-aux'): A proposition is rightly classified as a theorem if and only if it is known to have a proof.

(i-aux'), however, implies (i-aux) only on a constructive understanding of existence, and such an understanding is not part of the traditional conception of proof.

---

[16] A recent well-known example of such a joint undertaking is the classification of the finite simple groups.

Even if it were, though, problems would remain. For it's only the stricter forms of constructivism that require actual exhibition (or survey) of an object as adequate justification of its existence. More liberal varieties allow existence to be established by the provision of suitably clear descriptions of methods the (perhaps idealized) execution of which can be seen to guarantee an exhibition of a thing of the type claimed to exist.

(i-aux) thus has little to recommend it as a traditional condition of theoremhood. Related remarks apply to (i-supp) and condition (i). (i-supp) must give way to

(i-supp'): A proposition known to have a proof is known to have an *a priori* justification[17],

and (i) to

(i') A proposition rightly classified as a theorem is known to have an *a priori* justification.[18]

The question then becomes whether the compressed argument for the 4CT comports with (i'), and it seems that it does. It (the compressed argument) can reasonably be taken to show that there is a proof of the 4CT, and this together with (i-supp') implies that the 4CT has an *a priori* justification. This being so, the compressed argument would not seem to demand a revision of what is in truth the traditional view—namely, that whether or not they are humanly graspable, there nonetheless exist *a priori* justifications for mathematical theorems.

Let's now consider (iii). As stated, it's too vague to assess. Clarified in the way Tymoczko's suggests, however, it runs counter to the traditional conception of proof. This conflict is due to a strong property of self-sufficiency that Tymoczko attributes to the traditional conception of proof. A proof is, he says,

> . . . an exhibition, a derivation of the conclusion, and it needs nothing outside of itself to be convincing.   [Tymoczko 1979], p. 59

Is such self-sufficiency characteristic of the traditional conception of proof?

There are reasons to think that it isn't. Conviction often, perhaps typically, requires not only proofs *per se* but *reflections on* them. These may be as simple as the application of certain checking procedures or as complex as reflections on the meaning and/or plausibility of ideas and principles used in a proof. Whatever their particular character, reflections on proofs and their components are often vital to acceptance of their conclusions. At the same time, though, they do not strictly belong to the proofs themselves, at least not as proofs are ordinarily thought of these days.[19]

Paul Teller ([Teller 1980]) made a similar point in arguing that Tymoczko was wrong to regard surveyability as a necessary condition of proof. It's not a property of proofs *per se*, he said, but a property some proofs have and others lack. It signifies not the extent to which an argument actually *is* a proof, but the extent to which it can be verified as such. It's thus a matter of degrees. Some proofs are so simple and easy that little knowledge and training is required

---

[17] Briefly, a proposition *p* will be said to *have* an *a priori* justification if there is a warrant for it that does not depend on taking the contents of any experience as evidence.

[18] Later we'll see reason to think that this might even be strengthened to something like 'A proposition known to have a proof is known *a priori*'.

[19] This is in part due to the fact that modern understandings of the notion of *axiom* do not typically retain the classical requirement of self-evidence.

for their verification. Others 'are so complicated that only a few mathematicians' ([Teller 1980], p. 798) can verify their correctness. Still others are 'out of the reach of the best [verifiers]' (*loc. cit.*). Despite this variation in their verifiability, however, all may be genuine proofs. The AH proof of the 4CT thus represents at most an extension of our means of surveying proofs, not a change in our concept of proof.

All in all, it seems wrong to say that, on the traditional conception, proof is a unit of reasoning that needs nothing outside itself to be convincing. (iii) is therefore not a basic tenet of the traditional conception of proof. Conviction in mathematics often involves not only proofs but judgments *about* them—judgments which do not themselves belong to the proofs in question. Neither has conviction traditionally been seen as requiring proof, as the discussion of section 2 makes clear. In sum, proof has not traditionally been regarded as either necessary or sufficient for conviction.

Tymoczko and his claims aside, the AH proof of the 4CT raises other interesting questions regarding proof and mathematical knowledge. One of these concerns what if any difference there might be between actually having an *a priori* warrant for a proposition and simply having evidence that one exists.

Suppose that I become convinced, on evidence I know to be reliable, that a certain statement κ is provable from certain *a priori* warranted statements $\pi_1, \ldots, \pi_n$ whose warrants I know to be reliable. Suppose, in addition, that I have good reason to believe that the shortest proof of κ from these statements is not humanly surveyable.

Under such circumstances, it would not be humanly possible to have a proof of κ from the statements mentioned, and I might even know or believe this to be so. This notwithstanding, I might still know that *there is* such a proof. More exactly, under the circumstances described, my knowledge that the statements in $\pi_1, \ldots, \pi_n$ are *a priori* warranted, and my knowledge that κ is provable from $\pi_1, \ldots, \pi_n$ could assure me that κ cannot be empirically falsified. This being so, my attitude towards κ would at least be similar to what it would be were I actually to have a proof of κ. Applying this to the case of the AH argument, we see that though it may not itself be a proof of the 4CT, it might still provide a warrant similar in *a priori* character to that provided by a proof.[20]

But does it provide mathematical knowledge? This is a subtler and more difficult question. I can read about a proof in a newspaper or the announcements section of a journal and learn that a certain theorem (e.g., Fermat's Last Theorem, FLT, for short) has been proved. When the publication and my reading of it are both properly judged to be reliable, this learning can amount to knowledge. I can thus know that FLT is provable, and, supposing I know that the methods of proof used are reliable, I can also know FLT.

I can know all these things—I and many people in fact do—and still not have *mathematical* knowledge of FLT. The reason, roughly, is that in knowing what I know, I don't know the *mathematical reasons* for FLT. Neither do I know how (i.e. by what reasoning) they guarantee it. So, even though I might know that certain propositions (say, the axioms of second-order Peano Arithmetic) are true and that certain inferences are sound, and know that from these propositions and inferences a proof of FLT can be fashioned, I do not thereby gain mathematical knowledge of FLT.

---

[20] See Peressini [Peressini 2003], Fallis [Fallis 1996] and [Fallis 1997] for related discussions.

This raises a question concerning the AH proof of the 4CT—namely, whether it provides enough insight into the mathematical reasons for the 4CT to give genuinely mathematical knowledge of it—either to those who designed it (i.e. Appel, Haken and Koch) or to others with similar substantial knowledge of the program believed to be implemented by the computing device. I would estimate that it does, but my main point is that the mere fact that at various points the proof is turned over to a computer would not prohibit its being an adequate statement of mathematical reasons for its conclusion.[21]

A key distinction here is that between *program* and *implementation of* program. Knowledge of the computer-assisted proof of the 4CT can give sufficient knowledge of reasons for the 4CT only if two conditions are satisfied. The first is that the knower have extensive enough knowledge of the reducibility program to allow him properly to judge that it is an adequate program for doing what it's intended and/or believed to do. The second is that the knower should have sufficient knowledge of the implementation to warrant judgment that the aforementioned program is indeed the program executed by the machine whose output is used.

It seems the principal novelty of the AH proof concerns the second condition. In ordinary computations, knowledge that a computation implements a program comes through survey of the computation. In the case of the AH proof this is not and perhaps can not be the case. Whether this introduces an empirical element into the AH proof that is fundamentally unlike that which figures in more ordinary proofs may be doubted. This notwithstanding, the question of whether the type of knowledge an informed knower can have of the program that figures in the AH proof and of its implementation can amount to proper knowledge of the *reasons* for the 4CT is an important one, and one that deserves more careful discussion than I can give it here.[22]

4 Formalization and Rigor

The prevailing view of proof sees rigor as a necessary feature of proof and formalizability as a necessary condition of rigor. On this view, a rigorous proof is one that can be known not to conceal substantive (i.e. non-logical) information. Its inferences can be seen to be valid solely by appeal to logical relations between concepts and not to their 'senses' or 'contents'. As Pasch put it:

> ... the process of inferring must always be independent of the sense of ... concepts just as it must be independent of diagrams. It is only relations between ... concepts that should be taken into account in the propositions and definitions that are dealt with. In the course of the deduction, it is certainly legitimate and useful, though by no means necessary, to think of the reference of the concepts involved. If it is indeed necessary to so think, the defectiveness of the deduction and the inadequacy of the ... proof is thereby revealed unless it is possible to remove the gaps by modification of the reasoning used.
>
> [Pasch 1912], p. 98[23]

---

[21] See Fallis [Fallis 2003] for an interesting discussion of the phenomenon of "gaps" in the statements of "reasons" offered by proofs.

[22] For an indication of one direction such further discussion might take, see Fallis [Fallis 2002] where broad questions concerning the fit between the goals of mathematics and its methods are considered.

[23] My translation. The same basic idea was defended in the middle of the eighteenth century by J. H. Lambert (cf. [Lambert 1766], p. 162).

Rigorous proof, on this view, is reasoning all of whose inferences track purely logical relations between concepts. In the late nineteenth and early twentieth centuries, syntactical criteria for such relations were developed and these have become the basis for the currently prevailing view of formalization.

The reasoning behind this view is straightforward: (i) proper proofs are proofs that either are or can readily be made rigorous; (ii) proofs that are or can readily be made rigorous are formalizable; therefore (iii) all proper proofs are formalizable. Call this argument the *common argument* and its conclusion the *common view*.

Both the view and its argument seem dubious. Mathematical proofs are not commonly formalized, either at the time they're presented or afterwards. Neither are they generally presented in a way that makes their formalizations either apparent or routine. This notwithstanding, they are commonly presented in a way that *does* make their *rigor* clear—if not at the start, then at least by the time they're widely circulated among peers and/or students. There are thus indications that rigor and formalization are independent concerns.

This is not the common view, however. On that view, non-formalized proofs are typically close enough to formalized proofs to make the fact of formalizability clear and the remaining work of formalization routine. Saunders Mac Lane maintained such a view.

> A mathematical proof is rigorous when it is (or could be) written out in the first order predicate language $L(\in)$ as a sequence of inferences from the axioms ZFC, each inference made according to one of the stated rules.... practically no one actually bothers to write out... formal proofs. In practice, a proof is a sketch, in sufficient detail to make possible a routine translation of this sketch into a formal proof.
>
> [Mac Lane 1986], p. 377

This, as I said, is the common view. But common or not, not everyone agrees, and the dissenters include some who have great experience in the work of formalization. The artificial intelligence researcher John A. Robinson (an expert in automated theorem-proving) is a case in point. His experience with formalization causes him to remark that it is often "surprisingly difficult" and only occasionally a routine matter (cf. Robinson [Robinson 1997], p. 54).

> In most cases it requires considerable ingenuity, and has the feel of a fresh and separate mathematical problem in itself. In some cases... formalization is so elusive as to seem to be impossible. [Robinson 1997], p. 54

Still more importantly he sees *standard* formalization—that is, formalization of the usual reduction-to-syntactically-presented-rudimentary-logical-inference variety—as often undesirable even if manageable. The reason is that a prime goal of proof is explanation (cf. [Mac Lane 1986], pp. 378–79 and [Robinson 2000], p. 277) and standard formalization can obscure explanatory connections. Indeed, Robinson believes that it "typically destroys all traces of the explanatory power of the informal proof" (Robinson [Robinson 1997], p. 56; see also Robinson [Robinson 2000], pp. 293–94).

That this is so is due to the fact that standard formalization breaks an informal proof down into many artificially small steps of reasoning. Explanation, on the other hand, is typically carried by "large-scale, high-level" ([Robinson 2000], p. 279) "architectural" patterns of reasoning, patterns which may be obscured when embedded in a mass of rudimentary logical inferences.

> Too much detail causes difficulty in viewing the big picture. One cannot see the forest for the trees.   [Robinson 2000], p. 479[24]

Robinson believes that real mathematical proofs are essentially *performances*, and not "structured static texts" ([Robinson 2000], p. 281) in which nothing *happens*. To focus on formalized proofs is to view proofs as such texts, and to do this is like experiencing music only by reading musical scores (cf. *loc. cit.*). Scores are important, but there is more to a 'living' piece of music than its score.

In the same way, there is more to a real proof than its formalization. A formal proof is "only the score, only the script, only the instructions for producing the real proof" (*loc. cit.*). Indeed, it's typically not even that, since it's generally an afterthought rather than a guide to proof.[25]

Robinson supports this view by outlining a form of performative experiment—an introspective experiment in which he looks for theorems that he (a) understands, but which he (b) finds incredible and for which he (c) possesses a proof that is within his power to understand with a reasonable effort. Once a selection is made, the "experiment" consists in learning how to 'perform' the given proof while monitoring the process to detect the "crucial moments in the proof" (*op. cit.*, p. 282) when his attitude turns from incredulity to acceptance. Robinson illustrates these ideas with an example from number theory, namely, Erdős's proof of Bertrand's Conjecture that for every positive integer $n > 1$, there is at least one prime p such that $n < p < 2n$.[26]

The result of such an experiment, Robinson suggests, is that a proof comes to be stored as "a collection of relatively few leading ideas dealing with interesting . . . phenomena" ([Robinson 2000], p. 291). Increasing familiarity with these phenomena eventually gives them "an aura of certainty" and they become "established resources" which can be "triggered at will". The mind, Robinson says, is "hungry" (*op. cit.*, p. 292) for such "key ideas" that capture the gist of a proof. Fixing our attention on them provides a better grasp of the overall plan of a proof and, with the overall plan before it, the mind can then "understand" (*loc. cit.*) the proof and not get lost in its details. Details can thus be the enemy of understanding and blind us to the overall architecture of a proof (cf. Robinson [Robinson 2000], p. 292). The central work to be done, then, is to identify those larger patterns of inference that guide actual mathematical practice and understanding and make a formal protocol (or different local formal protocols) of them.

---

[24] There is an unmistakable parallel here with Poincaré, who campaigned against the "logicization" of mathematical reasoning for similar reasons. See Poincaré [Poincaré 1905], ch. I. See also Detlefsen [Detlefsen 1992] for a fuller discussion of these ideas of Poincaré's.

[25] For related though in certain respects broader discussions of mathematical *activities*, see Giaquinto [Giaquinto 2005a] and Rota [Rota 1997]. In addition to proof, the former identifies discovery, justification and explanation as other key epistemic activities. The latter describes mathematical practice as concerned with such things as investigations, intuitions, conjectures and verifications, all of which are taken to be different from, albeit related to, proof. Rota also discusses the axiomatic method and how it can sometimes conceal explanatory connections, and offers a few observations concerning the processes through which mathematical reasoning is refined.

[26] The conjecture was first proved in 1850 by Chebychev. Erdős's proof is more picturesque than Chebychev's, however, and (therefore?) more memorable, or so Robinson argues (cf. Robinson [Robinson 2000], pp. 283–89).

Robinson notes certain difficulties involved in attempting to do this (cf. Robinson [Robinson 2000], pp. 293–294). There is, though, a larger possibility that Robinson seems to overlook—namely, that there may simply be no family of perceivable entailments that (a) are individually "larger" than those of rudimentary logic and (b) offer adequate protection from the admission of dangerous gaps in mathematical reasoning. This, at any rate, is a central problem confronting Robinson's and similar proposals.

There is also a question concerning a possible deeper relation between explanatory content/character and rigor. Traditionally at least, mathematical reasoning has been taken to be at its most rigorous when it is also at its most potently explanatory. We're most certain to avoid gaps in reasoning when premises *explain* conclusions.

Hilbert suggested such a view when he wrote

It is an error to believe that rigor in proof is an enemy of simplicity. On the contrary we find it confirmed by numerous examples that the rigorous method is at the same time the simpler and the more easily comprehended. The very effort for rigor forces us to find out simpler methods of proof.   [Hilbert 1902], p. 441

In any event, it seems at least possible to think of the rigor as linked to explanatory transparency—an inference being rigorous to the extent that its premises can be seen to *explain* its conclusion.[27] The greater such explanatory transparency, the more confident we can be that unrecognized information has not been used to connect a conclusion to premises in ways that matter. To the extent, then, that formalization decreases explanatory transparency, it also decreases rigor. A reexamination of the commonly presumed connection(s) between rigor and formalization would thus seem to be in order.

5 *Visualization and Diagrammatic Reasoning in Mathematics*

The common view of diagrams and their role in proof has for some time been that they are merely heuristic devices, useful instruments to aid the discovery, formulation and/or the intuitive comprehension of proofs, but lacking any genuinely justificative role in proof. Leibniz stated the essentials of this view as follows.

. . . geometers do not derive their proofs from diagrams, though the expository approach makes it seem so. The cogency of demonstration is independent of the diagram, whose only role is to make it easier to understand what is meant and to fix one's attention. It is universal propositions, i.e. definitions and axioms and theorems which have already been demonstrated, that make up the reasoning, and they would sustain it even if there were no diagram.   [Leibniz 1981], p. 360[28]

---

[27] For further recent discussions of the role of explanation in mathematics, see Mancosu [Mancosu 2000] and [Mancosu 2001], Tappenden [Tappenden 2005] and Mancosu, Jorgensen *et al.* [Hafner/Mancosu 2005].

[28] The *New Essays* were published posthumously in 1765. It was written over an extended period of time and completed sometime between 1709 and Leibniz' death in 1716.

Many have advocated similar views,[29] although some have disagreed, the preeminent example being Kant. In Kant's view, diagrammatic reasoning (or something like it) was not only to be admitted into genuine proof, it was generally necessary for it ([Kant 1781–87], A716–17/B744–45; A713–14/B741–42.)

Less radical, but still supportive of the use of diagrams in geometrical reasoning were Hobbes ([Hobbes 1655], [Hobbes 1656]), Newton ([Newton 1720], appendix, pp. 229–230), Locke ([Locke 1697], p. 58) and such lesser figures as Francis Maseres ([Maseres 1758], pp. ii–iii). Some, indeed, went beyond sympathy. C. S. Peirce, for example, maintained that virtually all reasoning—logical as well as mathematical—is either diagrammatic overall or has essential diagrammatic aspects ([Peirce 1898]).

Historically, there have been two main reasons for denying a genuinely justificative role to diagrams. One is unreliability (see [Hahn 1933] for a summary statement), the other their *particularity* ([Locke 1689–90], Bk IV, ch 1, sect 9; [Hume 1748], sect XII, part I; [Berkeley 1709], in [Berkeley 1948–57], vol. 1, p. 221, [Berkeley 1710], Bk IV, ch 7, sects 7–13). In geometry, there are well-known examples used to support the charge of unreliability. A widely used example is the famous diagrammatic "proof" that all triangles are equilateral. This was a favorite of Hilbert's which he repeated in various of his lecture courses on the foundations of geometry.

The other, perhaps more fundamental reason for denying justificative status to diagrammatic reasoning is their *particularity*. Mathematical truths are typically general truths while diagrams are *particular* figures. Since deductive reasoning concerning a particular figure can not establish a general truth, diagrammatic reasoning can not deductively justify a typical mathematical truth. It can only do so by some sort of analogical or broadly inductive extension. So, at any rate, the traditional reasoning goes.

In recent decades there has been renewed interest in diagrammatic reasoning in logic and mathematics. One influential example is the investigation and defense initiated by Jon Barwise and John Etchemendy, and pursued by various of their students and others. Barwise and Etchemendy argue that diagrams can and often do play a genuine epistemic role in proof: "we claim that visual forms of representation can be important, not just as heuristic and pedagogic tools, but as legitimate elements of mathematical proofs" ([Barwise/Etchemendy 1991], p. 9). Later they strengthen this by saying that "diagrams and other forms of visual representation can be *essential* and legitimate components in valid deductive reasoning" (*op. cit.*, p. 16, emphasis added).

They offer two responses to the charge that diagrammatic (or broadly visual) reasoning is unreliable. The first is the basic logical point that the existence of fallacious instances of reasoning of a given broad type does not impugn *all* instances of that type. Accordingly, even though some diagrammatic reasoning is fallacious, not all of it need be.

The second is a set of specific examples of what Barwise and Etchemendy characterize as 'perfectly valid proofs' that use diagrammatic (or other visual) reasoning in justificative ways ([Barwise/Etchemendy 1991], p. 12). Their mathematical example is an argument for

---

[29] We quoted a remark from Pasch earlier that expressed such a view. Hilbert too said such things (cf. lecture notes on geometry of the summer semester of 1894, the winter semester of 1898/99 and the summer semester of 1927). It is more difficult to determine his final view, however, because of the emphasis he elsewhere placed on 'intuitive grasp' (*anschaulichen Erfassen*) in geometric thinking ([Hilbert/Cohn-Vossen 1932], V, VI).

Russell too held such views ([Russell 1901], pp. 88–89, [Russell 1919], p. 145), as did Hans Hahn ([Hahn 1933]). More recent examples include Dieudonné [Dieudonné 1960], p. v and Tennant [Tennant 1986], p. 304.

the Pythagorean theorem ([Barwise/Etchemendy 1991], p. 12–13, Example 3) that combines diagrammatic reasoning and algebraic reasoning. The basic diagram is as follows:

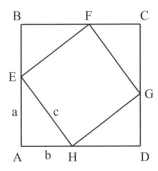

The focal triangle is $\triangle EAH$, and the claim to be proved is that $a^2 + b^2 = c^2$. The argument begins by constructing a square on EH and replicating $\triangle EAH$ three times as indicated in the diagram. One then determines that each side of $ABCD$ is a straight line by appealing to the theorem that the sum of the angles of a triangle is a straight line. From this, we're told, "one easily sees" that $ABCD$ is itself a square, and a square whose area can be computed in two ways: $(a+b)^2$ and $c^2 + 4(\frac{ab}{2})$. Equating these two and doing the obvious calculations, we arrive at $a^2 + b^2 = c^2$, which is what was to be proved.

Commenting on this argument, Barwise and Etchemendy ([Barwise/Etchemendy 1991], p. 12) make five claims.

1. It is (clearly) a "legitimate proof of the Pythagorean theorem."
2. It is a "combination of geometric manipulation of a diagram and algebraic manipulation of nondiagrammatic symbols."
3. The diagrammatic elements "play a crucial role in the proof."
4. The diagrammatic elements are primary (and typical of many traditional geometric proofs) in two related ways.
   (a) They make the algebraic steps of the argument "almost transparent." Once the diagrammatic steps are in place, the algebraic steps are easy to devise.
   (b) An analogous 'linguistic' proof would be both difficult to discover and difficult to remember without the use of diagrams. (This may give a sense for the suggestion noted above that the diagrammatic elements of the argument are "essential" to it.)
5. The proof (clearly) does not make use of "accidental features" of the diagrams involved.

Barwise and Etchemendy do not argue for these claims and, to my mind, none are evident. I'll briefly state some of my reservations below. Before doing that, though, I'd like to mention that both 4(a) and 4(b) are consonant with the traditional view of diagrams—namely, that they can be heuristically valuable, even though they play no legitimate justificative role in proof. Accepting 4(a) and 4(b) would thus not commit one to a justificative role for diagrams.

Regarding 1, I note two points. First, since the argument does not clearly identify all the different propositions it depends upon, it seems wrong to say that it is clearly a legitimate proof. In fact, it contains gaps, which legitimate proofs are not supposed to have. As an example of a gap, consider the step where Barwise and Etchemendy call for the "replication" of the original triangle three times as "shown" in the diagram. What goes into such "replication"? And what justifies it? There are plausible answers to these questions, but they need to be added to the argument if it is to deserve the title of 'proof'.

Nor does the Barwise-Etchemendy argument seem to support the suggestion in 4(a). In particular, its diagrammatic starting point does not seem to determine its algebraic details or make them 'transparent'. Indeed, as Yanney and Calderhead ([Yanney/Calderhead 1898]) argued long ago, there are at least four significantly different ways to carry a proof of the Pythagorean Theorem forward from the initial diagrammatic starting point of constructing a square on the hypotenuse of $\triangle EAH$. There may be (and perhaps typically is) some point in the development of a diagrammatic argument where a set of algebraic steps capable of completing the argument becomes "transparent". This would not be enough to show, though, that it's the diagrams that produce the transparency. It might instead be sheer accumulation of information, be it diagrammatically or non-diagrammatically supplied. Is this what is happening in the Barwise-Etchemendy proof? Or is there some distinctive gain in transparency that is specifically due to their use of diagrams?

Regarding claim 2, my main concern is clarity: specifically, the clarity of the key notion of 'geometrical manipulation'. On some level, the claim is uncontroversial. There are surely elements of the Barwise-Etchemendy argument that in some sense(s) are *geometrical manipulations*. A more difficult question, though, is what role specifically *visual* information plays in these operations. The answer to this is not, I think, clear, and this lack of clarity mounts when we consider claim 2 in conjunction with claims 3 and 5.

What are the 'geometrical manipulations' that supposedly both reflect essentially visual information and are clearly not accidental? It's hard to see what they might be. The likely candidate is the so-called 'replication' of the original triangle on the four sides of the square constructed on its hypotenuse. But it should be noted that this cannot be regarded as non-accidental unless we're able to establish non-accidentally that the sides of $ABCD$ are straight lines.

Barwise and Etchemendy rightly recognize that there needs to be a proof that the sides of $ABCD$ are straight lines. What is not clear from their argument is that this relies in any significant way on the visual information in their diagrams. Their reasoning is basically that the sides of $ABCD$ are straight lines because the interior angles of a triangle sum to a straight line. This is non-accidentally true of the sides of $ABCD$, however, only because of the similarity (in the geometrical sense) of triangles AEH, BFE, CGF and DHG. It does not derive from the visual qualities of their diagram. We know that AB, for example, is a straight line because: (i) the sum of the interior angles of a triangle is a straight angle and (ii) $\angle AEB$ is equal to such a sum. That (ii) is non-accidentally true follows from the additional facts that (a) $\angle AEB$ is composed of $\angle HEA$, $\angle HEF$ and $\angle FEB$, that (b) $\triangle AEH$ is composed of $\angle HAE$, $\angle AHE$ and $\angle HEA$, and that (c) $\angle HEA = \angle HEA$, $\angle HEF = \angle HAE$ and $\angle FEB = \angle AHE$. Similarly for the other sides of $ABCD$.

Where is the appeal to visual information in all of this? Nothing suggests that it's in (i). We are thus left with (ii). But where in (ii)? Not from (b), since that comes from the definition of $\triangle AEH$. Not from (c) either, though, since we get that from logic ($\angle HEA = \angle HEA$) and the knowledge that $\triangle AEH$ and $\triangle BFE$ are similar. This latter knowledge is not genuinely visual since it comes from knowledge that $\angle HAE$ and $\angle EBF$ are both right angles (because of the way $ABCD$ is constructed), that the angles in a triangle sum to two right angles ((i) again), that $\angle HEF$ is a right angle (due to $EFGH$'s being constructed as a square), and that $\angle HEA$ and $\angle FEB$ are therefore complementary. From this it follows that $\angle FEB = \angle AHE$ and $\angle AEH = \angle BFE$.

There are, of course, various heuristic roles that visual information may play in this reasoning—for example, in first *suggesting* that $\angle AEB = \angle AEH + \angle HEF + \angle FEB$. What is not so clear, however, is what justificative role this information might play.

Suppose, for the sake of argument, that knowledge that $\angle AEB = \angle AEH + \angle HEF + \angle FEB$ *does* come from the visual experience of the diagram. The *content* of that experience—$\angle AEB = \angle AEH + \angle HEF + \angle FEB$—plays a logical role in the argument. This notwithstanding, this content, and its logical relations to the contents of the other elements of the argument, are not the *only* things that affect the type of warrant the argument provides for its conclusion.

Traditionally, proofs have been intended to support belief in the necessity of their conclusions. Basing belief that $\angle AEB = \angle AEH + \angle HEF + \angle FEB$ (or any other premise of the Barwise-Etchemendy argument) on visual information would not provide for the realization of this intention. If our only reason for believing that $\angle AEB = \angle AEH + \angle HEF + \angle FEB$ is that it visually appears to be the case, we will not be in a position at the end of the Barwise-Etchemendy argument to know that it's necessarily the case that $a^2 + b^2 = c^2$. This, it seems to me, is similar to Berkeley's and Hume's objections to Locke's view of diagrams.

There are other questions and concerns raised by the Barwise-Etchemendy proposal, but in the space that remains I'll consider other recent work concerning the use of diagrams and visual information in mathematics. One example is Norman ([Norman 2006]), which argues for a broadly Kantian viewpoint according to which diagrammatic reasoning of the type found in classical geometry can contribute to *a priori* justification. That this is so is due to the fact that the reasoner typically forms concepts of (types of) geometrical objects, that she reasons with diagrams by taking them to represent instances of these concepts and that she then infers a general conclusion by taking the diagram to represent not merely a particular picture or image itself but the full set of such images producible by the same essential process of construction by which the given diagram was constructed. Having argued this, Norman nonetheless concedes that, in the end, diagrammatic reasoning lacks the rigor generally required of proofs. He thus suggests, in the end, that proof is but one means of attaining mathematical knowledge, an idea not unlike that suggested by Wallis, Tymoczko and others.

James Robert Brown has defended a similar view (cf. Brown [Brown 1999], pp. 24–43), arguing that diagrams and pictures can provide evidential grounds for propositions concerning mathematical objects we do not see. Since, however, the propositions supported by diagrams are often more general than the diagrams themselves are, the justificative role of diagrams can not generally be due to their *depiction* of the subject-matters of the propositions they support. They're not so much pictures, says Brown, as 'windows into Plato's heaven' ([Brown 1997], p. 174).

What these 'windows' are and how they're supposed to work is not clear. Brown mentions a "structural similarity" ([Brown 1997], p. 173) between diagrams and what they depict, and maintains that this somehow unites the items that belong to the justificative range of a diagram. This structural similarity is, however, presumably different from what Barwise and Etchemendy had in mind when they claimed that a 'good diagram is isomorphic, or at least homomorphic, to the situation it represents' ([Barwise/Etchemendy 1991], p. 22). Brown at any rate emphasizes that diagrams are not generally either isomorphic or homomorphic to what they represent ([Brown 1997], p. 173).

In the end, Brown doesn't show what he claims to show—namely, that diagrams either constitute proofs or play justificative roles in them. Indeed, some of what he says goes against this. He notes, in particular, that the main epistemic function of diagrammatic reasoning is to provide rational conviction rather than understanding (cf. Brown [Brown 1999], pp. 42–43), the latter being typically reserved for the more conventional 'propositional' proofs. But one must then wonder whether diagrammatic reasoning supports the higher forms of mathematical knowledge. Brown has nothing convincing to say on this point.

Marcus Giaquinto offers yet another account of visual reasoning in [Giaquinto 1994] and [Giaquinto 2005]. Where Brown emphasizes the legitimate evidential force of visual reasoning in establishing various theorems of number theory and analysis (e.g. Bolzano's Intermediate Value Theorem), Giaquinto emphasizes the differences between geometry and analysis, and argues for a much less extensive role for visual reasoning in analysis than in geometry.

Giaquinto's account of the epistemic role of visual reasoning also differs from Brown's. He maintains that visual reasoning is a legitimate means of discovery, where 'discovery' for him has a justificative aspect.[30] "One *discovers* a truth (which one does not already believe) by coming to believe it *independently* in an *epistemically acceptable* way." ([Giaquinto 1994], p. 790, emphases added). By requiring independence he means to rule out mere reliance on testimony. By *epistemically acceptable* ways he means reliable ways that are not undermined by an agent's other beliefs.

Giaquinto expressly denies that visual reasoning can be used as evidence for various theorems of analysis (e.g. the Intermediate Value Theorem, cf. Giaquinto [Giaquinto 1994], p. 793). He denies, in particular, that it can serve as a legitimate means of discovery, in the sense of the term described above. The reason is that the theorems mentioned exhibit a type of generality that defies discovery (in the above sense) by visual reasoning. In visual reasoning in geometry, generality is achieved because the visual reasoning typically 'brings to mind' a reliable general '*form* of thinking'. In the analytic cases mentioned, similarly general reasoning is unreliable.

In the case of Bolzano's Intermediate Value Theorem, he argues, in particular, for the falsity of the following:

(i) Any continuous function that changes signs on an interval has an uninterrupted curve from a point above the x-axis to a point below it.
(ii) Any function whose curve meets the x-axis has a zero value.

The latter is false because, judged according to visual criteria, a curve with a single point gap at the x-axis will nonetheless *look* like it intersects the x-axis. The two parts of such a line "could not appear to be separated by just one point, as a point has zero breadth" (*op. cit.*, p. 800). Visualization therefore cannot be a way of discovering (in Giaquinto's sense) Bolzano's theorem.

Assumption (i) seems even less defensible. The assumption that every continuous function has an uninterrupted curve is false because not every continuous function has any (visualizable) curve at all. Giaquinto offers Weierstrass' everywhere continuous but nowhere differentiable function as an example. At every stage of visualization (assuming the stages to follow the imagined steps of magnification) of this function there is a smooth part and the function is thus

---

[30] He distinguishes two different types of justification, however. One (*demonstrative* justification) requires both the absence of any violation of basic standards of rationality and an ability to explicitly give a reason for one's belief. The other (*default* justification) requires only the former (cf. [Giaquinto 1994], p. 791).

differentiable. On the other hand, a curve which is non-differentiable at a point "makes a sharp turn at that point, and a curve consisting of sharp turns at every point, without any smooth segments between sharp points, is unvisualizable" (*op. cit.*, p. 801). Some continuous functions thus have no curves (i.e. no visualizable curves) at all. This being so, theorems pertaining to such functions cannot be discovered through visualization. Or so Giaquinto reasons.

He also discusses what he sees as a signal difference between visual reasoning in geometry and visual reasoning in analysis (*op. cit.*, pp. 804–805). Certain geometrical concepts (e.g. those of a circle and a straight line) can have visual representations because some physical (or visually imagined) figures can appear to be *perfect* exemplars of their geometrical type. Thus, some physical or imagined circles can appear to be *perfectly* circular and some physical or imagined straight lines can appear to be *perfectly* straight. Geometrical concepts thus amount to idealizations or perfections, and can be visually represented by exemplars that are near enough to being perfect that their defects are not visually detectable.

The same is not true of such analytic concepts as continuous function, differentiable function and the integral. The first cannot be visualized by an uninterrupted curve since this both excludes some continuous functions and includes come non-continuous ones. The second cannot be visualized as a function with a smooth curve. The third cannot be visualized because of demands that analysis places on the concept of area. Hence, there are significant differences between geometry and analysis as regards the discoverability of theorems via visualization. Visualization may often be an aid to understanding and a stimulus or "trigger" to discovery in analysis, but it is only rarely a *mode* of discovery (cf. *op. cit.*, p. 811). This notwithstanding, it may still be a valuable tool to the analyst, and one whose value can be expected to grow with increased experience in analysis (*op. cit.*, p. 812).[31]

Thus far, I've not questioned the traditional assumption that diagrammatic reasoning is useful. I'll now briefly consider this question and the growing body of literature concerned with it. It contains some of the most interesting recent work concerning diagrammatic reasoning.

An important earlier study was Jill Larkin & Herbert Simon's "Why a Diagram is (Sometimes) Worth Ten Thousand Words" ([Larkin/Simon 1987]). A key difference between diagrammatic and linguistic (what they and others term 'sentential') reasoning, they claimed, is the degree to which information explicit in the one is implicit in the other. Diagrams characteristically display information explicitly that is only implicit in their linguistic counterparts. Since implicit information has to be computed in order to be used, linguistic reasoning typically involves more computation than diagrammatic reasoning, and this means that it's less easy.[32] Larkin and Simon argue that this is due in large part to the fact that linguistic representation is sequential or linear, while diagrammatic representation is planar. In these planar representations, spatially adjacent parts of the diagram often carry inferentially adjacent information. A process of diagrammatic reasoning is thus commonly driven by visual traversal or survey of a diagram and requires relatively little extraction of (i.e., search for) tacit elements.

---

[31] For more on the idea that diagrams and/or other types of pictures can act causally as "triggers" for belief-formation see Giaquinto [Giaquinto 2005].

[32] This assumes, of course, that we're talking about linguistic and diagrammatic expressions of the same information. Representations are treated as informationally equivalent when the information in each is inferrable from the information in the other. Informationally equivalent representations are then said to be computationally equivalent when, roughly, every inference in the one is as easy as the parallel inference in the other.

A different explanation of the relative efficiency of diagrammatic over linguistic reasoning is pursued by Stenning and Lemon ([Stenning/Lemon 2001]). They argue that it is typically due to diagrams' having a lower capacity for expression, in particular, a lower capacity for expressing abstractions. They argue further that restricted capacity to express abstractions generally makes for tractability of inference, while enhanced such capacity makes for intractability. The authors broadly attribute these differences to the planar character of diagrammatic representations.

> ... the expressive restrictions on DRs [diagrammatic representations] arise from an interaction between topological and geometrical constraints on plane surfaces, and the ways in which diagrams are interpreted.
>
> [Stenning/Lemon 2001], p. 30, brackets added

The topological constraints mentioned stem from a theorem of Helly's which limits the number of convex regions the mutual inclusion/exclusion relationships and emptiness/non-emptiness features that can be accurately presented in a planar array (cf. Stenning and Lemon [Stenning/Lemon 2001], pp. 45–46). Diagrams that exceed this limit are not generally trustworthy as regards the inclusion/exclusion, emptiness/non-emptiness information they convey.

Stenning and Lemon offer general characterizations of diagrammatic represensensentation systems and efficacious diagrammatic representation systems. Roughly, a representation system functions diagrammatically to the extent that its interpretation can be directly read off its spatial characteristics. A little more exactly, a diagrammatic representation is "a plane structure in which representing tokens are objects whose mutual spatial and graphical relations are directly interpreted as relations in the target structure" (*op. cit.*, p. 36).

The directness mentioned plays a key role in the efficiency of diagrammatic reasoning. As the interpretation of a representational system grows in abstractness (i.e., becomes less direct) its diagrammatic character decreases and the need for *extractive* interpretation (hence complexity) increases. Roughly, then, what makes diagrammatic reasoning efficient, when it is efficient, is the *directness* of its interpretation—the relatively great capacity a user has to read off key features of the target structure from the appearance of the diagram. It becomes useful to the extent that the features of the target system that can be directly read off the diagram comprise important features of the target system. See *op. cit.*, pp. 47–48 for a general characterization of diagrammatic effectiveness.

Unfortunately, the examples treated in this paper, as in most other recent work on diagrams, deal mainly with the use of diagrams in purely logical reasoning. Little attention is given to more complicated cases such as the use of diagrams in geometrical reasoning.[33] It may be that the general characterizations of diagrammatic reasoning and effective diagrammatic reasoning that Stenning and Lemon offer can be extended to such cases, but they offer little to support such a view. Nor, finally, do they engage the question of the general relationship between diagrammatic reasoning and proof.

These limitations notwithstanding, I commend the work for its attempt to provide a psychologically plausible explanation of why diagrammatic reasoning seems so often useful. Any serious account of the role of diagrams in proof will ultimately have to come to grips with the

---

[33] For examples of what mathematicians count as diagrammatic reasoning, see the continuing series of *Proofs without Words* in the *Mathematics Magazine* and also the two books by Nelsen ([Nelsen 1997], [Nelsen 2001]).

issues these authors address. I might also mention that the references in the paper provide the interested reader with valuable suggestions for continued study of these questions.

6 Concluding Thoughts

I've focused on three preoccupations of recent writings on proof:

1. **The role and possible effects of empirical reasoning in mathematics.** Do recent developments (specifically, the computer-assisted proof of the 4CT) point to something essentially new as regards the need for and/or effects of using broadly empirical and inductive reasoning in mathematics? In particular, should we see such things as the computer-assisted proof of the 4CT as pointing to the existence of mathematical truths of which we cannot have *a priori* knowledge?

2. **The role of formalization in proof.** What are the patterns of inference according to which mathematical reasoning naturally proceeds? Are they of 'local' character (i.e. sensitive to the subject-matter of the reasoning concerned) or 'global' character (i.e. invariant across all subject-matters)? Finally, what if any relationship is there (a) between the patterns of inference manifest in a proof and its explanatory capacity and (b) between explanatory capacity and rigor?

3. **Diagrams and their role in mathematical reasoning.** What essentially *is* diagrammatic reasoning, and what is the nature and basis of its usefulness? Can it play a justificative role in the development of mathematical knowledge and, more particularly, in genuine proof? Finally, does the use of diagrammatic reasoning force an adjustment either in our conception of rigor or in our view of its importance?

Concerning 1, I've urged caution as regards the suggestion by Tymoczko (and others) that the computer-assisted proof of the 4CT calls for fundamental changes in our understanding of mathematical method and proof. Its chief novelty, in my view, is the adjustment it suggests in our views of how we may come to know that proofs exist. It offers a concrete illustration of a proof that may defy human surveyal but nonetheless admits of survey by a computational routine designed and verified by humans.

The broader proposal of Jaffe and Quinn to "institutionalize" the use of empirical methods in mathematics does not challenge our understanding of the nature of proof so much as our use of it as a justificative standard in mathematics. It joins questions regarding proper method in mathematics to larger questions of morality and social practice.

The questions raised in 2 remain largely open. Robinson's work emphasizes the importance of finding the patterns that carry the flow of information in mathematical proof, and presents reasons for thinking they're often determined by 'local' topic and are not of a topic-neutral logical character.

The questions raised in 3 remain similarly open. This notwithstanding, insightful cases have been made for the significance of diagrammatic reasoning as justificative (as distinct from purely heuristic). At the same time, our understanding of possible limits on justificative uses of diagrammatic reasoning have been similarly advanced.

As regards the broad questions identified at the beginning of this paper, I've argued that little has been done to challenge the traditional view that proof has a distinctive role to play in the development of mathematical knowledge. In particular, I've argued that there is nothing new in the view that broadly empirical methods can play a role in mathematical investigation.

The challenges by Robinson and others to the traditional view of formalizability as an ideal of proof are of greater interest. They suggest that the level of detail required by certain types of formalization may actually interfere with the recognition of larger-scale structures in proofs upon which their explanatory potential depends.

Finally, the growing body of work on diagrammatic reasoning is of similarly great interest and potential. It challenges traditional ideas concerning the role of diagrammatic reasoning in proof and in the development of mathematical knowledge more generally. It suggests, in particular, that diagrammatic reasoning has a justificative and not merely a heuristic role to play in proof. Much interesting work has already been done in this direction, and more is sure to follow.

References and Bibliography

[Appel/Haken 1977] Appel, K., and Haken, W., "Every planar map is four colorable. I. Discharging", *Illinois Journal of Mathematics* 21 (1977), pp. 429–490.

[Appel et al. 1977] Appel, K., Haken, W. & J. Koch, "Every planar map is four colorable. II. Reducibility", *Illinois Journal of Mathematics* 21 (1977), pp. 1–251.

[Archimedes] Archimedes, *The Method*, preface. Heiberg ed. text with an English translation in [Thomas 2000], pp. 221–223.

[Aristotle 1908] Aristotle, *Posterior Analytics*, Oxford Translations, W. D. Ross and J. A. Smith eds., Oxford University Press, Oxford, 1908–1954.

[Aristotle 2000] ——, *Nichomachean Ethics*, English trans. and ed. by Roger Crisp, Cambridge University Press, Cambridge &New York, 2000.

[Barwise/Etchemendy 1991] Barwise, J. and J. Etchemendy, "Visual information and valid reasoning", in *Visualization in teaching and learning mathematics*, W. Zimmermann & S. Cunningham eds., Mathematical Association of America, Washington, DC, 1991.

[Berkeley 1948–57] Berkeley, G., *The Works of George Berkeley, Bishop of Cloyne*.

[Berkeley 1709] ——, *An Essay Towards A New Theory Of Vision*, in [Berkeley 1948–57], vol. 1.

[Berkeley 1710] ——, *A Treatise concerning the Principles of Human Knowledge*, in [Berkeley 1948–57], vol. 2.

[Borwein 1992] Borwein, J., "Some observations on computer assisted analysis", *Notices of the American Mathematical Society* 39 (1992), pp. 825–829.

[Brown 1997] Brown, J. R., "Pictures and Proofs", *British Journal for the Philosophy of Science* 48 (1997), pp. 161–80.

[Brown 1999] ——, *Philosophy of Mathematics:An introduction to the world of proofs and pictures*, Routledge, London & New York, 1999.

[Cicero 1894] Cicero, M. T., *The orations of Marcus Tullius Cicero*, 4 vols., London, G. Bell & sons, 1894–1903.

[Descartes 1620–28] Descartes, R., "Rules for the Direction of the Mind", in *The Philosophical Writings of Descartes*, vol. I, trans. by J. Cottingham, R. Stoothoff and D. Murdoch, Cambridge University Press, Cambridge, England, 1985.

[Descartes 1637] ——, *La Géométrie*, 1637. First published as an appendix to his *Discourse on Method*. Page references are to the English translation by D. Smith and L. Latham, Dover, New York, 1954.

[Detlefsen/Luker 1980] Detlefsen, M. & M. Luker, "The Four-Color theorem and mathematical proof", *Journal of Philosophy* 77 (1980), pp. 803–820.

[Detlefsen 1992] ——, "Poincaré Against the Logicians", *Synthese* 90 (1992), pp. 349–378.

[Detlefsen 2004] ——, "Formalism", ch. 8 of *The Oxford Handbook of Philosophy of Mathematics and Logic*, S. Shapiro (ed.), Oxford University Press, Oxford, 2004.

[Dieudonné 1960] Dieudonné, J., *Foundations of Modern Analysis*, Academic Press, New York, 1960.

[Euclid 1956] Euclid, *The Elements*, English trans. by Sir Thomas Heath, second revised ed., 3 vols., Dover, New York, 1956.

[Ewald 1996] Ewald, W., *From Kant to Hilbert: A source book in the foundations of mathematics*, two volumes, W. Ewald (ed.), Oxford University Press, Oxford, New York, etc., 1996.

[Fallis 1996] Fallis, D., "Mathematical proof and the reliability of DNA evidence", *American Mathematical Monthly* 103 (1996), pp. 491–497.

[Fallis 1997] ——, "The Epistemic Status of Probabilistic Proof", *Journal of Philosophy* 94 (1997), pp. 165–186.

[Fallis 2002] ——, "What Do Mathematicians Want?: Probabilistic Proofs and the Epistemic Goals of Mathematicians", *Logique et Analyse* 45 (2002), pp. 373–388.

[Fallis 2003] ——, "Intentional gaps in mathematical proofs", *Synthese* 134 (2003), pp. 45–69.

[Giaquinto 1992] Giaquinto, M., "Visualizing as a Means of Geometrical Discovery", *Mind and Language* 7 (1992), pp. 382–401.

[Giaquinto 1993] ——, "Visualizing in Arithmetic", *Philosophy and Phenomenological Research* 53 (1993), pp. 385–396.

[Giaquinto 1994] ——, "Epistemology of Visual Thinking in Elementary Real Analysis", *British Journal for the Philosophy of Science* 45 (1994), pp. 789–813.

[Giaquinto 2005] ——, "From Symmetry Perception to Basic Geometry", in [Mancosu et al. 2005]

[Giaquinto 2005a] ——, "Mathematical Activity", in [Mancosu et al. 2005].

[Hafner/Mancosu 2005] Hafner, J. and P. Mancosu, "The Varieties of Mathematical Explanation", in [Mancosu et al. 2005].

[Hahn 1933] Hahn, H., "The Crisis in Intuition", in H. Hahn *Empiricism, Logic, and Mathematics: Philosophical Papers*, B. McGuinness ed., D. Reidel Publishing Co., Dordrecht, 1980.

[Hilbert 1902] Hilbert, D., "Mathematical Problems", *Bulletin of the American Mathematical Society* **8** (1902), pp. 437–479. English translation of "Mathematische Probleme", *Archiv der Mathematik und Physik* (3rd series) **1** (1901), pp. 44–63, 213–237. The latter is a transcription of Hilbert's 1900 address to the World Congress of Mathematicians.

[Hilbert/Cohn-Vossen 1932] Hilbert, D. & S. Cohn-Vossen, *Anschauliche Geometrie*, Julius Springer, Berlin, 1932.

[Hobbes 1655] Hobbes, T., *De Corpore*, English trans. in [Hobbes 1839–45], vol. 1.

[Hobbes 1656] ——, *Six Lessons to the Professors of the Mathematics, one of Geometry, the other of Astronomy, in the chairs set up by the noble and learned Sir Henry Savile, in the University of Oxford*, in [Hobbes 1839–45], vol. 7.

[Hobbes 1839–45] ——, *The English works of Thomas Hobbes of Malmesbury*. W. Molesworth (ed.). Eleven volumes. J. Bohn, London, 1839–1845.

[Hume 1748] Hume, D., *An Enquiry concerning Human Understanding*, first published as *Philosophical Essays concerning Human Understanding*, London. Reprinted as *Enquiries Concerning Human Understanding and Concerning the Principles of Morals*, ed. L.A. Selby-Bigge and P.H. Nidditch, Clarendon Press, Oxford, 1978.

[Hutton 1795] Hutton, C., *A Mathematical and Philosophical Dictionary*, J. Johnson, and G.G. and J. Robinson, London, 1795–96. Two volumes. Reprinted by G. Olms Verlag, Hildesheim and New York, 1973. Reprinted in four volumes by Thoemmes Press, Bristol, 2000.

[Jaffe/Quinn 1993] Jaffe, A. & F. Quinn, "'Theoretical mathematics': Towards a cultural synthesis of mathematics and theoretical physics", *Bulletin of the American Mathematical Society* 29 (1993), pp. 1–13.

[Jesseph 1999] Jesseph, D., *Squaring the circle: The War between Hobbes and Wallis*, University of Chicago Press, Chicago, 1999.

[Kant 1781–87] Kant, I., *Critique of Pure Reason*, English trans. and ed. by P. Guyer and A. Wood, Cambridge University Press, Cambridge, 1998. The first edition of this work appeared in 1781, the second in 1787.

[Lambert 1766] Lambert, J., "Theorie der Parallelinien". This was written in 1766 but not published until 1786, when J. Bernoulli prepared an edition of Lambert's work. It is reprinted in *Die Theorie der Parallelinien von Euklid bis auf Gauss*, F. Engel and P. St'ackel eds., Teubner, Leipzig, 1895. Page references are to this reprinting.

[Larkin/Simon 1987] Larkin, J. and H. Simon, "Why a Diagram is (Sometimes) Worth Ten Thousand Words", *Cognitive Science* 11 (1987), pp. 65–100.

[Leibniz 1707] Leibniz, G. W. F., *Opera philosophica quae exstant latina, gallica, germanica omnia*, J. E. Erdmann (ed.), Scientia, Aalen, 1959.

[Leibniz 1981] ——, *New Essays Concerning Human Understanding*, P. Remnant and J. Bennett trans., Cambridge University Press, Cambridge, 1981.

[Locke 1689–90] Locke, J., *An Essay concerning Human Understanding*, P. H. Nidditch ed., Oxford University Press, Oxford, 1975.

[Locke 1697] ——, "Letter to Edward, the Bishop of Worcester", vol. 4 *Philosophical Works & Selected Correspondence of John Locke*, InteLex, Charlottesville, 1997.

[Mac Lane 1986] Mac Lane, S., *Mathematics: Form and Function*, Springer, New York, 1986.

[MacLaurin 1742] MacLaurin, C., *A treatise of fluxions*, T.W. & T. Ruddimans, Edinburgh, 1742. Excerpts reprinted in [Ewald 1996], volume 1. Imprint New York, Johnson Reprint Corp., 1964-67. Page references are to this reprinting.

[Mancosu 2000] Mancosu, P., "On Mathematical Explanation", in *The growth of mathematical knowledge*, E. Grosholz and H. Breger eds., Synthese Library volume 289, Kluwer, Dordrecht, 2000.

[Mancosu 2001] ——, "Mathematical Explanation: Problems and Prospects", *Topoi* 20 (2001), pp. 97–117.

[Mancosu et al. 2005] Mancosu *et al.*, *Visualization, Explanation and Reasoning Styles in Mathematics*, P. Mancosu, Klaus Frovin Jørgensen & Stig Andur Pedersen eds., Synthese Library, volume 327, Springer, Dordrecht, 2005.

[Mancosu 2005a] Mancosu, P., "Visualization in Logic and Mathematics" in [Mancosu et al. 2005].

[Marx 1844] Marx, K., *Comments on James Mill, Éléments D'économie Politique*. English trans. by C. Dutt in *Karl Marx, Frederick Engels: Collected Works*, volume 3. International Publishers, New York, 1975.

[Maseres 1758] Maseres, F., *A dissertation on the use of the negative sign in algebra: containing a demonstration of the rules usually given concerning it*, Printed, by S. Richardson, London, 1758.

[Nelsen 1997] Nelsen, R., *Proofs Without Words: Exercises in Visual Thinking*, Math. Assoc. Amer., Washington, DC, 1997.

[Nelsen 2001] ——, *Proofs Without Words II: More Exercises in Visual Thinking*, Math. Assoc. Amer., Washington, DC, 2001.

[Newton 1720] Newton, I., *Universal Arithmetick, or a Treatise of Arithmetical Composition and Resolution*. Translated from the Latin by J. Raphson. London, 1720.

[Norman 2006] Norman, J., *After Euclid: Visual Reasoning & the Epistemology of Diagrams*, CSLI, Stanford, 2006.

[Pappus 2000] Pappus, *The Treasury of Analysis*, in [Thomas 2000].

[Pasch 1912] Pasch, M., *Vorlesungen über neuere Geometrie*, 2nd, ed., Teubner, Leipzig, 1912. First ed. 1882.

[Peirce 1898] Peirce, C. S., "The Logic of Mathematics in Relation to Education", *Educational Review* 8 (1898), pp. 209–216. Reprinted in [Ewald 1996], vol. 1.

[Peressini 2003] Peressini, A., "Proof, Reliability, and Mathematical Knowledge", *Theoria* 69 (2003), pp. 211–232.

[Poincaré 1902] Poincaré, H., *Science and Hypothesis*, English trans. of 1905 French original. Dover, New York, 1952.

[Poincaré 1905] ——, *The Value of Science*, English trans. by G. B. Halsted of 1905 French original. Dover, New York, 1958.

[Robinson 1997] Robinson, J., "Informal Rigor and Mathematical Understanding", in *Computational Logic and Proof Theory*. Proceedings of the 5th annual Kurt Gödel Colloquium, August 25–29, 1997. Springer, Heidelberg & New York, 1997.

[Robinson 2000] ——, "Proof = Guarantee + Explanation", in S. Hölldobler (ed.), *Intellectics and computational logic: papers in honor of Wolfgang Bibel*, Kluwer Academic Publishers, Dordrecht (the Netherlands) and Boston, 2000.

[Robinson 2004] ——, "Logic is not the whole story", in *First-order logic revisited*, V. Hendricks et al. (eds.), Logos, Berlin, 2004.

[Rota 1997] Rota, G.-C., "The Phenomenology of Mathematical Proof", in *Proof and Progress in Mathematics*, A. Kanamori ed., special issue, *Synthese* 111(2) (1997), pp. 183–196.

[Russell 1901] Russell, B., "Mathematicians and Metaphysicians", in *Mysticism and Logic*, Doubleday Anchor, 1957.

[Russell 1919] ——, *Introduction to Mathematical Philosophy*, George Allen and Unwin Ltd., London, 1919.

[Smith 1776] Smith, A., *An inquiry into the nature and causes of the wealth of nations*, printed for W. Strahan and T. Cadell, London, 1776.

[Stenning/Lemon 2001] Stenning, K. & O. Lemon, "Aligning logical and psychological perspectives on diagrammatic reasoning", *Artificial Intelligence Review* 15 (2001), pp. 29–62.

[Sturm 1700] Sturm, J., *Mathesis enucleata, or, The elements of the mathematicks*, printed for Robert Knaplock and Dan. Midwinter and Tho. Leigh, London, 1700.

[Tappenden 2005] Tappenden, J., "Proof Style and Understanding in Mathematics I: Visualization, Unification and Axiom Choice", in [Mancosu et al. 2005].

[Teller 1980] Teller, P., "Computer proof", *Journal of Philosophy* 77 (1980), pp. 797–803.

[Tennant 1986] Tennant, N., "The withering away of formal semantics?", *Mind and Language* 1 (1986), pp. 302–318.

[Thomas 2000] Thomas, I., *Selections illustrating the history of Greek mathematics*, vol. II, Aristarchus to Pappus, Loeb Classical Library, Harvard University Press, Cambridge, MA, 2000.

[Tymoczko 1979] Tymoczko, T., "The Four-Color problem and its philosophical significance", *Journal of Philosophy* 76 (1979), pp. 57–83.

[Tymoczko 1980] ——, "Computers, proofs and mathematicians: a philosophical investigation of the four-color proof", *Mathematics Magazine* 53 (1980), pp. 131–138.

[Tymoczko 1986] ——, *New directions in the philosophy of mathematics*, T. Tymoczko (ed.), Birkhäuser, Boston, 1986.

[Viète 1591] Viète, F., *Introduction to the Analytic Art*, ed. and trans. by T. R. Witmer, Kent State University Press, Kent, Ohio, 1983.

[Wallis 1685] Wallis, J., *A Treatise of Algebra, both historical and practical: shewing the original, progress, and advancement thereof, from time to time, and by what steps it hath attained to the height at which it now is.* Printed for Richard Davis by John Playford, London, 1685.

[Wallis 1693] ——, *Opera Mathematica*, Oxoni: E Theatro Sheldoniano, 1693–1699. Reprinted (with a foreword by C. Scriba) by Georg Olms, Hildesheim, New York, 1972.

[Wallis 1656] ——, *The Mathematics of Infinitesimals: John Wallis 1656.* Translated from the Latin by J. Steadall. Springer, New York, 2004.

[Wolff 1739] Wolff, C., *A Treatise of algebra: with the application of it to a variety of problems in arithmetic, to geometry, trigonometry, and conic sections: with the several methods of solving and constructing equations of the higher kind*, printed for A. Bettesworth and C. Hitch, London, 1739.

[Yanney/Calderhead 1898] Yanney, B. & J. Calderhead "New and old proofs of the Pythagorean Theorem", *American Mathematical Monthly* 5 (1898), pp. 73–74.

2

Implications of Experimental Mathematics for the Philosophy of Mathematics[1]

Jonathan Borwein[2]
Faculty of Computer Science
Dalhousie University

From the Editors

When computers were first introduced, they were much more a tool for the other sciences than for mathematics. It was many years before more than a very small subset of mathematicians used them for anything beyond word-processing. Today, however, more and more mathematicians are using computers to actively assist their mathematical research in a range of ways. In this chapter, Jonathan Borwein, one of the leaders in this trend, discusses ways that computers can be used in the development of mathematics, both to assist in the discovery of mathematical facts and to assist in the development of their proofs. He suggests that what mathematics requires is secure knowledge that mathematical claims are true, and an understanding of why they are true, and that proofs are not necessarily the only route to this security. For teachers of mathematics, computers are a very helpful, if not essential, component of a constructivist approach to the mathematics curriculum.

Jonathan Borwein holds a Canada Research Chair in the Faculty of Computer Science at Dalhousie University (users.cs.dal.ca/ jborwein/). His research interests include scientific computation, numerical optimization, image reconstruction, computational number theory, experimental mathematics, and collaborative technology. He was the founding Director of the Centre for Experimental and Constructive Mathematics, a Simon Fraser University research center within the Departments of Mathematics and Statistics and Actuarial Science, established in 1993. He has received numerous awards including the Chauvenet Prize of the MAA in 1993 (with

---

[1] The companion web site is at **www.experimentalmath.info**

[2] Canada Research Chair, Faculty of Computer Science, 6050 University Ave, Dalhousie University, Nova Scotia, B3H 1W5 Canada. Email: `jborwein@cs.dal.ca`

P.B. Borwein and D.H. Bailey) for "Ramanujan, Modular Equations and Pi or How to Compute a Billion Digits of Pi," (Monthly *1989), Fellowship in the Royal Society of Canada (1994), and Fellowship in the American Association for the Advancement of Science (2002). Jointly with David Bailey he operates the Experimental Mathematics Website, www.experimentalmath.info. He is the author of several hundred papers, and the co-author of numerous books, including, with L. Berggren and P.B. Borwein,* Pi: a Source Book *(Springer-Verlag 1997); with David Bailey,* Mathematics by Experiment: Plausible Reasoning in the 21st Century *(AK Peters 2003); with David Bailey and Roland Girgensohn,* Experiments in Mathematics CD *(AK Peters 2006); with these same co-authors,* Experimentation in Mathematics: Computational Paths to Discovery *(AK Peters 2004); with David Bailey, Neil Calkin, Roland Girgensohn, D. Luke, and Victor Moll,* Experimental Mathematics in Action *(AK Peters 2007[3]); and he has just completed a related book with Keith Devlin,* The Computer as Crucible, *currently in press with AK Peters. Borwein and Bailey have also developed a number of software packages for experimental mathematics (crd.lbl.gov/ dhbailey/expmath/software/).*

Christopher Koch [Koch 2004] accurately captures a great scientific distaste for philosophizing:

> "Whether we scientists are inspired, bored, or infuriated by philosophy, all our theorizing and experimentation depends on particular philosophical background assumptions. This hidden influence is an acute embarrassment to many researchers, and it is therefore not often acknowledged."   (Christopher Koch, 2004)

That acknowledged, I am of the opinion that mathematical philosophy matters more now than it has in nearly a century. The power of modern computers matched with that of modern mathematical software and the sophistication of current mathematics is changing the way we do mathematics.

In my view it is now both necessary and possible to admit quasi-empirical inductive methods fully into mathematical argument. In doing so carefully we will enrich mathematics and yet preserve the mathematical literature's deserved reputation for reliability—even as the methods and criteria change. What do I mean by reliability? Well, research mathematicians still consult Euler or Riemann to be informed, anatomists only consult Harvey[4] for historical reasons. Mathematicians happily quote old papers as core steps of arguments, physical scientists expect to have to confirm results with another experiment.

1 Mathematical Knowledge as I View It

Somewhat unusually, I can exactly place the day at registration that I became a mathematician and I recall the reason why. I was about to deposit my punch cards in the 'honours history bin'. I remember thinking

---

[3] An earlier version of this chapter was taught in this short-course based book.

[4] William Harvey published the first accurate description of circulation, "An Anatomical Study of the Motion of the Heart and of the Blood in Animals," in 1628.

"If I do study history, in ten years I shall have forgotten how to use the calculus properly. If I take mathematics, I shall still be able to read competently about the War of 1812 or the Papal schism." (Jonathan Borwein, 1968)

The inescapable reality of objective mathematical knowledge is still with me. Nonetheless, my view then of the edifice I was entering is not that close to my view of the one I inhabit forty years later.

I also know when I became a computer-assisted fallibilist. Reading Imre Lakatos' *Proofs and Refutations*, [Lakatos 1976], a few years later while a very new faculty member, I was suddenly absolved from the grave sin of error, as I began to understand that missteps, mistakes and errors are the grist of all creative work.[5] The book, his doctorate posthumously published in 1976, is a student conversation about the Euler characteristic. The students are of various philosophical stripes and the discourse benefits from his early work on Hegel with the Stalinist Lukács in Hungary and from later study with Karl Popper at the London School of Economics. I had been prepared for this dispensation by the opportunity to learn a variety of subjects from Michael Dummett. Dummett was at that time completing his study rehabilitating Frege's status, [Dummett 1973].

A decade later the appearance of the first 'portable' computers happily coincided with my desire to decode Srinivasa Ramanujan's (1887–1920) cryptic assertions about theta functions and elliptic integrals, [Borwein et al. 1989]. I realized that by coding his formulae and my own in the *APL* programming language[6], I was able to rapidly confirm and refute identities and conjectures and to travel much more rapidly and fearlessly down potential blind alleys. I had become a computer-assisted fallibilist, at first somewhat falteringly, but twenty years have certainly honed my abilities.

Today, while I appreciate fine proofs and aim to produce them when possible, I no longer view proof as the royal road to secure mathematical knowledge.

2 Introduction

I first discuss my views, and those of others, on the nature of mathematics, and then illustrate these views in a variety of mathematical contexts. A considerably more detailed treatment of many of these topics is to be found in my book with Dave Bailey entitled *Mathematics by Experiment: Plausible Reasoning in the 21st Century*—especially in Chapters One, Two and Seven, [Borwein/Bailey 2003]. Additionally, [Bailey et al. 2007] contains several pertinent case studies as well as a version of this current chapter.

Kurt Gödel may well have overturned the mathematical apple cart entirely deductively, but nonetheless he could hold quite different ideas about legitimate forms of mathematical reasoning, [Gödel 1995]:

"If mathematics describes an objective world just like physics, there is no reason why inductive methods should not be applied in mathematics just the same as in physics." (Kurt Gödel[7], 1951)

---

[5] Gila Hanna [Hanna 2006] takes a more critical view placing more emphasis on the role of proof and certainty in mathematics; I do not disagree, so much as I place more value on the role of computer-assisted refutation. Also 'certainty' usually arrives late in the development of a proof.

[6] Known as a 'write only' very high level language, APL was a fine tool, albeit with a steep learning curve whose code is almost impossible to read later.

[7] Taken from a previously unpublished work, [Gödel 1995] originally given as the 1951 Gibbs lecture.

While we mathematicians have often separated ourselves from the sciences, they have tended to be more ecumenical. For example, a recent review of *Models. The Third Dimension of Science*, [Brown 2004], chose a mathematical plaster model of a Clebsch diagonal surface as its only illustration. Similarly, authors seeking examples of the aesthetic in science often choose iconic mathematics formulae such as $E = MC^2$.

Let me begin by fixing a few concepts before starting work in earnest. Above all, I hope to persuade you of the power of mathematical experimentation—it is also fun—and that the traditional accounting of mathematical learning and research is largely an ahistorical caricature. I recall three terms.

mathematics, n. *a group of related subjects, including algebra, geometry, trigonometry and calculus, concerned with the study of number, quantity, shape, and space, and their inter-relationships, applications, generalizations and abstractions.*

This definition—taken from my Collins Dictionary [Borowski/Borwein 2006]—makes no immediate mention of proof, nor of the means of reasoning to be allowed. The Webster's Dictionary [Webster's 1999] contrasts:

induction, n. *any form of reasoning in which the conclusion, though supported by the premises, does not follow from them necessarily.*; and

deduction, n. *a process of reasoning in which a conclusion follows necessarily from the premises presented, so that the conclusion cannot be false if the premises are true.*
b. a conclusion reached by this process.

Like Gödel, I suggest that both should be entertained in mathematics. This is certainly compatible with the general view of mathematicians that in some sense "mathematical stuff is out there" to be discovered. In this paper, I shall talk broadly about experimental and heuristic mathematics, giving accessible, primarily visual and symbolic, examples.

3 Philosophy of Experimental Mathematics

"The computer has in turn changed the very nature of mathematical experience, suggesting for the first time that mathematics, like physics, may yet become an empirical discipline, a place where things are discovered because they are seen."
(David Berlinski, [Berlinski 1997], p. 39)

The shift from *typographic* to *digital culture* is vexing for mathematicians. For example, there is still no truly satisfactory way of displaying mathematics on the web—and certainly not of asking mathematical questions. Also, we respect *authority*, [Grabiner 2004], but value *authorship* deeply—however much the two values are in conflict, [Borwein/Stanway 2005]. For example, the more I recast someone else's ideas in my own words, the more I enhance my authorship while undermining the original authority of the notions. Medieval scribes had the opposite concern and so took care to attribute their ideas to such as Aristotle or Plato.

And we care more about the *reliability* of our literature than does any other science. Indeed I would argue that we have over-subscribed to this notion and often pay lip-service, not real attention, to our older literature. How often does one see original sources sprinkled like holy water in papers that make no real use of them—the references offering a false sense of scholarship?

The traditional central role of proof in mathematics is arguably and perhaps appropriately under siege. Via examples, I intend to pose and answer various questions. I shall conclude with a variety of quotations from our progenitors and even contemporaries:

My Questions. What constitutes secure mathematical knowledge? When is computation convincing? Are humans less fallible? What tools are available? What methodologies? What of the 'law of the small numbers'? Who cares for certainty? What is the role of proof? How is mathematics actually done? How should it be? I mean these questions both about the apprehension (discovery) and the establishment (proving) of mathematics. This is presumably more controversial in the formal proof phase.

My Answers. To misquote D'Arcy Thompson (1860–1948) 'form follows function', [Thompson 1992]: rigour (proof) follows reason (discovery); indeed, excessive focus on rigour has driven us away from our wellsprings. Many good ideas are wrong. Not all truths are provable, and not all provable truths are worth proving. Gödel's incompleteness results certainly showed us the first two of these assertions while the third is the bane of editors who are frequently presented with correct but unexceptional and unmotivated generalizations of results in the literature. Moreover, near certainty is often as good as it gets—intellectual context (community) matters. Recent complex human proofs are often very long, extraordinarily subtle and fraught with error—consider Fermat's last theorem, the Poincaré conjecture, the classification of finite simple groups, presumably any proof of the Riemann hypothesis, [Economist 2005]. So while we mathematicians publicly talk of certainty we really settle for security.

In all these settings, modern computational tools dramatically change the nature and scale of available evidence. Given an interesting identity buried in a long and complicated paper on an unfamiliar subject, which would give you more confidence in its correctness: staring at the proof, or confirming computationally that it is correct to 10,000 decimal places?

Here is such a formula ([Bailey/Borwein 2005], p. 20):

$$\frac{24}{7\sqrt{7}} \int_{\pi/3}^{\pi/2} \log \left| \frac{\tan t + \sqrt{7}}{\tan t - \sqrt{7}} \right| dt \overset{?}{=} L_{-7}(2)$$

$$= \sum_{n=0}^{\infty} \left[\frac{1}{(7n+1)^2} + \frac{1}{(7n+2)^2} - \frac{1}{(7n+3)^2} + \frac{1}{(7n+4)^2} - \frac{1}{(7n+5)^2} - \frac{1}{(7n+6)^2} \right]. \quad (1)$$

This identity links a volume (the integral) to an arithmetic quantity (the sum). It arose out of some studies in quantum field theory, in analysis of the volumes of ideal tetrahedra in hyperbolic space. The question mark is used because, while no hint of a path to a formal proof is yet known, it has been verified numerically to 20,000 digit precision—using 45 minutes on 1024 processors at Virginia Tech.

A more inductive approach can have significant benefits. For example, as there is still some doubt about the proof of the classification of finite simple groups it is important to ask whether the result is true but the proof flawed, or rather if there is still perhaps an 'ogre' sporadic group even larger than the 'monster.' What heuristic, probabilistic or computational tools can increase our confidence that the ogre does or does not exist? Likewise, there are experts who still believe the *Riemann hypothesis*[8] (RH) may be false and that the billions of zeroes found so far are much too small to be representative.[9] In any event, our understanding of the complexity of various crypto-systems relies on (RH) and we should like secure knowledge that any counter-example is enormous.

Peter Medawar (1915–87)—a Nobel prize winning oncologist and a great expositor of science— writing in *Advice to a Young Scientist*, [Medawar 1979], identifies four forms of scientific experiment:

1. *The Kantian experiment: generating "the classical non-Euclidean geometries (hyperbolic, elliptic) by replacing Euclid's axiom of parallels (or something equivalent to it) with alternative forms."* All mathematicians perform such experiments while the majority of computer explorations are of the following Baconian form.
2. *The Baconian experiment is a contrived as opposed to a natural happening, it "is the consequence of 'trying things out' or even of merely messing about."* Baconian experiments are the explorations of a happy if disorganized beachcomber and carry little predictive power.
3. *Aristotelian demonstrations: "apply electrodes to a frog's sciatic nerve, and lo, the leg kicks; always precede the presentation of the dog's dinner with the ringing of a bell, and lo, the bell alone will soon make the dog dribble."* Arguably our 'Corollaries' and 'Examples' are Aristotelian, they reinforce but do not predict. Medawar then says the most important form of experiment is:
4. *The Galilean experiment is "a critical experiment—one that discriminates between possibilities and, in doing so, either gives us confidence in the view we are taking or makes us think it in need of correction."* The Galilean is the only form of experiment which stands to make Experimental Mathematics a serious enterprise. Performing careful, replicable Galilean experiments requires work and care.

Reuben Hersh's arguments for a humanist philosophy of mathematics, especially ([Hersh 1995], pp. 590–591), and ([Hersh 1999], p. 22), as paraphrased below, become even more convincing in our highly computational setting.

1. Mathematics is human. *It is part of and fits into human culture. It does not match Frege's concept of an abstract, timeless, tenseless, objective reality.*[10]
2. Mathematical knowledge is fallible. *As in science, mathematics can advance by making mistakes and then correcting or even re-correcting them. The "fallibilism" of mathematics is brilliantly argued in Lakatos' Proofs and Refutations.*
3. There are different versions of proof or rigor. *Standards of rigor can vary depending on time, place, and other things. The use of computers in formal proofs, exemplified by the*

---

[8] All non-trivial zeroes—not negative even integers—of the zeta function lie on the line with real part 1/2.

[9] See [Odlyzko 2001] and various of Andrew Odlyzko's unpublished but widely circulated works.

[10] That Frege's view of mathematics is wrong, for Hersh as for me, does not diminish its historical importance.

computer-assisted proof of the four color theorem in 1977,[11] *is just one example of an emerging nontraditional standard of rigor.*

4. Empirical evidence, numerical experimentation and probabilistic proof all can help us decide what to believe in mathematics. *Aristotelian logic isn't necessarily always the best way of deciding.*

5. Mathematical objects are a special variety of a social-cultural-historical object. *Contrary to the assertions of certain post-modern detractors, mathematics cannot be dismissed as merely a new form of literature or religion. Nevertheless, many mathematical objects can be seen as shared ideas, like Moby Dick in literature, or the Immaculate Conception in religion.*

I entirely subscribe to points 2., 3., 4., and with certain caveats about objective knowledge[12] to points 1. and 5. In any event mathematics is and will remain a uniquely human undertaking.

This version of humanism sits *fairly* comfortably along-side current versions of **social-constructivism** as described next.

"The social constructivist thesis is that mathematics is a social construction, a cultural product, fallible like any other branch of knowledge." (Paul Ernest, [Ernest 1990], §3)

But only if I qualify this with *"Yes, but much-much less fallible than most branches of knowledge."* Associated most notably with the writings of Paul Ernest—an English Mathematician and Professor in the Philosophy of Mathematics Education who in [Ernest 1998] traces the intellectual pedigree for his thesis, a pedigree that encompasses the writings of Wittgenstein, Lakatos, Davis, and Hersh among others—social constructivism seeks to define mathematical knowledge and epistemology through the social structure and interactions of the mathematical community and society as a whole.

This interaction often takes place over very long periods. Many of the ideas our students—and some colleagues—take for granted took a great deal of time to gel. The Greeks suspected the impossibility of the three *classical construction problems*[13] and the irrationality of the golden mean was well known to the Pythagoreans.

While concerns about potential and completed infinities are very old, until the advent of the calculus with Newton and Leibnitz and the need to handle fluxions or infinitesimals, the level of need for rigour remained modest. Certainly Euclid is in its geometric domain generally a model of rigour, while also Archimedes' numerical analysis was not equalled until the 19th century.

The need for rigour arrived in full force in the time of Cauchy and Fourier. The treacherous countably infinite processes of analysis and the limitations of formal manipulation came to the fore. It is difficult with a modern sensibility to understand how Cauchy's proof of the continuity

---

[11] Especially since a new implementation by Seymour, Robertson and Thomas in 1997 has produced a simpler, clearer and less troubling implementation.

[12] While it is not Hersh's intention, a superficial reading of point 5. hints at a cultural relativism to which I certainly do not subscribe.

[13] Trisection, circle squaring and cube doubling were taken by the educated to be impossible in antiquity. Already in 414 BCE, in his play *The Birds*, Aristophanes uses 'circle-squarers' as a term for those who attempt the impossible. Similarly, the French Academy stopped accepting claimed proofs a full two centuries before the 19th century achieved proofs of their impossibility.

of pointwise-limits could coexist in texts for a generation with clear counter-examples originating in Fourier's theory of heat.[14]

By the end of the 19th century Frege's (1848–1925) attempt to base mathematics in a linguistically based *logicism* had foundered on Russell and other's discoveries of the paradoxes of naive set theory. Within thirty five years Gödel—and then Turing's more algorithmic treatment[15]—had similarly damaged both Russell and Whitehead's and Hilbert's programs.

Throughout the twentieth century, bolstered by the armor of abstraction, the great ship Mathematics has sailed on largely unperturbed. During the last decade of the 19th and first few decades of the 20th century the following main streams of philosophy emerged explicitly within mathematics to replace logicism, but primarily as the domain of philosophers and logicians.

- *Platonism.* Everyman's idealist philosophy—stuff exists and we must find it. Despite being the oldest mathematical philosophy, Platonism—still predominant among working mathematicians—was only christened in 1934 by Paul Bernays.[16]
- *Formalism.* Associated mostly with Hilbert—it asserts that mathematics is invented and is best viewed as formal symbolic games without intrinsic meaning.
- *Intuitionism.* Invented by Brouwer and championed by Heyting, intuitionism asks for inarguable monadic components that can be fully analyzed and has many variants; this has interesting overlaps with recent work in cognitive psychology such as Lakoff and Nunez' work, [Lakoff/Nunez 2001], on 'embodied cognition'.[17]
- *Constructivism.* Originating with Markoff and especially Kronecker (1823–1891), and refined by Bishop it finds fault with significant parts of classical mathematics. Its 'I'm from Missouri, tell me how big it is' sensibility is not to be confused with Paul Ernest's 'social constructivism', [Ernest 1998].

The last two philosophies deny the principle of the *excluded middle*, "*A* or not *A*," and resonate with computer science—as does some of formalism. It is hard after all to run a deterministic program which does not know which disjunctive logic-gate to follow. By contrast the battle between a Platonic idealism (a 'deductive absolutism') and various forms of 'fallibilism' (a quasi-empirical 'relativism') plays out across all four, but fallibilism perhaps lives most easily within a restrained version of intuitionism which looks for 'intuitive arguments' and is willing to accept that 'a proof is what convinces'. As Lakatos shows, an argument that was convincing a hundred years ago may well now be viewed as inadequate. And one today trusted may be challenged in the next century.

---

[14] Cauchy's proof appeared in his 1821 text on analysis. While counterexamples were pointed out almost immediately, Stokes and Seidel were still refining the missing uniformity conditions in the late 1840s.

[15] The modern treatment of incompleteness leans heavily on Turing's analysis of the *Halting problem* for so-called Turing machines.

[16] See Karlis Podnieks, "Platonism, Intuition and the Nature on Mathematics," available at www.ltn.lv/podnieks/gt1.html

[17] The cognate views of Henri Poincaré (1854–1912) ([Poincaré 2004], p. 23) on the role of the *subliminal* are reflected in "The mathematical facts that are worthy of study are those that, by their analogy with other facts are susceptible of leading us to knowledge of a mathematical law, in the same way that physical facts lead us to a physical law." He also wrote "It is by logic we prove, it is by intuition that we invent," [Poincaré 1904].

As we illustrate in the next section or two, it is only perhaps in the last twenty five years, with the emergence of powerful mathematical platforms, that any approach other than a largely undigested Platonism and a reliance on proof and abstraction has had the tools[18] to give it traction with working mathematicians.

In this light, Hales' proof of Kepler's conjecture that *the densest way to stack spheres is in a pyramid* resolves the oldest problem in discrete geometry. It also supplies the most interesting recent example of intensively computer-assisted proof, and after five years with the review process was published in the *Annals of Mathematics*—with an "only 99% checked" disclaimer, withdrawn very late in the process and after being widely reported.

This process has triggered very varied reactions [Kolata 2004] and has provoked Thomas Hales to attempt a formal computational proof which he expects to complete by 2011, [Economist 2005]. Famous earlier examples of fundamentally computer-assisted proof include the *Four color theorem* and proof of the *Non-existence of a projective plane of order 10*. The three raise and answer quite distinct questions about computer-assisted proof—both real and specious. For example, there were real concerns about the completeness of the search in the 1976 proof of the Four color theorem but there should be none about the 1997 reworking by Seymour, Robertson and Thomas.[19] Correspondingly, Lam deservedly won the 1992 *Lester R. Ford award* for his compelling explanation of why to trust his computer when it announced there was no plane of order ten, [Lam 1991]. Finally, while it is reasonable to be concerned about the certainty of Hales' conclusion, was it really the *Annal's* purpose to suggest all other articles have been more than 99% certified?

To make the case as to how far mathematical computation has come we trace the changes over the past half century. The 1949 computation of π to 2,037 places suggested by von Neumann, took 70 hours. A billion digits may now be computed in much less time on a laptop. Strikingly, it would have taken roughly 100,000 ENIAC's to store the Smithsonian's picture—as is possible thanks to *40 years of Moore's law* in action.[20]

This is an astounding record of sustained exponential progress without peer in the history of technology. Additionally, mathematical tools are now being implemented on parallel platforms, providing *much* greater power to the research mathematician. Amassing huge amounts of processing power will not alone solve many mathematical problems. There are very few mathematical 'Grand-challenge problems', [JBorwein/PBorwein 2001] where, as in the physical sciences, a few more orders of computational power will resolve a problem.

For example, an order of magnitude improvement in computational power currently translates into one more day of accurate weather forecasting, while it is now common for biomedical researchers to design experiments today whose outcome is predicated on 'peta-scale' computation being available by say 2010, [Rowe et al. 2005]. There is, however, much more value in *very rapid 'Aha's'* as can be obtained through "micro-parallelism;" that is, where we benefit by being able to compute many simultaneous answers on a neurologically-rapid scale and so can hold many parts of a problem in our mind at one time.

---

[18] That is, to broadly implement Hersh's central points (2.-4.).

[19] See www.math.gatech.edu/thomas/FC/fourcolor.html.

[20] **Moore's Law** is now taken to be the assertion that *semiconductor technology approximately doubles in capacity and performance roughly every 18 to 24 months.*

To sum up, in light of the discussion and terms above, I now describe myself a sort-of social-constructivist, and as a computer-assisted fallibilist with constructivist leanings. I believe that more-and-more of the interesting parts of mathematics will be less-and-less susceptible to classical deductive analysis and that Hersh's 'non-traditional standard of rigor' must come to the fore.

4 Our Experimental Methodology

Despite Picasso's complaint that "computers are useless, they only give answers," the main goal of computation in pure mathematics is arguably to yield *insight*. This demands speed or, equivalently, substantial *micro-parallelism* to provide answers on a cognitively relevant scale; so that we may ask and answer more questions while they remain in our consciousness. This is relevant for rapid verification; for validation; for *proofs* and *especially for refutations* which includes what Lakatos calls "monster barring," [Lakatos 1976]. Most of this goes on in the daily small-scale accretive level of mathematical discovery but insight is gained even in cases like the proof of the Four color theorem or the Non-existence of a plane of order ten. Such insight is not found in the case-enumeration of the proof, but rather in the algorithmic reasons for believing that one has at hand a tractable unavoidable set of configurations or another effective algorithmic strategy. For instance, Lam [Lam 1991] ran his algorithms on known cases in various subtle ways, and also explained why built-in redundancy made the probability of machine-generated error negligible. More generally, the act of programming—if well performed—always leads to more insight about the structure of the problem.

In this setting it is enough to equate *parallelism* with access to requisite *more* space and speed of computation. Also, we should be willing to consider all computations as 'exact' which provide truly reliable answers.[21] This now usually requires a careful *hybrid* of symbolic and numeric methods, such as achieved by *Maple*'s liaison with the *Numerical Algorithms Group* (NAG) Library[22], see [Bornemann et al. 2004], [Borwein 2005b]. There are now excellent tools for such purposes throughout analysis, algebra, geometry and topology, see [Borwein/Bailey 2003], [Borwein et al. 2004], [Bornemann et al. 2004], [JBorwein/PBorwein 2001], [Borwein/Corless 1999].

Along the way questions required by—or just made natural by—computing start to force out older questions and possibilities in the way beautifully described a century ago by Dewey regarding evolution.

> "Old ideas give way slowly; for they are more than abstract logical forms and categories. They are habits, predispositions, deeply engrained attitudes of aversion and preference. Moreover, the conviction persists—though history shows it to be a hallucination—that all the questions that the human mind has asked are questions that can be answered in terms of the alternatives that the questions themselves present. But in fact intellectual progress usually occurs through sheer abandonment of questions together with both of the alternatives they assume; an abandonment that results from their decreasing vitality and a change of urgent interest. We do not solve them: we get over them. Old questions

---

[21] If careful interval analysis can certify that a number known to be integer is larger than 2.5 and less than 3.5, this constitutes an exact computational proof that it is 3.

[22] See www.nag.co.uk/.

are solved by disappearing, evaporating, while new questions corresponding to the changed attitude of endeavor and preference take their place. Doubtless the greatest dissolvent in contemporary thought of old questions, the greatest precipitant of new methods, new intentions, new problems, is the one effected by the scientific revolution that found its climax in the 'Origin of Species.'"   (John Dewey, [Dewey 1997])

Lest one think this a feature of the humanities and the human sciences, consider the artisanal chemical processes that have been lost as they were replaced by cheaper industrial versions. And mathematics is far from immune. Felix Klein, quoted at length in the introduction to [JBorwein/PBorwein 1987], laments that "now the younger generation hardly knows abelian functions." He goes on to explain that:

"In mathematics as in the other sciences, the same processes can be observed again and again. First, new questions arise, for internal or external reasons, and draw researchers away from the old questions. And the old questions, just because they have been worked on so much, need ever more comprehensive study for their mastery. This is unpleasant, and so one is glad to turn to problems that have been less developed and therefore require less foreknowledge—even if it is only a matter of axiomatics, or set theory, or some such thing."   (Felix Klein, [Klein 1928], p. 294)

Freeman Dyson has likewise gracefully described how taste changes:

"I see some parallels between the shifts of fashion in mathematics and in music. In music, the popular new styles of jazz and rock became fashionable a little earlier than the new mathematical styles of chaos and complexity theory. Jazz and rock were long despised by classical musicians, but have emerged as art-forms more accessible than classical music to a wide section of the public. Jazz and rock are no longer to be despised as passing fads. Neither are chaos and complexity theory. But still, classical music and classical mathematics are not dead. Mozart lives, and so does Euler. When the wheel of fashion turns once more, quantum mechanics and hard analysis will once again be in style."   (Freeman Dyson, [Dyson 1996])

For example recursively defined objects were once anathema—Ramanujan worked very hard to replace lovely iterations by sometimes-obscure closed-form approximations. Additionally, what is "easy" changes: high performance computing and networking are blurring, merging disciplines and collaborators. This is democratizing mathematics but further challenging authentication—consider how easy it is to find information on *Wikipedia*[23] and how hard it is to validate it.

Moving towards a well articulated Experimental *Mathodology*—both in theory and practice—will take much effort. The need is premised on the assertions that intuition is acquired—we can and must better mesh computation and mathematics, and that visualization is of growing importance—in many settings even three is a lot of dimensions.

---

[23] *Wikipedia* is an open source project at `en.wikipedia.org/wiki/Main_Page`; "wiki-wiki" is Hawaiian for "quickly."

"Monster-barring" (Lakatos's term, [Lakatos 1976], for refining hypotheses to rule out nasty counter-examples[24]) and "caging" (Nathalie Sinclair tells me this is my own term for imposing needed restrictions in a conjecture) are often easy to enhance computationally, as for example with randomized checks of equations, linear algebra, and primality or graphic checks of equalities, inequalities, areas, etc. Moreover, our mathodology fits well with the kind of pedagogy espoused at a more elementary level (and without the computer) by John Mason in [Mason 2006].

4.1 Eight Roles for Computation

I next recapitulate eight roles for computation that Bailey and I discuss in our two recent books [Borwein/Bailey 2003], [Borwein et al. 2004]:

#1. Gaining insight and intuition or just knowledge. Working algorithmically with mathematical objects almost inevitably adds insight to the processes one is studying. At some point even just the careful aggregation of data leads to better understanding.

#2. Discovering new facts, patterns and relationships. The number of *additive partitions* of a positive integer n, $p(n)$, is *generated* by

$$P(q) := 1 + \sum_{n \geq 1} p(n)q^n = \frac{1}{\prod_{n=1}^{\infty}(1 - q^n)}. \tag{2}$$

Thus, $p(5) = 7$ since

$$5 = 4 + 1 = 3 + 2 = 3 + 1 + 1 = 2 + 2 + 1 = 2 + 1 + 1 + 1 = 1 + 1 + 1 + 1 + 1.$$

Developing (2) is a fine introduction to enumeration via *generating functions*. Additive partitions are harder to handle than multiplicative factorizations, but they are very interesting ([Borwein et al. 2004], Chapter 4). Ramanujan used Major MacMahon's table of $p(n)$ to intuit remarkable deep congruences such as

$$p(\mathbf{5n + 4}) \equiv 0 \mod \mathbf{5}, \quad p(\mathbf{7n + 5}) \equiv 0 \mod \mathbf{7}, \quad p(\mathbf{11n + 6}) \equiv 0 \mod \mathbf{11},$$

from relatively limited data like

$$\begin{aligned}
P(q) = {} & 1 + q + 2\,q^2 + 3\,q^3 + \underline{5}\,q^4 + \overline{7}\,q^5 + 11\,q^6 + 15\,q^7 \\
& + 22\,q^8 + \underline{30}\,q^9 + 42\,q^{10} + 56\,q^{11} + \overline{77}\,q^{12} + 101\,q^{13} + \underline{135}\,q^{14} \\
& + 176\,q^{15} + 231\,q^{16} + 297\,q^{17} + 385\,q^{18} + \overline{490}\,q^{19} \\
& + 627\,q^{20}b + 792\,q^{21} + 1002\,q^{22} + \cdots + p(200)q^{200} + \cdots
\end{aligned} \tag{3}$$

Cases $5n + 4$ and $7n + 5$ are flagged in (3). Of course, it is markedly easier to (heuristically) confirm than find these fine examples of *Mathematics: the science of patterns*.[25] The study of such congruences—much assisted by symbolic computation—is very active today.

---

[24] Is, for example, a polyhedron always convex? Is a curve intended to be simple? Is a topology assumed Hausdorff, a group commutative?

[25] The title of Keith Devlin's 1996 book, [Devlin 1996].

#3. Graphing to expose mathematical facts, structures or principles. Consider Nick Trefethen's fourth challenge problem as described in [Bornemann et al. 2004], [Borwein 2005b]. It requires one to find ten good digits of:

4. What is the global minimum of the function

$$\exp(\sin(50x)) + \sin(60e^y) + \sin(70\sin x) + \sin(\sin(80y))$$
$$- \sin(10(x+y)) + (x^2 + y^2)/4?$$

As a foretaste of future graphic tools, one can solve this problem graphically and interactively using current *adaptive 3-D plotting* routines which can catch all the bumps. This does admittedly rely on trusting a good deal of software.

#4. Rigourously testing and especially falsifying conjectures. I hew to the Popperian scientific view that we primarily falsify; but that as we perform more and more testing experiments without such falsification we draw closer to firm belief in the truth of a conjecture such as: *the polynomial $P(n) = n^2 - n + p$ has prime values for all $n = 0, 1, \ldots, p-2$, exactly for* Euler's lucky prime numbers, *that is, $p = 2, 3, 5, 11, 17,$ and 41.*[26]

#5. Exploring a possible result to see if it *merits* formal proof. A conventional deductive approach to a hard multi-step problem really requires establishing all the subordinate lemmas and propositions needed along the way—especially if they are highly technical and un-intuitive. Now some may be independently interesting or useful, but many are only worth proving if the entire expedition pans out. Computational experimental mathematics provides tools to survey the landscape with little risk of error: only if the view from the summit is worthwhile, does one lay out the route carefully. I discuss this further at the end of the next Section.

#6. Suggesting approaches for formal proof. The proof of the *cubic theta function identity* discussed in ([Borwein et al. 2004], p. 210ff), shows how a fully intelligible human proof can be obtained entirely by careful symbolic computation.

#7. Computing replacing lengthy hand derivations. Who would wish to verify the following prime factorization by hand?

$$6422607578676942838792549775208734746307$$
$$= (2140992015395526641)(1963506722254397)(1527791).$$

Surely, what we value is understanding the underlying algorithm, not the human work?

#8. Confirming analytically derived results. This is a wonderful and frequently accessible way of confirming results. Even if the result itself is not computationally checkable, there is often an accessible corollary. An assertion about bounded operators on Hilbert space may have a useful consequence for three-by-three matrices. It is also an excellent way to error correct, or to check calculus examples before giving a class.

5 Finding Things versus Proving Things

I now illuminate these eight roles with eight mathematical examples. At the end of each I note some of the roles illustrated.

---

[26] See [Weisstein WWW] for the answer.

Figure 2.1 (**Ex. 1.**): **Graphical comparison of** $-x^2 \ln(x)$ (lower local maximum in both graphs) with $x - x^2$ (left graph) and $x^2 - x^4$ (right graph)

1. **Pictorial comparison** of $y - y^2$ and $y^2 - y^4$ to $-y^2 \ln(y)$, when y lies in the unit interval, is a much more rapid way to divine which function is larger than by using traditional analytic methods.

   Figure 2.1 below shows that it is clear in the latter case that the functions cross, and so it is futile to try to prove one majorizes the other. In the first case, evidence is provided to motivate attempting a proof and often the picture serves to guide such a proof—by showing monotonicity or convexity or some other salient property. ∎

   This certainly illustrates roles #3 and #4, and perhaps role #5.

2. **A proof and a disproof.** Any modern computer algebra can tell one that

$$0 < \int_0^1 \frac{(1-x)^4 x^4}{1+x^2} \, dx = \frac{22}{7} - \pi, \tag{4}$$

   since the integral may be interpreted as the area under a positive curve. We are however no wiser as to why! If however we ask the same system to compute the indefinite integral, we are likely to be told that

$$\int_0^t \cdot = \frac{1}{7} t^7 - \frac{2}{3} t^6 + t^5 - \frac{4}{3} t^3 + 4t - 4 \arctan(t).$$

   Then (4) is now rigorously established by differentiation and an appeal to the Fundamental theorem of calculus. ∎

   This illustrates roles #1 and #6. It also falsifies the bad conjecture that $\pi = 22/7$ and so illustrates #4 again. Finally, the computer's proof is easier (#7) and very nice, though probably it is not the one we would have developed by ourselves. The fact that $22/7$ is a continued fraction approximation to π has led to many hunts for generalizations of (4), see [Borwein et al. 2004], Chapter 1. None so far are entirely successful.

3. **A computer discovery and a 'proof' of the series for** $\arcsin^2(x)$. We compute a few coefficients and observe that there is a regular power of 4 in the numerator, and integers in the denominator; or equivalently we look at $\arcsin(x/2)^2$. The generating function package 'gfun' in *Maple*, then predicts a recursion, r, for the denominators and solves it, as R.

```
>with(gfun):
>s:=[seq(1/coeff(series(arcsin(x/2)^2,x,25),x,2*n),n=1..6)]:
>R:=unapply(rsolve(op(1, listtorec(s,r(m))),r(m)),m);[seq(R(m),m=0..8)];
```

yields, $s := [4, 48, 360, 2240, 12600, 66528]$,

$$R := m \mapsto 8 \frac{4^m \, \Gamma(3/2 + m)(m + 1)}{\pi^{1/2}\Gamma(1 + m)},$$

where Γ is the Gamma function, and then returns the sequence of values

$$[4, 48, 360, 2240, 12600, 66528, 336336, 1647360, 7876440].$$

We may now use Sloane's *Online Encyclopedia of Integer Sequences*[27] to reveal that the coefficients are $R(n) = 2n^2 \binom{2n}{n}$. More precisely, sequence A002544 identifies $R(n + 1)/4 = \binom{2n+1}{n}(n + 1)^2$.

```
> [seq(2*n^2*binomial(2*n,n),n=1..8)];
```

confirms this with

$$[4, 48, 360, 2240, 12600, 66528, 336336, 1647360].$$

Next we write

```
> S:=Sum((2*x)^(2*n)/(2*n^2*binomial(2*n,n)),n=1..infinity):S=values(S);
```

which returns

$$\frac{1}{2} \sum_{n=1}^{\infty} \frac{(2x)^{2n}}{n^2 \binom{2n}{n}} = \arcsin^2(x).$$

That is, we have discovered—and proven if we trust or verify *Maple*'s summation algorithm—the desired Maclaurin series.

   As prefigured by Ramanujan, it transpires that there is a beautiful closed form for $\arcsin^{2m}(x)$ for all $m = 1, 2, \ldots$. In [Borwein/Chamberland 2007] there is a discussion of the use of *integer relation methods*, [Borwein/Bailey 2003], Chapter 6, to find this closed form and associated proofs are presented.   ∎

   Here we see an admixture of all of the roles save #3, but above all #2 and #5.

4. Discovery without proof. Donald Knuth[28] asked for a closed form evaluation of:

$$\sum_{k=1}^{\infty} \left\{ \frac{k^k}{k! \, e^k} - \frac{1}{\sqrt{2\pi k}} \right\} = -0.084069508727655 \ldots \ldots \qquad (5)$$

Since about 2000 CE it has been easy to compute 20—or 200—digits of this sum in *Maple* or *Mathematica*; and then to use the 'smart lookup' facility in the *Inverse Symbolic Calculator*(ISC). The ISC at `oldweb.cecm.sfu.ca/projects/ISC` uses a variety of search algorithms and heuristics to predict what a number might actually be. Similar ideas are now implemented as 'identify' in *Maple* and (for algebraic numbers only) as 'Recognize' in *Mathematica*, and are described in [Borwein 2005b], [Borwein/Bailey 2003],

---

[27] At `www.research.att.com/~njas/sequences/index.html`.

[28] Posed as an MAA Problem [Knuth 2002].

[Borwein/Corless 1999], [Bailey/Borwein 2000]. In this case it *rapidly* returns

$$0.084069508727655 \approx \frac{2}{3} + \frac{\zeta(1/2)}{\sqrt{2\pi}}.$$

We thus have a prediction which *Maple* 9.5 on a 2004 laptop *confirms* to 100 places in under 6 seconds and to 500 in 40 seconds. Arguably we are done. After all we were asked to *evaluate* the series and we now know a closed-form answer.

Notice also that the 'divergent' $\zeta(1/2)$ term is formally to be expected in that while $\sum_{n=1}^{\infty} 1/n^{1/2} = \infty$, the *analytic continuation* of $\zeta(s) := \sum_{n=1}^{\infty} 1/n^s$ for $s > 1$ evaluated at $1/2$ does occur! ∎

We have discovered and tested the result and in so doing gained insight and knowledge while illustrating roles #1, #2 and #4. Moreover, as described in [Borwein et al. 2004], p. 15, one can also be led by the computer to a very satisfactory computer-assisted but also very human proof, thus illustrating role #6. Indeed, the first hint is that the computer algebra system returned the value in (5) very quickly even though the series is very slowly convergent. This suggests the program is doing something intelligent—and it is! Such a use of computing is termed "instrumental" in that the computer is fundamental to the process, see [Lagrange 2005].

5. A striking conjecture with no known proof strategy (as of spring 2007)[29] given in [Borwein et al. 2004], p. 162, is: for $n = 1, 2, 3 \ldots$

$$8^n \zeta\left(\{\bar{2}, 1\}_n\right) \stackrel{?}{=} \zeta\left(\{2, 1\}_n\right). \tag{6}$$

Explicitly, the first two cases are

$$8 \sum_{n>m>0} \frac{(-1)^n}{n^2 m} = \sum_{n>0} \frac{1}{n^3} \quad \text{and} \quad 64 \sum_{n>m>o>p>0} \frac{(-1)^{n+o}}{n^2 m \, o^2 p} = \sum_{n>m>0} \frac{1}{n^3 m^3}.$$

The notation should now be clear—we use the 'overbar' to denote an alternation. Such alternating sums are called *multi-zeta values* (MZV) and positive ones are called *Euler sums* after Euler who first studied them seriously. They arise naturally in a variety of modern fields from combinatorics to mathematical physics and knot theory.

There is abundant evidence amassed since 'identity' (6) was found in 1996. For example, very recently Petr Lisonek checked the first 85 cases to 1000 places in about 41 HP hours with only the *predicted round-off error*. And the case $n = 163$ was checked in about ten hours. These objects are very hard to compute naively and require substantial computation as a precursor to their analysis.

Formula (6) is the *only* identification of its type of an Euler sum with a distinct MZV and we have no idea why it is true. Any similar MZV proof has been both highly non-trivial and illuminating. To illustrate how far we are from proof: can just the case $n = 2$ be proven *symbolically* as has been the case for $n = 1$? ∎

This identity was discovered by the British quantum field theorist David Broadhurst and me during a large hunt for such objects in the mid-nineties. In this process we discovered and proved many lovely results (see [Borwein/Bailey 2003], Chapter 2, and [Borwein et al. 2004], Chapter 4), thereby illustrating #1,#2, #4, #5 and #7. In the case of 'identity' (6) we have failed with

---

[29] A quite subtle proof has now been found by Zhao and is described in the second edition of [Borwein/Bailey 2003].

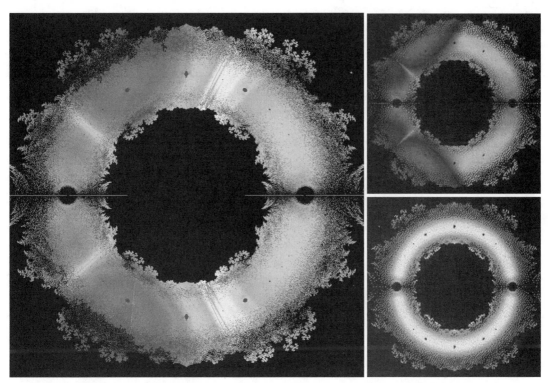

Figure 2.2 **(Ex. 6.): "The price of metaphor is eternal vigilance."** (Arturo Rosenblueth & Norbert Wiener, [Lewontin 2001])

#6, but we have ruled out many sterile approaches. It is one of many examples where we can now have (near) certainty without proof. Another was shown in equation (1) above.

6. **What you draw *is* what you see.** *Roots of polynomials with coefficients 1 or −1 up to degree 18.*

As the quote suggests, pictures are highly metaphorical. The shading in Figure 2.2 is determined by a normalized sensitivity of the coefficients of the polynomials to slight variations around the values of the zeros with red indicating low sensitivity and violet indicating high sensitivity.[30] It is hard to see how the structure revealed in the pictures above[31] would be seen other than through graphically data-mining. Note the different shapes—now proven by P. Borwein and colleagues—of the holes around the various roots of unity.

The striations are unexplained but all re-computations expose them! And the fractal structure is provably there. Nonetheless different ways of measuring the stability of the calculations reveal somewhat different features. This is very much analogous to a chemist discovering an unexplained but robust spectral line. ∎

This certainly illustrates #2 and #7, but also #1 and #3.

---

[30] Colour versions may be seen at `oldweb.cecm.sfu.ca/personal/loki/Projects/Roots/Book/` and on the cover of this book.

[31] We plot all complex zeroes of polynomials with only −1 and 1 as coefficients up to a given degree. As the degree increases some of the holes fill in—at different rates.

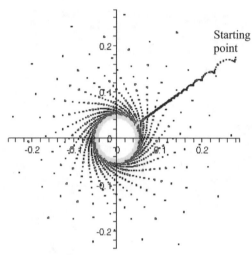

Figure 2.3 (Ex. 7.): "Visual convergence in the complex plane"

7. Visual Dynamics. In recent continued fraction work, Crandall and I needed to study the *dynamical system* $t_0 := t_1 := 1$:

$$t_n := \frac{1}{n} t_{n-1} + \omega_{n-1} \left(1 - \frac{1}{n} \right) t_{n-2},$$

where $\omega_n = a^2, b^2$ for n even, odd respectively, are two unit vectors. Think of this as a **black box** which we wish to examine scientifically. Numerically, all one *sees* is $t_n \to 0$ slowly. Pictorially, in Figure 2.3, we *learn* significantly more.[32] If the iterates are plotted with colour changing after every few hundred iterates,[33] it is clear that they spiral roman-candle like in to the origin:

Scaling by $\sqrt{n}$, and distinguishing even and odd iterates, *fine structure* appears in Figure 2.4. We now observe, predict and validate that the outcomes depend on whether or not one or both of a and b are roots of unity (that is, rational multiples of π). Input a pth root of unity and out come p spirals, input a non-root of unity and we see a circle. ∎

This forceably illustrates role #2 but also roles #1, #3, #4. It took my coauthors and me, over a year and 100 pages to convert this intuition into a rigorous formal proof, [Bailey/Borwein 2005]. Indeed, the results are technical and delicate enough that I have more faith in the facts than in the finished argument. In this sentiment, I am not entirely alone.

Carl Friedrich Gauss, who drew (carefully) and computed a great deal, is said to have noted, *I **have** the result, but I do not yet know how to get it.*[34] An excited young Gauss writes: "A new field of analysis has appeared to us, self-evidently, in the study of functions etc." (October 1798,

---

[32] ... "Then felt I like a watcher of the skies, when a new planet swims into his ken." From John Keats (1795–1821) poem *On first looking into Chapman's Homer.*

[33] A colour version may be seen on the cover of [Bailey et al. 2007].

[34] Like so many attributions, the quote has so far escaped exact isolation!

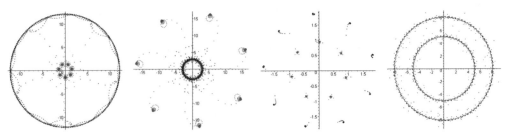

Figure 2.4 (Ex. 7.): The attractors for various $|a| = |b| = 1$

reproduced in [Borwein/Bailey 2003], Fig. 1.2, p.15). It had and the consequent proofs pried open the doors of much modern elliptic function and number theory.

My penultimate and more comprehensive example is more sophisticated and I beg the less-expert analyst's indulgence. Please consider its structure and not the details.

8. A full run. Consider the *unsolved* **Problem 10738** from the 1999 American Mathematical Monthly, [Borwein et al. 2004]:

Problem: For $t > 0$ let

$$m_n(t) = \sum_{k=0}^{\infty} k^n \exp(-t) \frac{t^k}{k!}$$

be the nth moment of a *Poisson distribution* with parameter t. Let $c_n(t) = m_n(t)/n!$. Show

a) $\{m_n(t)\}_{n=0}^{\infty}$ is log-convex[35] for all $t > 0$.
b) $\{c_n(t)\}_{n=0}^{\infty}$ is not log-concave for $t < 1$.
c*) $\{c_n(t)\}_{n=0}^{\infty}$ is log-concave for $t \geq 1$.

Solution. (a) Neglecting the factor of $\exp(-t)$ as we may, this reduces to

$$\sum_{k,j \geq 0} \frac{(jk)^{n+1} t^{k+j}}{k! j!} \leq \sum_{k,j \geq 0} \frac{(jk)^n t^{k+j}}{k! j!} k^2 = \sum_{k,j \geq 0} \frac{(jk)^n t^{k+j}}{k! j!} \frac{k^2 + j^2}{2},$$

and this now follows from $2jk \leq k^2 + j^2$.

(b) As

$$m_{n+1}(t) = t \sum_{k=0}^{\infty} (k+1)^n \exp(-t) \frac{t^k}{k!},$$

on applying the binomial theorem to $(k+1)^n$, we see that $m_n(t)$ satisfies the recurrence

$$m_{n+1}(t) = t \sum_{k=0}^{n} \binom{n}{k} m_k(t), \qquad m_0(t) = 1.$$

In particular for $t = 1$, we computationally obtain as many terms of the sequence

$$1, 1, 2, 5, 15, 52, 203, 877, 4140 \ldots$$

---

[35] A sequence $\{a_n\}$ is *log-convex* if $a_{n+1}a_{n-1} \geq a_n^2$, for $n \geq 1$ and log-concave when the inequality is reversed.

as we wish. These are the *Bell numbers* as was discovered again by consulting *Sloane's Encyclopedia* which can also tell us that, for $t = 2$, we have the *generalized Bell numbers*, and gives the exponential generating functions.[36] Inter alia, an explicit computation shows that

$$t \frac{1+t}{2} = c_0(t)\, c_2(t) \le c_1(t)^2 = t^2$$

exactly if $t \ge 1$, which completes (b).

Also, preparatory to the next part, a simple calculation shows that

$$\sum_{n \ge 0} c_n u^n = \exp\left(t(e^u - 1)\right). \tag{7}$$

(c^*)[37] We appeal to a recent theorem, [Borwein et al. 2004], p. 42, due to E. Rodney Canfield which proves the lovely and quite difficult result below. A self-contained proof would be very fine.

Theorem 1: *If a sequence* $1, b_1, b_2, \ldots$ *is non-negative and log-concave then so is the sequence* $1, c_1, c_2, \ldots$ *determined by the generating function equation*

$$\sum_{n \ge 0} c_n u^n = \exp\left(\sum_{j \ge 1} b_j \frac{u^j}{j}\right).$$

Using equation (7) above, we apply this to the sequence $\mathbf{b_j} = \mathbf{t}/(\mathbf{j} - \mathbf{1})!$ which is log-concave exactly for $t \ge 1$. ∎

A search in 2001 on *MathSciNet* for "Bell numbers" since 1995 turned up 18 items. Canfield's paper showed up as number 10. Later, *Google* found it immediately!

Quite unusually, the given solution to (c) was the only one received by the Monthly. The reason might well be that it relied on the following sequence of steps:

| A **(Question Posed)** ⇒ Computer Algebra System ⇒ Interface ⇒ |
| Search Engine ⇒ Digital Library ⇒ Hard New Paper ⇒ **(Answer)** |

Without going into detail, we have visited most of the points elaborated in Section 4.1. Now if only we could already automate this process!

Jacques Hadamard, describes the role of proof as well as anyone—and most persuasively given that his 1896 proof of the Prime number theorem is an inarguable apex of rigorous analysis.

"The object of mathematical rigor is to sanction and legitimize the conquests of intuition, and there was never any other object for it."   (Jacques Hadamard[38])

Of the eight uses of computers instanced above, let me reiterate the central importance of heuristic methods for determining what is true and whether it merits proof. I tentatively offer the

---

[36] Bell numbers were known earlier to Ramanujan—an example of *Stigler's Law of Eponymy*, [Borwein et al. 2004], p. 60. Combinatorially they count the number of nonempty subsets of a finite set.

[37] The '*' indicates this was the unsolved component.

[38] J. Hadamard, in E. Borel, Lecons sur la theorie des fonctions, 3rd ed. 1928, quoted in ([Polya 1981](2), p. 127). See also [Poincaré 2004].

following surprising example which is very very likely to be true, offers no suggestion of a proof and indeed may have no reasonable proof.

9. **Conjecture.** *Consider*

$$x_n = \left\{ 16x_{n-1} + \frac{120n^2 - 89n + 16}{512n^4 - 1024n^3 + 712n^2 - 206n + 21} \right\}. \tag{8}$$

The sequence $\beta_n = (\lfloor 16x_n \rfloor)$, where (x_n) is the sequence of iterates defined in equation (8), precisely generates the hexadecimal expansion of $\pi - 3$.

(Here $\{\cdot\}$ denotes the fractional part and $(\lfloor \cdot \rfloor)$ denotes the integer part.) In fact, we know from [Borwein/Bailey 2003], Chapter 4, that the first million iterates are correct and in consequence:

$$\sum_{n=1}^{\infty} \|x_n - \{16^n \pi\}\| \leq 1.46 \times 10^{-8} \ldots. \tag{9}$$

where $\|a\| = \min(a, 1 - a)$. By the first Borel-Cantelli lemma this shows that the hexadecimal expansion of π only finitely differs from (β_n). Heuristically, the probability of any error is very low. ∎

6 Conclusions

To summarize, I do argue that reimposing the primacy of mathematical knowledge over proof is appropriate. So I return to the matter of what it takes to persuade an individual to adopt new methods and drop time honoured ones. Aptly, we may start by consulting Kuhn on the matter of paradigm shift:

> "The issue of paradigm choice can never be unequivocally settled by logic and experiment alone. . . . in these matters neither proof nor error is at issue. The transfer of allegiance from paradigm to paradigm is a conversion experience that cannot be forced."
>
> (Thomas Kuhn[39])

As we have seen, the pragmatist philosopher John Dewey eloquently agrees, while Max Planck, [Planck 1949], has also famously remarked on the difficulty of such paradigm shifts. This is Kuhn's version[40]:

> "And Max Planck, surveying his own career in his Scientific Autobiography, sadly remarked that 'a new scientific truth does not triumph by convincing its opponents and making them see the light, but rather because its opponents eventually die, and a new generation grows up that is familiar with it.'"
>
> (Albert Einstein, [Kuhn 1996], [Planck 1949])

This transition is certainly already apparent. It is certainly rarer to find a mathematician under thirty who is unfamiliar with at least one of *Maple*, *Mathematica* or *MatLab*, than it is to one

---

[39] In [Regis 1988], *Who Got Einstein's Office?* The answer is Arne Beurling.

[40] Kuhn is quoting Einstein quoting Planck. There are various renderings of this second-hand German quotation.

over sixty five who is really fluent. As such fluency becomes ubiquitous, I expect a re-balancing of our community's valuing of deductive proof over inductive knowledge.

In his famous lecture to the Paris International Congress in 1900, Hilbert writes[41]

"Moreover a mathematical problem should be difficult in order to entice us, yet not completely inaccessible, lest it mock our efforts. It should be to us a guidepost on the mazy path to hidden truths, and ultimately a reminder of our pleasure in the successful solution."   (David Hilbert, [Yandell 2002])

Note the primacy given by a most exacting researcher to discovery and to truth over proof and rigor. More controversially and most of a century later, Greg Chaitin invites us to be bolder and act more like physicists.

"I believe that elementary number theory and the rest of mathematics should be pursued more in the spirit of experimental science, and that you should be willing to adopt new principles. . . . *And the Riemann Hypothesis isn't self-evident either, but it's very useful.* A physicist would say that there is ample experimental evidence for the Riemann Hypothesis and would go ahead and take it as a working assumption. . . . We may want to introduce it formally into our mathematical system."

(Greg Chaitin, [Borwein/Bailey 2003], p. 254)

Ten years later:

"[Chaitin's] "Opinion" article proposes that the Riemann hypothesis (RH) be adopted as a new axiom for mathematics. Normally one could only countenance such a suggestion if one were assured that the RH was undecidable. However, a proof of undecidability is a logical impossibility in this case, since if RH is false it is provably false. Thus, the author contends, one may either wait for a proof, or disproof, of RH—both of which could be impossible—or one may take the bull by the horns and accept the RH as an axiom. He prefers this latter course as the more positive one."   (Roger Heath Brown[42])

Much as I admire the challenge of Greg Chaitin's statements, I am not yet convinced that it is helpful to add axioms as opposed to proving conditional results that start "Assuming the continuum hypothesis" or emphasize that "without assuming the Riemann hypothesis we are able to show. . . . " Most important is that we lay our cards on the table. We should explicitly and honestly indicate when we believe our tools to be heuristic, we should carefully indicate why we have confidence in our computations—and where our uncertainty lies—and the like.

On that note, Hardy is supposed to have commented—somewhat dismissively—that Landau, a great German number theorist, would never be the first to prove the Riemann Hypothesis, but that if someone else did so then Landau would have the best possible proof shortly after. I certainly hope that a more experimental methodology will better value independent replication and honour

---

[41] See the late Ben Yandell's fine account of the twenty-three "*Mathematische Probleme*" lecture, Hilbert Problems and their solvers, [Yandell 2002]. The written lecture (given in [Yandell 2002]) is considerably longer and further ranging that the one delivered in person.

[42] Roger Heath-Brown's *Mathematical Review* of [Chaitin 2004], 2004.

the first transparent proof[43] of Fermat's last theorem as much as Andrew Wiles' monumental proof. Hardy also commented that he did his best work past forty. Inductive, accretive, tool-assisted mathematics certainly allows brilliance to be supplemented by experience and—as in my case—stands to further undermine the notion that one necessarily does one's best mathematics young.

6.1 As for Education

The main consequence for me is that a *constructivist educational curriculum*—supported by both good technology and reliable content—is both possible and highly desirable. In a traditional instructivist mathematics classroom there are few opportunities for realistic discovery. The current sophistication of dynamic geometry software such as *Geometer's Sketchpad*, *Cabri* or *Cinderella*, of many fine web-interfaces, and of broad mathematical computation platforms like *Maple* and *Mathematica* has changed this greatly—though in my opinion both *Maple* and *Mathematica* are unsuitable until late in high-school, as they presume too much of both the student and the teacher. A thoughtful and detailed discussion of many of the central issues can be found in J.P. Lagrange's article [Lagrange 2005] on teaching functions in such a milieu.

Another important lesson is that we need to teach procedural or *algorithmic thinking*. Although some vague notion of a computer program as a repeated procedure is probably ubiquitous today, this does not carry much water in practice. For example, five years or so ago, while teaching future elementary school teachers (in their final year), I introduced only one topic not in the text: extraction of roots by Newton's method. I taught this in class, tested it on an assignment and repeated it during the review period. About half of the students participated in both sessions. On the final exam, I asked the students to compute $\sqrt{3}$ using Newton's method starting at $x_0 = 3$ to estimate $\sqrt{3} = \underline{1.732}050808 \ldots$ so that the first three digits after the decimal point were correct. I hoped to see $x_1 = 2, x_2 = 7/4$ and $x_3 = 97/56 = \underline{1.732}142857\ldots$. I gave the students the exact iteration in the form

$$x_{\text{NEW}} = \frac{x + 3/x_{\text{OLD}}}{2}, \tag{10}$$

and some other details. The half of the class that had been taught the method had no trouble with the question. The rest almost without exception "guessed and checked." They tried $x_{\text{OLD}} = 3$ and then rather randomly substituted many other values in (10). If they were lucky they found some x_{OLD} such that x_{NEW} did the job.

My own recent experiences with technology-mediated curriculum are described in Jen Chang's 2006 MPub, [Chang 2006]. There is a concurrent commercial implementation of such a middle-school *Interactive School Mathematics* currently being completed by *MathResources*.[44] Many of the examples I have given, or similar ones more tailored to school [Borwein 2005a], are easily introduced into the curriculum, but only if the teacher is not left alone to do so. Technology also allows the same teacher to provide enriched material (say, on fractions, binomials, irrationality, fractals or chaos) to the brightest in the class while allowing more practice for those still

---

[43] Should such exist and as you prefer be discovered or invented.

[44] See www.mathresources.com/products/ism/index.html. I am a co-founder of this ten-year old company. Such a venture is very expensive and thus relies on commercial underpinning.

struggling with the basics. That said, successful mathematical education relies on active participation of the learner and the teacher and my own goal has been to produce technological resources to support not supplant this process; and I hope to make learning or teaching mathematics more rewarding and often more fun.

6.2 Last Words

To reprise, I hope to have made convincing arguments that the traditional deductive accounting of Mathematics is a largely ahistorical caricature—Euclid's millennial sway not withstanding.[45] Above all, mathematics is primarily about *secure knowledge* not proof, and that while the aesthetic is central, we must put much more emphasis on notions of supporting evidence and attend more closely to the reliability of witnesses.

Proofs are often out of reach—but understanding, even certainty, is not. Clearly, computer packages can make concepts more accessible. A short list includes linear relation algorithms, Galois theory, Groebner bases, etc. While progress is made *"one funeral at a time,"*[46] in Thomas Wolfe's words *"you can't go home again"* and as the co-inventor of the Fast Fourier transform properly observed, in [Tukey 1962][47]

> "Far better an approximate answer to the right question, which is often vague, than the exact answer to the wrong question, which can always be made precise."

Acknowledgements My gratitude is due to many colleagues who offered thoughtful and challenging comments during the completion of this work, and especially to David Bailey, Neil Calkin and Nathalie Sinclair. Equal gratitude is owed to the editors, Bonnie Gold and Roger Simons, for their careful and appropriately critical readings of several earlier drafts of this chapter.

References

[Bailey/Borwein 2000] D.H. Bailey and J.M. Borwein, "Experimental Mathematics: Recent Developments and Future Outlook," pp. 51–66 in Vol. I of *Mathematics Unlimited—2001 and Beyond,* B. Engquist & W. Schmid (Eds.), Springer-Verlag, 2000.

[Bailey/Borwein 2005] ———, "Experimental Mathematics: Examples, Methods and Implications," *Notices Amer. Math. Soc.*, **52** No. 5 (2005), pp. 502–514.

[Bailey et al. 2007] D. Bailey, J. Borwein, N. Calkin, R. Girgensohn, R. Luke, and V. Moll, *Experimental Mathematics in Action,* A.K. Peters, Ltd., 2007.

[Berlinski 1997] David Berlinski, "Ground Zero: A Review of The Pleasures of Counting, by T. W. Koerner," by David Berlinski, *The Sciences*, July/August 1997, pp. 37–41.

[Bornemann et al. 2004] F. Bornemann, D. Laurie, S. Wagon, and J. Waldvogel, *The SIAM 100 Digit Challenge: A Study in High-Accuracy Numerical Computing*, SIAM, Philadelphia, 2004.

---

[45] Most of the cited quotations are stored at `jborwein/quotations.html`

[46] This grim version of Planck's comment is sometimes attributed to Niels Bohr but this seems specious. It is also spuriously attributed on the web to Michael Milken, and I imagine many others.

[47] Ironically, despite often being cited as in that article, I can not locate it!

[Borowski/Borwein 2006] E.J. Borowski and J.M. Borwein, *Dictionary of Mathematics*, Smithsonian/ Collins Edition, 2006.

[Borwein 2005a] J.M. Borwein "The Experimental Mathematician: The Pleasure of Discovery and the Role of Proof," *International Journal of Computers for Mathematical Learning*, **10** (2005), pp. 75–108.

[Borwein 2005b] ——, "The 100 Digit Challenge: an Extended Review," *Math Intelligencer*, **27** (4) (2005), pp. 40–48. Available at `users.cs.dal.ca/~jborwein/digits.pdf`.

[Borwein/Bailey 2003] J.M. Borwein and D.H. Bailey, *Mathematics by Experiment: Plausible Reasoning in the 21st Century,* AK Peters Ltd, 2003, second expanded edition, 2008.

[Borwein et al. 2004] J.M. Borwein, D.H. Bailey and R. Girgensohn, *Experimentation in Mathematics: Computational Paths to Discovery,* AK Peters Ltd, 2004.

[JBorwein/PBorwein 1987] J.M. Borwein and P.B. Borwein, *Pi and the AGM,* CMS Monographs and Advanced Texts, John Wiley, 1987.

[JBorwein/PBorwein 2001] ——, "Challenges for Mathematical Computing," *Computing in Science & Engineering*, **3** (2001), pp. 48–53.

[Borwein et al. 1989] J.M. Borwein, P.B. Borwein, and D.A. Bailey, "Ramanujan, modular equations and pi or how to compute a billion digits of pi," *MAA Monthly*, **96** (1989), pp. 201–219. Reprinted in *Organic Mathematics Proceedings*, (`www.cecm.sfu.ca/organics`), April 12, 1996. Print version: *CMS/AMS Conference Proceedings*, **20** (1997), ISSN: 0731-1036.

[Borwein/Chamberland 2007] Jonathan Borwein and Marc Chamberland, "Integer powers of Arcsin," *Int. J. Math. & Math. Sci.*, 10 Pages, Art. ID 19381, June 2007. [D-drive preprint 288].

[Borwein/Corless 1999] Jonathan M. Borwein and Robert Corless, "Emerging Tools for Experimental Mathematics," *MAA Monthly*, **106** (1999), pp. 889–909.

[Borwein/Stanway 2005] J.M. Borwein and T.S. Stanway, "Knowledge and Community in Mathematics," *The Mathematical Intelligencer*, **27** (2005), pp. 7–16.

[Brown 2004] Julie K. Brown, "Solid Tools for Visualizing Science," *Science,* November 19, 2004, pp. 1136–37.

[Chaitin 2004] G.J. Chaitin, "Thoughts on the Riemann hypothesis," *Math. Intelligencer,* **26** (2004), no. 1, pp. 4–7. (MR2034034)

[Chang 2006] Jen Chang, "The SAMPLE Experience: the Development of a Rich Media Online Mathematics Learning Environment," MPub Project Report, Simon Fraser University, 2006. Available at `locutus.cs.dal.ca:8088/archive/00000327/`.

[Dewey 1997] John Dewey, *Influence of Darwin on Philosophy and Other Essays*, Prometheus Books, 1997.

[Devlin 1996] Keith Devlin, *Mathematics the Science of Patterns,* Owl Books, 1996.

[Dongarra/Sullivan 2000] J. Dongarra, F. Sullivan, "The top 10 algorithms," *Computing in Science & Engineering,* **2** (2000), pp. 22–23. (See `www.cecm.sfu.ca/personal/jborwein/algorithms.html`.)

[Dummett 1973] Michael Dummett, *Frege: Philosophy of Language,* Harvard University Press, 1973.

[Dyson 1996] Freeman Dyson, Review of *Nature's Numbers* by Ian Stewart (Basic Books, 1995). *American Mathematical Monthly*, August–September 1996, p. 612.

[Economist 2005] "Proof and Beauty," *The Economist*, March 31, 2005. (See `www.economist.com/science/displayStory.cfm?story_id=3809661`.)

[Ernest 1990] Paul Ernest, "Social Constructivism As a Philosophy of Mathematics. Radical Constructivism Rehabilitated?" A 'historical paper' available at `www.people.ex.ac.uk/PErnest/`.

[Ernest 1998] ——, *Social Constructivism As a Philosophy of Mathematics*, State University of New York Press, 1998.

[Gödel 1995] Kurt Gödel, "Some Basic Theorems on the Foundations," p. 313 in *Collected Works, Vol. III. Unpublished essays and lectures.* Oxford University Press, New York, 1995.

[Grabiner 2004] Judith Grabiner, "Newton, Maclaurin, and the Authority of Mathematics," *MAA Monthly*, December 2004, pp. 841–852.

[Hanna 2006] Gila Hanna, "The Influence of Lakatos," preprint, 2006.

[Hersh 1995] Reuben Hersh, "Fresh Breezes in the Philosophy of Mathematics", *MAA Monthly*, August 1995, pp. 589–594.

[Hersh 1999] ——, *What is Mathematics Really?* Oxford University Press, 1999.

[Klein 1928] Felix Klein, *Development of Mathematics in the 19th Century*, 1928, Trans Math. Sci. Press, R. Hermann Ed. (Brookline, MA, 1979).

[Kolata 2004] Gina Kolata, "In Math, Computers Don't Lie. Or Do They?" *NY Times*, April 6th, 2004.

[Koch 2004] Christopher Koch, "Thinking About the Conscious Mind," a review of John R. Searle's *Mind. A Brief Introduction,* Oxford University Press, 2004. *Science*, November 5, 2004, pp. 979–980.

[Knuth 2002] Donald Knuth, *American Mathematical Monthly Problem* 10832, November 2002.

[Kuhn 1996] T.S. Kuhn, *The Structure of Scientific Revolutions,* 3rd ed., U. of Chicago Press, 1996.

[Lakatos 1976] Imre Lakatos, *Proofs and Refutations*, Cambridge Univ. Press, 1976.

[Lakoff/Nunez 2001] George Lakoff and Rafael E. Nunez, *Where Mathematics Comes From: How the Embodied Mind Brings Mathematics into Being,* Basic Books, 2001.

[Lam 1991] Clement W.H. Lam, "The Search for a Finite Projective Plane of Order 10," *Amer. Math. Monthly* **98** (1991), pp. 305–318.

[Lagrange 2005] J.B. Lagrange, "Curriculum, Classroom Practices, and Tool Design in the Learning of Functions through Technology-aided Experimental Approaches," *International Journal of Computers in Math Learning,* **10** (2005), pp. 143–189.

[Lewontin 2001] R. C. Lewontin, "In the Beginning Was the Word," (*Human Genome Issue*), *Science*, February 16, 2001, pp. 1263–1264.

[Mason 2006] John Mason, *Learning and Doing Mathematics*, QED Press; 2nd revised ed.b, 2006.

[Medawar 1979] Peter B. Medawar, *Advice to a Young Scientist,* HarperCollins, 1979.

[Odlyzko 2001] Andrew Odlyzko, "The 10^{22}-nd zero of the Riemann zeta function. Dynamical, spectral, and arithmetic zeta functions," *Contemp. Math.*, **290** (2001), pp. 139–144.

[Planck 1949] Max Planck, *Scientific Autobiography and Other Papers,* trans. F. Gaynor (New York, 1949), pp. 33–34.

[Poincaré 2004] Henri Poincaré, "Mathematical Invention," pp. 20–30 in *Musing's of the Masters*, Raymond Ayoub editor, MAA, 2004.

[Poincaré 1904] ——, *Mathematical Definitions in Education*, (1904).

[Polya 1981] George Pólya, *Mathematical Discovery: On Understanding, Learning, and Teaching Problem Solving* (Combined Edition), New York, Wiley & Sons, 1981.

[Regis 1988] Ed Regis, *Who got Einstein's office?* Addison Wesley, 1988.

[Rowe et al. 2005] Kerry Rowe et al., *Engines of Discovery: The 21st Century Revolution.* The Long Range Plan for HPC in Canada, NRC Press, Ottawa, 2005.

[Thompson 1992] D'Arcy Thompson, *On Growth and Form,* Dover Publications, 1992.

[Tukey 1962] J.W. Tukey, "The future of data analysis," *Ann. Math. Statist.* **33**, (1962), pp. 1–67.

[Webster's 1999] *Random House Webster's Unabridged Dictionary*, Random House, 1999.

[Weisstein WWW] Eric W. Weisstein, "Lucky Number of Euler," from *MathWorld–A Wolfram Web Resource.* `mathworld.wolfram.com/LuckyNumberofEuler.html`.

[Yandell 2002] Benjamin Yandell, *The Honors Class*, AK Peters, 2002.

3

On the Roles of Proof in Mathematics

Joseph Auslander
Department of Mathematics
University of Maryland

From the Editors

This third perspective on proof comes from a mathematician with a more traditional perspective than Borwein's. The author brings his considerable experience both in developing his own proofs and in reviewing others' to questions about the roles of proof. His discussion on the roles of proof contains some interesting new ideas, such as proof as exploration and proof as justification of definitions—ideas that are relevant to us as we think about how we teach mathematics. At the end he offers some extended illustrations of his main points, from his experience working in topological dynamics and ergodic theory.

Joseph Auslander is a Professor Emeritus of Mathematics at the University of Maryland. He has published extensively in topological dynamics and ergodic theory. He is the author of Minimal Flows and Their Extensions *(1988) and co-editor, with Walter H. Gottschalk, of* Topological Dynamics, an international symposium *(1968). He has published two reviews of books in the philosophy of mathematics:* What is Mathematics, Really? *by Reuben Hersh,* Where Mathematics Comes From: How the Embodied Mind Brings Mathematics into Being *by George Lakoff and Rafael E. Núñez. Those reviews appeared in* SIAM Review *(2000) and* American Scientist *(2001), respectively. With Bonnie Gold, he organized a panel for the winter 2001 joint mathematics meetings in New Orleans on "The Philosophy of Mathematics: That Which is of Interest to Mathematicians," which led to the founding of POMSIGMAA. He was the first Secretary of POMSIGMAA, and gave a talk, "When is a Proof a Proof?" at the POMSIGMAA contributed paper session in January 2004.*

In this article, I will make, and try to justify, the following points.

Deductive proof is almost the defining feature of mathematics. Mathematics without proof
would not be mathematics. This is so although mathematics consists of more than proof,
and proof occurs in other disciplines.

Proof is necessary for validation of a mathematical result. But there are other, equally
compelling reasons for proof.

Standards of proof vary over time, and even among different mathematicians at a given time.

The question of "when is a proof a proof?" is a complex one. This has always been an issue,
but it is particularly so now in the light of computer assisted proofs and very long proofs.

1 Proof as a Defining Feature of Mathematics

I am writing as a working mathematician, not as a philosopher. My approach to proof is consistent
with the viewpoint, cogently put forth by Reuben Hersh [1997] and Paul Ernest [1998], that
mathematics is socially constructed. That is, it has been constructed by humans, and is part of
human culture. Therefore I will focus on what mathematicians actually do. This is what Hersh
calls "practical proof—the argument that convinces the qualified skeptical expert" rather than
formal proof.

Thomas Hales clarifies this distinction well [Hales www.]:

"Traditional mathematical proofs are written in a way to make them easily understood
by mathematicians. Routine logical steps are omitted. An enormous amount of context
is assumed on the part of the reader. Proofs, especially in topology and geometry, rely
on intuitive arguments in situations where a trained mathematician would be capable
of translating those intuitive arguments into a more rigorous argument." This is distinguished from formal proof where "all the intermediate logical steps are supplied" and
"no appeal is made to intuition."

I will not try to give a precise definition of mathematics; the definitions I've seen are either
too restrictive or too inclusive, but certainly the use of deductive proof is an essential feature.
Mathematics is not just about "results." (One might refer to the belief in the primacy of results to
the exclusion of anything else as the "Vince Lombardi" approach, after the football coach who
said that "winning is the only thing.")

Mathematics is a process, which includes definitions, conjectures, examples, numerical evidence, statements of theorems, modelling, algorithms, and proofs, as well as heuristic arguments
which fall short of proof. These are all woven together. In particular the proof is inextricably
bound up with the result; indeed one can't really separate them. This is part of the aesthetics of
mathematics, but it also has "practical" consequences. Proofs often contain "subresults," as well
as implicit or explicit lemmas, which are of interest in themselves. These would be lost if one just
catalogued "results." Moreover, often a proof yields more than is explicitly stated, and it may
point the way to new theorems. This is illustrated by Hillel Furstenberg's proof of the Szemeredi
theorem, which will be discussed below.

As John Franks [1989] eloquently puts it "a proof is not some kind of super spell checker
that merely validates mathematical facts . . . Proofs (are) the central content of mathematical
knowledge . . . Who would be satisfied if God were to announce that the Riemann hypothesis is
true, but deny us the proof?" (Regarding the last point, we might ask if we would be satisfied if

a computer "announced" that a theorem had been proved, but we couldn't see the proof. See the section on "Proof as Certification" for more about this.)

Another gloss on this topic was stated by the biologist Richard Lewontin [2005] writing in the New York Review of Books: "Science, indeed scholarship in general, is a domain in which the integrity of the process is more important than . . . any particular result. This is . . . a question of the very survival of the process of investigation." Lewontin in this passage was concerned with issues of honesty and fraud in science, but the point holds in a more general context.

As was mentioned above, mathematics is not only about proof. Moreover, the notion of proof also occurs in other areas (in other sciences of course—physical, biological, and social—and also such disciplines as law and history) but it has a somewhat different meaning, and different methods are used to attain it. These are characterized by a mixture of deductive reasoning and empirical evidence. Debates about the relation of these are at the heart of the philosophy of science.

I should say at the outset that I am definitely not asserting that proofs in mathematics are in some sense more "valid" than those in other disciplines. Rather, there are different methods of arriving at conclusions, and that deductive proof is central to mathematics to a much higher degree than in other areas. This is in spite of some challenges to this central role and even some predictions of the "death of proof."

We might accept as a provisional definition of proof a (valid) sequence of deductions, starting with the hypothesis, and arriving at the conclusion. Somewhat more formally [Kitcher 1984, p. 38] "We can now define a proof as a sequence of statements such that every member of the sequence is either a basic *a priori* statement or a statement which follows from previous members of the sequence in accordance with some apriority-preserving rule of inference."

This is somewhat at variance with our earlier emphasis on "practical proof," and in fact this tension is one of the things that makes the issue interesting. Nevertheless, mathematicians do feel that their proofs essentially accomplish what Kitcher describes (as the quotation from Hales in Section 1 points out). In fact, it's fair to say that this is a necessary and sufficient condition for a proof. That is, if this is achieved, we have a proof, and if it isn't there is no proof.

In a sense, that's all there is to it. As Gian-Carlo Rota [1996] puts it, "Mathematical proof does not admit degrees. A sequence of steps in an argument is either a proof, or it is meaningless. . . . The mathematical notion of proof is strikingly at variance with notions of proof in . . . law, everyday conversation, and physics."

However, I will argue that the situation is more complex than Rota makes it out to be. While any two mathematicians will agree in the abstract what a proof is, it's when one gets down to cases that problems may arise. Many of these can be reduced to "how do we know that a theorem has in fact been proved?" For example, what about "Proof: Obvious," or "Proof: This follows from the previous lemma?" At the other extreme, how do we evaluate a 15000 page proof, which may itself rely on papers the author hasn't read? Or a proof dependent on an unpublished or unobtainable paper? Or, a hot topic these days, a proof making use of computer calculations?

The issue of proofs in elementary and secondary school mathematics has been much discussed. Many (this writer included) lament the lack of emphasis on proofs in today's high school geometry classes, in contrast to what occurred in previous generations. The following quotation of Ken Ross [1998] addresses this point.

"While science verifies through observation, mathematics verifies through logical reasoning. Thus the essence of mathematics lies in proofs. . . . It should be emphasized that results in mathematics follow from hypotheses. . . . Moreover, beginning in the 8th grade, students should distinguish between inductive and deductive reasoning, be able to identify the hypothesis and conclusion in a deduction, test an assertion with examples, realize that one counterexample is enough to show that an assertion is false, and recognize whether something is being proved or merely given a plausibility argument."

2 The Roles of Proof

Mathematicians have a range of views on the role of proof in mathematics. Several of these views are illustrated by the following quotations, in which I have italicized words that emphasize the role of proof being mentioned.

> Hyman Bass [2003]: "The characteristic that distinguishes mathematics from all other sciences is the nature of mathematical knowledge and its *certification* by means of mathematical proof . . . it is the only science that thus pretends to claims of absolute certainty."
>
> Gian-Carlo Rota [1993, p. 93]: "Mathematicians cannot afford to behave like physicists, who take experimental verification as *confirmation* of the truth."
>
> In fact, the physicist Steven Weinberg [2001] makes essentially the same point as Rota: "You give up worrying about *certainty* when you make that turn in your career that makes you a physicist rather than a mathematician."
>
> David Gale [1990]: "The main goal of science is to observe and then to explain phenomena. In mathematics the *explanation* is the proof . . . the theorem-proof methodology . . . (is) the only *methodology* we have."
>
> Philip Davis and Reuben Hersh [1981, p. 151]: "Proof serves many purposes simultaneously . . . (It is) subject to a constant process of criticism and revalidation. Errors, ambiguities, and misunderstandings are cleared up by constant exposure. Proof is respectability. Proof is the seal of authority . . . (It) increases understanding by revealing the heart of the matter. Proof suggests new mathematics. Proof is mathematical power, the electric voltage of the subject which vitalizes the static assertions of the theorems."
>
> Saunders Mac Lane [Responses 1994, p.190]: "Intuition is glorious, but the heaven of mathematics requires much more. . . . Mathematics rests on proof—and proof is eternal."

There is no doubt that the overwhelming majority of mathematicians is committed to proofs in the traditional sense, and endorses the sentiments, if not the exact wording of the above quotations. Later, I'll express reservations about some of the assertions.

I would like to single out several (not unrelated) roles of proof, including certification (or validation), explanation, and exploration.

2.1 Proof as Certification

We accept that a purported result is correct when we hear that it has been proved by a mathematician we trust and "validated" by experts in the author's mathematical specialty. This is the case even if we haven't read the proof, or more frequently when we don't have the background to follow the proof. As an extreme, perhaps hackneyed, example, mathematicians accept Wiles' proof of Fermat's last theorem because number theorists have "certified" it to be correct. While certification is the most "primitive" or "elementary" aspect of proof, it is worthwhile looking at this role more closely. It is an indication that we are part of a community whose members trust one another. In fact, mathematics could not be a coherent discipline, as opposed to a random collection of techniques and results, without the process of certification.

Usually, certification of a result is a consequence of its appearance as a paper in a refereed journal. In fact, we might agree that this is a necessary condition for certification. In this case it is generally accepted that the "burden of proof" (the pun is inevitable) has shifted, and the result is presumed correct, unless there is a compelling reason to believe otherwise. It should be emphasized that it's necessary that one is convinced that a competent mathematician has worked out the proof, rather than it being "announced" by "God" (as in the quotation from John Franks earlier).

However, this process is far from perfect, and should be regarded as provisional. For one thing, it is well known that standards of refereeing vary widely. Some papers—for example, the proof of Fermat's last theorem, and Hales' proof of the Kepler conjecture discussed below—concern famous problems, and thus have received intense scrutiny. Other papers receive more routine treatment. Ralph Boas, who was for many years the editor of *Mathematical Reviews*, is said to have remarked that of the new results in papers reviewed most are true but the corresponding proofs are perhaps half the time wrong.

An interesting example was the published assertion by Waraskiewicz [1937], that a homogeneous plane continuum is necessarily a simple closed curve. This "result" was generally accepted, and in fact a more general assertion was published by Choquet [1944]. However, a counterexample was provided by Bing [1948]. (Another example will be discussed in the section "Four Examples.")

Also, referees are generally told that it is not their job to determine whether a paper is correct—this is the responsibility of the author—although the referee should be reasonably convinced. The referee is typically asked to determine whether the paper is worthwhile. Of course this begs the question somewhat. If the result is not correct, then the paper is not worthwhile. In the case of very long papers, referees usually don't try to check every line. Robert MacPherson, an editor of the *Annals of Mathematics* says "I try to understand the internal logic of the proof and do consistency checks." [Szpiro 2003, p. 208] Moreover, there are (presumably refereed) papers in respectable journals where the claimed result is false (in some cases not so noted for many years).

The issue of the refereeing process—real and ideal—in mathematics is fascinating and largely unexplored. Gossip on this topic abounds but I know of no systematic study.

Certification of a result allows us to use it in further research. In theory, one just checks the hypotheses, and if they are appropriate to the given situation, applies the result and goes on from there. This may be necessary (one can't develop all of mathematics each time one writes out a proof) but it brings along certain dangers. For reasons which aren't entirely clear, applying a result mechanically, without an understanding of the proof, can lead to errors. For example, sometimes one is fooled by notation. (This is borne out by my own experience. In fact, on one occasion I was attempting to apply something I had proved earlier without thinking it through carefully, and I made an elementary error.)

The point is that a mathematician is not absolved from understanding the proof, even when the result in question has been accepted by the mathematical community. When one uses a result in one's own research or teaching, the stakes are higher. It then becomes necessary to understand at least the basic outlines of the proof. One requires a higher degree of certainty for the use of a result than is obtained by the passive acceptance of it.

This was put well by Daniel Biss [2004]: "No honest mathematician uses a result simply because it has been published. Rather we use results we trust are true . . . the defining threshold for this notion is . . . a complex mélange of what has been published, what has been accepted as true by a larger community, and . . . what we believe ourselves to understand."

There is a recent tendency for (some) mathematicians to post their papers on preprint servers. Frequently this is preliminary to the submission of these papers to a journal (in which case it's not particularly different from the former practice of the distribution of preprints, allowing access of the results to researchers in the field prior to publication), but in some cases there is no intention of submission to a journal. Even given the imperfect process of refereeing, this somewhat undermines the certification of the results in question.

2.2 Proof as Explanation

Our second role of proof is explanation. This is what concerns most mathematicians. One should be able to follow at least the broad outlines of the argument, and be confident that one can fill in the details. As Andrew Gleason [Yandell 2001, p. 150] points out, "Proofs really aren't there to convince you that something is true ... they're there to show you why it is true."

Ideally this is what proof is all about. Almost by definition, a proof is supposed to explain the result. Now, it must be admitted that not all proofs meet this standard. To some extent this is in the eye of the beholder. Indeed sometimes the conviction that a result is correct may arise not from the proof, but from (say) numerical evidence, illuminating examples, or visual representation. Such considerations have often led to the development of new, more understandable, proofs.

The great mathematician Paul Erdős spoke of "The Book" in which "God" maintained the "perfect" proofs of theorems. In fact there is a real book, appropriately titled *Proofs from THE BOOK* by Martin Aigner and Gunter Ziegler [1999] which presents many proofs in this spirit. Erdős collaborated on this book shortly before his death, and many of the proofs are due to him.

The first chapter consists of six different proofs of the infinity of primes, starting with the familiar proof due to Euclid. The sixth proof, due to Erdős, proves more, namely that the sum of the reciprocals of the primes diverges. (The first proof of this fact was given by Euler.) Erdős' proof is by contradiction—suppose the sum converges. If $p_1, p_2, \ldots$ is the sequence of primes written in increasing order, then there is a k such that $\sum_{i \geq k+1} \frac{1}{p_i} < \frac{1}{2}$. Call $p_1, \ldots, p_k$ the small primes, and the others the big primes. For a fixed $N > 0$ let N_b be the number of $n \leq N$ which are divisible by at least one big prime, and N_s the number of such integers with only small prime divisors. Clearly $N = N_b + N_s$. On the other hand, Erdős shows, by an intricate combinatorial argument, that for a suitable N (in fact 2^{k+2}), $N_b + N_s < N$, which gives the contradiction.

2.3 Proof as Exploration

The above proof is also an example of the third role of proof, that of exploration. Every mathematician knows that when he/she writes out a proof, new insights, ideas, and questions emerge. Moreover, the proof requires techniques which may then be applied to the consideration of new problems. What makes this topic interesting, and somewhat complex, is that there is not always a hard line between explanation and exploration. Often the hallmark of a good proof is that it proves more than the statement of the theorem, as the Erdős proof illustrates.

A fascinating example of proof as exploration is the story of the proof of the alternating sign matrix conjecture, a topic on the boundary of algebra and combinatorics. An alternating sign matrix (ASM) is a square matrix of 0s, 1s, and –1s such that the sum of the entries in each row and each column is 1 and the nonzero entries in each row and each column alternate in sign. These are generalizations of permutation matrices. The ASM conjecture (now the ASM theorem) concerns the number A_n of such $n \times n$ matrices, which is given by $A_n = \prod_{0 \le j \le n-1} \frac{(3j+1)!}{(n+j)!}$. (In contrast, there are $n!$ permutation matrices.)

The history of the proof is brilliantly developed in David Bressoud's book *Proofs and Confirmations* [Bressoud 1999]. (The title was inspired by Imre Lakatos' book *Proofs and Refutations* [Lakatos 1976] which in turn was adapted from Karl Popper's *Conjectures and Refutations* [Popper 1963].) Bressoud presents the proof as an exploration, and in fact the chapter containing the proof is entitled "Explorations." He is referring to the development of the proof of the ASM conjecture, which he's presenting the way it developed historically.

Woven into the narrative are classical antecedents of the ASM conjecture, including an algorithm for the evaluation of determinants due to Charles Dodgson (Lewis Carroll), the appearance of many participants (including Mills, Robbins, Rumsey, Stanley, Andrews, and Zeilberger) as well as other results and conjectures. In fact, the ASM conjecture is one of fourteen related conjectures, two of which are still unproved. (One of these was "checked by one of the largest army of reviewers any paper has seen: 88 referees and one computer.")

Bressoud writes that the ASM proof "lay in unexpected territory and revealed a host of new insights and engaging problems." The unexpected territory included plane partitions, symmetric functions, and hypergeometric series. Indeed, it turns out that physicists were interested in ASMs, but they called them six vertex models or square ice.

The strategy of the proof was to try to find a one-to-one correspondence between $n \times n$ ASMs and descending plane partitions with largest part less than or equal to n. (Plane partitions are partitions of integers arranged as a two dimensional array, with certain restrictions. Some of the other conjectures concern generating functions - namely power series whose coefficients count the number of certain plane partitions.)

2.4 Proof as Justification of Definitions

Still another reason for proof, closely connected to teaching, is the justification for mathematical definitions. (I am indebted to my colleague Paul Green for this observation.) For example, one proves that the sum and product of continuous functions is continuous to confirm that the $\varepsilon - \delta$ definition is successful in capturing the intuitive idea of continuity. Similarly, the proof of the intermediate value theorem justifies the definition of the real number system. Yet another example is the use of the fundamental theorem of calculus to show that there is a real valued function whose derivative is e^{-x^2}. What is involved here is the very definition of a function. It demonstrates that a function need not be given by a simple formula, which is something we want to drive home to students. Only a formal proof can guarantee its existence and allow it to be studied.

Proofs also develop and underscore connections between different branches of mathematics, frequently to the benefit of both areas. Well known instances of this phenomenon are combinations of algebra and topology, and of combinatorics and number theory. In "Four Examples," we'll discuss in detail Furstenberg's proof on the Szemeredi theorem, which combines ergodic theory and combinatorial number theory.

2.5 The Dieudonné-Katznelson Encounter

It is frequently asserted that one can fill in the details of an informal argument to obtain a formally correct proof. To quote Bourbaki [1968, p. 8]: "In general [a mathematician] is content to bring the exposition to a point where his experience and mathematical flair tell him that translation into a formal language would be no more than an exercise of patience (though doubtless a very tedious one)." As if in reply, Hersh [1997, p. 52] says "It may be true. It's a matter of faith."

In this connection, let me turn to a personal recollection. In 1971, the distinguished mathematicians Yitzhak Katznelson and Jean Dieudonné visited the University of Maryland for a semester. Katznelson gave a course in ergodic theory, to which Dieudonné was a faithful attendee (as was I). Katznelson's lectures were well organized, although somewhat informal. Dieudonné (who had been a member of Bourbaki) didn't give Katznelson a moment's peace. He kept saying "That is not a proof" or sometimes "That's a nice presentation of the idea—now let's see the proof" and made Katznelson go over the argument until it was accomplished to his (Dieudonné's) satisfaction.

I'm certainly not saying that Dieudonné was more "rigorous" than Katznelson. Katznelson's proofs definitely met the standards of mathematical discourse. There are many acceptable styles of proof. (One might imagine Dieudonné lecturing, with Alonzo Church in the audience, who would say that Dieudonné's arguments were not proofs.)

Dieudonné had a high regard for Katznelson and the course (as he told me) and probably thought that the latter's arguments were essentially correct. But Dieudonné was not playing games. I'm sure he was serious in asserting that Katznelson's arguments fell short of proof, and felt that it was worth the class time for the development of one which was acceptable to him.

2.6 The Jaffe-Quinn Article

An extremely interesting discussion of various issues concerning proof was initiated by an article in the *Bulletin of the American Mathematical Society* by Arthur Jaffe and Frank Quinn [1993], and the responses it generated [Responses 1994]. The article (henceforth referred to as JQ) is entitled "Theoretical mathematics: towards a cultural synthesis of mathematics and theoretical physics." JQ use the term "theoretical mathematics" for "speculative and intuitive work" (this terminology was much criticized by a number of the respondents) and "rigorous mathematics" for "proof oriented work." While they agree that mathematics is "nearly characterized by the use of rigorous proofs" (which they unequivocally endorse) they call attention to "a trend towards basing mathematics on intuitive reasoning without proof" and say that this "may be the beginning of fundamental changes in the way mathematics is organized."

JQ contrast mathematics with physics. In the latter there is a "division of labor" between experimenters and theoreticians. But "the mathematical community has not undergone a bifurcation into theoretical and rigorous branches."

There is at least an implication by JQ that such a "bifurcation" would be desirable. But the lack of it is not accidental, and I doubt that it can be created by fiat. Of course there always has been a speculative and intuitive component to mathematics (and JQ correctly point to this as one of mathematics' "success stories") but I don't think there can be a division of mathematicians into two kinds, as there is in physics. That is, in general a mathematician's work is both intuitive *and* rigorous. Certainly there are individuals—Mandelbrot (one of the respondents) and Feigenbaum come to mind—whose main activity is "theoretical," but it's doubtful that there will be an entire community of such.

A year later the *Bulletin* printed a number of responses to JQ (by pure and applied mathematicians, physicists, and a historian of mathematics), as well as a separate article by Bill Thurston. These were in turn followed by a response by JQ.

While some of the responders are in substantial agreement with JQ, there are attacks from both the "right" and the "left." Mac Lane felt that physics is not a good model for mathematics. (The quote from Mac Lane in the section "The Roles of Proof," above, is part of his response.) Moe Hirsch (presumably tongue in cheek) suggests that "published mathematics... like good wine, should carry a date. If after ten years no errors have been found the theorem will be generally accepted" and that one should "attach a label to each proof, e.g., computer aided, mass collaboration, formal, informal, constructive, fuzzy, etc."

A particularly negative response was by Benoit Mandelbrot. He finds JQ "appalling" and refers to rigorous mathematicians as "Charles" mathematicians (since the AMS office in Providence is on Charles Street). He characterizes mathematical rigor as "besides the point and usually distracting, even where possible."

Richard Palais, the editor of the *Bulletin*, wrote that ("with mixed feelings") the *Bulletin* would no longer publish "controversial" articles. (Such would be restricted to the *Notices of the American Mathematical Society*.) One wonders about the subtext of this decision.

3 Computers and Proof

There is no question that computers are having a profound impact on mathematical practice. Perhaps their main role has been in experimentation, production of pictures, data, and the generation of conjectures. But computers have been used in some controversial proofs.

The relation between computers and proof is quite complex, and is still being sorted out. This paper will consider only a few such cases. It is interesting that Rota, in a passage following the quotation cited in the section "The Roles of Proof," above, says that it is *because* of computers that proof is "more indispensable than ever" (since "conjectures in number theory may fail for integers... beyond the reach of... computers"). There are some mathematicians, notably Paul Halmos and Pierre Deligne, who completely reject the use of computers. For example, Deligne has written "I don't believe in a proof done by a computer... I believe in a proof if I understand it." [Szpiro 2003, p. 21] In the same spirit, Eugene Wigner is reported to have said [Robertson 2003, p. 80] "It's nice to know that the computer understands the problem. But I would like to understand it, too." On the other hand, Thomas Hales says "I now feel that computer proofs are vital to the progress of mathematics." [Szpiro 2003, p. 212]

I take an intermediate point of view. Regardless of anyone's feelings (even Deligne's), one cannot wish away the use of computers in proofs. Mathematicians will use them if they find them necessary, or even convenient, and it's necessary to come to terms with this phenomenon. On the other hand, it's somewhat disingenuous to say that there is no difference between a calculation done by a computer and one done "by hand."

The issue is not whether one should "believe" a proof making use of a computer. Indeed, it may well be the case that a computer calculation is more reliable than a traditional one, especially if the latter is very long (witness the competing attempts at proving the Kepler conjecture, discussed below). Some of the same processes as in traditional proofs, for example modifications of the original argument, and repeated scrutiny, occur with computer proofs, and confirm the truth of the claimed assertion.

Moreover, there are certain proofs which just couldn't be accomplished without a computer. One such is the much discussed proof of the four color theorem, by Appel and Haken. The problem was reduced to several thousand cases, which were then checked by the computer.

The point is that it is necessary to recognize that there are tradeoffs involved here, namely the achievement of results versus the understanding of the reasons for their proofs. Even a rote computation in a traditional proof involves a certain amount of thinking. In the case of replication of a computer argument we cannot determine easily what hidden assumptions or errors lie in the shared bits of coding or hardware. At some point in the proof, a result is true because the computer "said so."

With regard to computers and proof, the story of the Kepler conjecture on sphere packing is particularly striking. (In my opinion, it is an order of magnitude more interesting than the four color theorem, although the latter was the first well known problem to make use of the computer for its solution).

The conjecture is that the densest way to pack spheres is the hexagonal close (or "greengrocers") packing. This is a four hundred year old problem, the oldest problem in discrete geometry, which was also part of Hilbert's 18th problem. There was a disputed proof (by Hsiang), and then a very long, computer assisted proof (by Hales), which is apparently correct. And the latter has led to conjectures and proofs of new results. All of this is recounted in detail in the excellent book *Kepler's Conjecture* by George C. Szpiro [2003].

A proposed proof, by Wu-Yi Hsiang [1993], was actually published. This proof made no use of the computer, just tools from (relatively) elementary geometry and calculus. The consensus of the mathematical community is that the attempted proof is incorrect, although Hsiang still stands by it. The proof that is now generally accepted is by Thomas Hales, with significant help from his student Samuel Ferguson. It consists of six papers, as well as a computer program. It was submitted to the *Annals of Mathematics* (in fact it was solicited by the *Annals*) and a team of 12 referees worked on it for four years. They returned a report saying that they were unable to completely certify the proof, although they were 99 percent certain of it.

In fact, the *Annals* has published Hales' proof [Hales 2005], although not the computer code on which it was based. The original plan was to publish it with a disclaimer, but after Hales reorganized it, it appeared as a single (more than one hundred page) paper, without a disclaimer. On the first page, Hales writes, "Here we describe the top-level outline of the proof and give sources of details of the proof. The latter are to appear as several papers in *Discrete and Computational Geometry*."

The Szpiro book has a chapter entitled "But is it really a proof?" There does seem to be a strong consensus that the Kepler conjecture is now proved—that it is "certified." There is considerably less agreement as to whether it meets the criterion of "explanation." For example, the mathematician and science writer Ian Stewart likens Hales' proof to a telephone directory, in contrast to Wiles' proof of Fermat's last theorem, which he compares to "War and Peace."

In this case, how are we to decide if the "telephone book" nature of Hales' proof is inherent to the problem?

This type of thing is unprecedented in mathematics. Regardless of one's feelings about the use of computers in proof, it must be recognized that Hales' work is a major scientific achievement. The story is not over; although the proof has appeared in print, there will very likely be simplifications that will really embed the result into mathematics.

The proofs of the four color theorem and the Kepler conjecture definitely fall within the traditional framework of proof as a sequence of deductions, although the computer plays an essential role. But there is another trend which in fact challenges the accepted dichotomy between a proof and an argument which falls short of proof. This is not concerned with the computer as an aid to proof, but rather envisions computer calculations as replacing proof.

One of the most provocative challenges to traditional proof was put forth by Doron Zeilberger in an article "Theorems for a price: tomorrow's semi-rigorous mathematical culture." [Zeilberger 1993] (As we'll see, "for a price" is meant literally.) The tone is set by Zeilberger's much quoted (and by now notorious) statement that in the future "rigorous old style mathematicians . . . may be viewed by mainstream mathematicians as a fringe sect of harmless eccentrics." He continues: "The computer has already started doing to mathematics what the telescope and microscope did to astronomy and biology. . . . In the future mathematicians will not care about absolute certainty, since there will be so many exciting new facts to discover." After presenting a number of identities which were proved by, or with the aid of, a computer, he envisions an abstract of a paper (c. 2100); "We show in a certain precise sense that the Goldbach conjecture is true with a probability larger than 0.9999 and that its complete truth could be determined with a budget of $10 billion." (Perhaps intentionally, there is no explanation of this assertion by the 2100-era mathematician.)

I should mention that Zeilberger is an outstanding mathematician, and in fact was one of the participants in the solution of the alternating sign matrix conjecture. But on this question I think he is quite wrongheaded.

Zeilberger's article was reprinted in the *Mathematical Intelligencer*, where it is followed by a response from his friend and collaborator George Andrews [1994]. Andrews' article is entitled (in part) "You've got to be kidding." He challenges Zeilberger's evaluation of the role of the computer in the discovery and proof of the various identities, and says moreover that Zeilberger "ignores the insight provided by proof" and "has produced exactly no evidence that his Brave New World is on the way."

As for Zeilberger's assertion that important theorems can be proved "for a price"—I don't believe it. Mathematics just doesn't work that way. Although mathematicians are no more immune to the lure of money than anyone else, one can't imagine a "crash program" to prove the Riemann hypothesis or the twin prime conjecture. It's true that there is now a well publicized monetary prize for such proofs, but there is no reason to think that the proofs will be attained any earlier on that account.

Another proponent of this trend is the geologist Douglas Robertson [2003]. Robertson's point of view is similar to Zeilberger's (he might be termed "Zeilberger lite"). Interestingly, he is extremely frank and explicit about what may be lost by this process. "Just as astronomers had to accept the idea that the telescope vastly extends the reach of the naked eye, mathematicians will have to accept the idea that the computer similarly extends the reach of the human mind." [Robertson 2003, p. 81] He also says that the understanding of the reasons behind such a computer proof "may not be attainable." Robertson asserts that computers will throw light on whether π is a normal number.[1] (This is very doubtful, in my opinion. No amount of computer calculation can settle this question.)

4 Four Examples

The examples which I'll discuss at some length are from topological dynamics and ergodic theory, areas of which I have some knowledge.

---

[1] A number is said to be normal (say to base 10) for which every finite sequence in the decimal expansion occurs with the "right" limiting frequency. For example, the occurrence of 57 has limiting frequency .01. Normal numbers have full Lebesgue measure, but are of first category.

4.1 The Birkhoff Ergodic Theorem

Even an absurdly naive idea can lead to a valid proof. Recall the statement of G.D. Birkhoff's (pointwise) ergodic theorem: Let T be a measure preserving transformation on a probability space X and let f be an integrable function on X. Then $\lim_{n\to\infty} \frac{1}{n} \sum_{i=0}^{n} f(T^i x)$ exists for almost all x.

(T^i denotes the i fold composition of the transformation T.)

In his book *Lectures on Ergodic Theory* [Halmos 1956], Halmos, after proving and obtaining some consequences of the ergodic theorem, concludes a chapter with what he calls an alternative "proof."

If f is a non-negative function on the positive integers, write

$$\int f(n)dn = \lim_{n\to\infty} \frac{1}{n} \sum_{i=0}^{n-1} f(i)$$

whenever the limit exists, and call such functions integrable. If T is a measure preserving transformation on a space X and f is an integrable function on X, then

$$\iint |f(T^n x)| dn\, dx = \iint |f(T^n x)| dx\, dn = \iint |f(x)| dx\, dn = \int |f(x)| dx < \infty.$$

Hence by "Fubini's theorem"(!) $f(T^n x)$ is an integrable function of its two arguments, and therefore, for almost every fixed x, it is an integrable function of n. That is,

$$\lim_{n\to\infty} \frac{1}{n} \sum_{i=0}^{n-1} f(T^i x)$$

exists for almost all x.

(This "proof" would work if there were a probability measure on the integers which assigned equal measure to each integer.)

One might think that this was just a joke on the part of Halmos, but he then asks, "Can any of this nonsense be made meaningful?" In fact, more than thirty years later, Ornstein obtained a proof based on this idea. This is the most conceptual proof of the ergodic theorem. The proof is by a delicate approximation argument. Essentially, one considers the product of the measure space with a finite space. This is a good example of the role of explanation in proof. Other proofs, some of which are shorter, depend on a trick. With this proof one really sees what is happening. The ideas of the proof led to theorems (by Ornstein and Weiss) on the actions of locally compact amenable groups.

4.2 Topological Entropy

Lakatos characterizes mathematical knowledge as proceeding by a sequence of proofs and refutations, and indeed sometimes erroneous "proofs" lead to refinements and clarifications. A case in point occurred in the early work on entropy, which is an important numerical invariant for dynamical systems.

Actually, there are two kinds of entropy. Measure theoretical entropy (defined by Kolmogorov and Sinai around 1958) applies to measure preserving transformations of probability spaces, and is defined in terms of measurable partitions. Topological entropy, defined several years later by Adler, Konheim, and McAndrew [Adler *et al.* 1965] concerns continuous self maps of compact metric spaces, and is defined in terms of open sets. In each case, the entropy is an extended non-negative real number. That is, if T is the transformation defined on the (probability or metric) space X, then the entropy $h(T)$ satisfies $0 \leq h(T) \leq \infty$. There are interesting connections between the two notions of entropy.

For measure theoretic entropy, there is a product theorem. If T_1 and T_2 are measure preserving transformations of X_1 and X_2 respectively, and $T_1 \times T_2$ is the product transformation, then $h(T_1 \times T_2) = h(T_1) + h(T_2)$. Moreover, the proof follows almost immediately from the definition.

When Adler, et al. introduced topological entropy, their paper included what they presented as a proof of the corresponding theorem for continuous transformations, which closely mimics the measure theoretic proof. This "proof," which was published, was erroneous—it slipped by a careless referee.[2]

This error (which was discovered by Kakutani) inspired an alternative (equivalent) definition of topological entropy by Bowen, which is in some ways more useful than the original definition. In particular, a correct proof of the product theorem can be given using this definition (although as a matter of fact Bowen's first proof was also incorrect).

4.3 The Szemeredi Theorem

Many important theorems have more than one proof. On an elementary level, there are several proofs of the infinity of primes, and several hundred proofs of the quadratic residue theorem (ten by Gauss). Also, as was mentioned above, there are a number of proofs of the ergodic theorem.

Of course, in terms of validation of a result, one correct proof is sufficient. If there is any question, the existence of multiple proofs provides some confirmation that a proposed result is correct. In this respect, mathematics is like an experimental science. An alternative proof is something like a replication of an experiment.

But I think what is even more important is that a new proof frequently connects with other branches of mathematics. As Michael Atiyah [2005] says "different proofs have different strengths and weaknesses, and they generalize in different directions—they are not just repetitions of each other."

A striking example is the Szemeredi theorem, which says that every set of integers of positive upper density contains arbitrarily long arithmetic progressions. Szemeredi's original proof was combinatorial and extremely long. Hillel Furstenberg gave another proof, which was accomplished by translating the problem to ergodic theory.

The main ergodic theoretic lemma is:

Let $T_1, \ldots, T_k$ be commuting measure preserving transformations of a probability space (X, μ), and let A be a set of positive measure. Then there is a positive integer n such that $\mu(A \cap T_1^{-n} A \cap \cdots \cap T_k^{-n} A) > 0$. To see how this measure theoretic result implies Szemeredi's theorem, we consider $\Omega = \{0, 1\}^Z$, the space of doubly infinite sequences of zeroes and ones, provided with the product topology. The shift transformation T on Ω is defined by $T\omega(n) = \omega(n+1)$.

Now let S be a subset of the integers of positive upper density, and let 1_S be the indicator function of S (that is $1_S(n) = 1$ if $n \in S$ and 0 otherwise); 1_S is a point of Ω. Let X be the orbit closure of 1_S under T, and let $A = \{\omega \in X \mid \omega(0) = 1\}$. It can be shown that there is a measure μ on X which is invariant under the shift for which $\mu(A) > 0$. (This fact depends on, and in fact is equivalent with the assumption that S has positive upper density.) Now apply the above lemma to the commuting transformations $T, T^2, \ldots, T^k$. It follows that there is a point $\omega \in X$ for which

---

[2] I was the careless referee.

$T^{jn}(\omega) \in A$ for some n and $j = 1, 2, \ldots, k$ from which one easily deduces that for some h, one has $h, h + n, \ldots, h + kn \in S$.

Byproducts of the proof include a general structure theorem for ergodic transformations, which in turn inspired an analogous structure theorem (by Veech) in topological dynamics, as well as "multidimensional" Szemeredi theorems. Furstenberg's proof initiated a fruitful connection between ergodic theory and combinatorial number theory. The ideas introduced played a role in the spectacular recent work of Tau and Green on the existence of arbitrarily long arithmetic progressions in the primes [Green/Tau to appear].

4.4 A Fixed Point Theorem

Standards of proof vary over time.[3] For example, it's well known that Euclid's proofs were incomplete (although apparently all of his theorems are correct). Also, the proofs of the Italian algebraic geometers of the early part of the last century are now found wanting.

An interesting more recent case is provided by a paper by Morton Brown and Walter Neumann [1977], which is related to two papers of G.D. Birkhoff, [1913] and [1925]. Birkhoff claimed to prove a conjecture of Poincaré ("Poincaré's last geometric theorem"), which asserted the existence of two fixed points for an area preserving homeomorphism of an annulus which rotates the boundary circles in opposite directions. Over the years there were questions as to whether Birkhoff's proof was correct. The paper of Brown and Neumann presents a proof which the authors generously say is essentially the same as Birkhoff's.

Be that as it may, the language of Brown and Neumann is quite different from Birkhoff's, reflecting the development of topology in the intervening years. Some of Birkhoff's statements lacked precision. For example a curve is defined to be the boundary of an open set. It isn't even clear what is meant by rotating the boundary curves in opposite directions. (A clockwise rotation of one degree can be regarded as a counterclockwise rotation of 359 degrees.) In the Brown-Neumann paper, this is made precise by passing to the universal covering space, a notion which was probably not known to Birkhoff. Another tool is the homotopy lifting property, which also was probably not known explicitly to Birkhoff.

One expects that fifty years from now, some of the proofs of mathematicians of 2006 will be thought to be in need of correction or modification.

While there are differences among the four proofs just discussed, a common thread is what might be called reinforcement (a new proof, or a correction, or a reworking of an earlier proof). There seems to be no doubt that the theorems in question have been proved.

We are confronted with a different situation with certain very long proofs. We conclude with a brief discussion of two current (possible) proofs of important results, on which the jury is still out.

One is the classification of finite simple groups, organized by the late Daniel Gorenstein, of which there is some doubt whether it has actually been accomplished. (Moe Hirsch, in his response to JQ irreverently asks "Who's in charge here, anyway?")

The other is the (apparent) proof by Perelman of Thurston's geometrization conjecture (which implies the Poincaré conjecture). John Morgan, in a survey article [Morgan 2005], writes

---

[3] A fascinating discussion of differing standards of proof over time is presented in [Kleiner/Movshovitz-Hadar 1997].

"The mathematical community is still trying to digest his argument and ascertain whether it is indeed . . . complete and correct."

In the latter case, the expectation is that a proof will in fact emerge. In spite of certain differences (in particular, the computer plays no role in Perelman's arguments) something like the "Kepler process" is occurring. As was discussed above, following a lengthy and elaborate process, Hales' proof is now generally accepted, and it's quite likely the same will hold for Perelman's.[4]

On the other hand, opinion is sharply divided in regard to the classification of finite simple groups, and it's anyone's guess as to how it will finally turn out.

Acknowledgements I would like to thank the following: Ethan Akin, Stuart Antman, Ken Berg, Mort Brown, Bill Byers, Marshall Cohen, Keith Devlin, Hillel Furstenberg, Bonnie Gold, Judy Green, Paul Green, Denny Gulick, Frances Gulick, Brian Hayes, Karl Hofmann, Michael Keane, Jonathan King, Barbara Meeker, Jonathan Rosenberg, Doris Schattschneider, Tom Ward.

References

[Adler *et al.* 1965] Roy Adler, A.G. Konheim, and M.H. McAndrew, "Topological entropy," *Trans. Amer. Math. Soc.* 114 (1965), pp. 309 – 319.

[Andrews 1994] George G. Andrews, "The death of proof? Semi-rigorous mathematics? You've got to be kidding!" *The Mathematical Intelligencer* 16 (1994), pp. 16–18.

[Atiyah 2005] "Interview with Michael Atiyah and Isadore Singer," *Notices Amer. Math. Soc.* 52 (2005), pp. 225–233.

[Aigner/Ziegler 1999] Martin Aigner and Gunter M. Ziegler, *Proofs from THE BOOK*, Springer, 1999.

[Bass 2003] Hyman Bass, "The Carnegie initiative on the doctorate: the case of mathematics," *Notices Amer. Math. Soc.* 50 [2003], pp. 767–776.

[Bing 1948] RH Bing, "A homogeneous plane continuum," *Duke Mathematics Journal* 15 (1948), pp. 729–742.

[Birkhoff 1913] G.D. Birkhoff, "Proof of Poincaré's last geometric theorem," *Trans. Amer. Math Soc.* 14 (1913), pp. 14–22.

[Birkhoff 1925] ——, "An extension of Poincaré's last geometric theorem," *Acta Mathematica* 47 (1925), pp. 297–311.

[Biss 2004] Daniel Biss, "The elephant in the internet," *Notices Amer. Math. Soc.* 51 (2004), pp. 1217–1219.

[Bourbaki 1968] Nicholas Bourbaki, *Elements of Mathematics, Theory of Sets*, Addison-Wesley, 1968.

[Bressoud 1999] David Bressoud, *Proofs and Confirmations, The Story of the Alternate Sign Matrix Conjecture*, Cambridge University Press, 1999.

[Brown/Neumann 1977] Morton Brown and Walter Neumann, "Proof of the Poincaré -Birkhoff fixed point theorem," *Michigan Math. Jour.* 24 (1977), pp. 21–31.

---

[4] The apparent confirmation of Perelman's proof has come too late for detailed consideration in this article. Perelman has posted his papers on his website, but has refused to submit them for publication. As we have noted, this is contrary to accepted scientific practice. In any case, there has been extensive discussion of the proof, including articles in the *New York Times* (August 15, 2006) and *The New Yorker* (August 25, 2006).

[Choquet 1944] G. Choquet, "Prolongement d'homéomorphes," *Comptes Rendus* 219 (1944), pp. 542–544.

[Davis/Hersh 1981] Philip Davis and Reuben Hersh, *The Mathematical Experience*, Houghton Mifflin, 1981.

[Ernest 1998] Paul Ernest, *Social Constructivisn as a Philosophy of Mathematics*, State University of New York 1998.

[Franks 1989] John Franks, "Comments on the responses to my review of *Chaos*," *Mathematical Intelligencer* 11 (1989), pp. 12–13.

[Furstenberg 1981] H. Furstenberg, *Recurrence in Ergodic Theory and Combinatorial Number Theory*, Princeton University Press, 1981.

[Gale 1990] David Gale, "Proof as explanation," *Mathematical Intelligencer* 12 (1990), p. 4.

[Green/Tau to appear] Ben Green and Terry Tau, "The primes contain arbitrarily long arithmetic progressions," *Annals of Mathematics*, to appear.

[Hales 2005] Thomas Hales, "A proof of the Kepler conjecture," *Annals of Mathematics* 162 (2005), pp. 1063–1183.

[Hales www.] ——, Flyspeck project fact sheet (www.math.pitt.edu/~thales/flyspeck/).

[Halmos 1956] Paul R. Halmos, *Lectures on Ergodic Theory*, Mathematical Society of Japan, 1956.

[Hersh 1997] Reuben Hersh, *What is Mathematics, Really?* Oxford University Press 1997.

[Hsiang 1993] Wu-Yi Hsiang, "On the sphere packing problem and the proof of Kepler's conjecture," *International Journal of Mathematics* 4 (1993), pp. 739–781.

[Jaffe/Quinn 1993] Arthur Jaffe and Frank Quinn, "Theoretical mathematics: towards a cultural synthesis of mathematics and theoretical physics," *Bull. Amer. Math. Soc.* 29 (1993), pp. 1–13.

[Kitcher 1984] Philip Kitcher, *The Nature of Mathematical Knowledge*, Oxford University Press, 1984.

[Kleiner/Movshovitz-Hadar 1997] Israel Kleiner and Nitsa Movshovitz-Hadar, "Proof: a many-splendored thing," *Mathematical Intelligencer* 19 (1997), pp. 16–26.

[Lakatos 1976] Imre Lakatos, *Proofs and Refutations: The Logic of Mathematical Discovery*, Cambridge University Press, 1976.

[Lewontin 2005] Richard Lewontin, "On fraud in science: an exchange," *New York Review of Books*, February 10, 2005, pp. 46–48.

[Morgan 2005] John Morgan, "Recent progress on the Poincaré conjecture and the classification of 3-manifolds," *Bull. Amer. Math. Soc.* 42 (2005), pp. 57–78.

[Popper 1963] Karl Popper, *Conjectures and Refutations: The Growth of Scientific Knowledge*, Routledge, 1963.

[Responses 1994] "Responses to 'Theoretical mathematics,'" *Bull. Amer. Math Soc.* 30 (1994), pp. 178–207.

[Robertson 2003] Douglas Robertson *Phase Change, the Computer Revolution in Science and Mathematics*, Oxford University Press, 2003.

[Ross 1998] Kenneth A. Ross, "The place of algorithms and proofs in school mathematics," *American Mathematical Monthly* 105 (1998), pp. 252–255.

[Rota 1993] Gian-Carlo Rota, "The concept of mathematical truth," pp. 91–96 in Alvin White, ed., *Essays in Humanistic Mathematics*, Mathematical Association of America, 1993.

[Rota 1996] ——, "The phenomenology of mathematical proof," pp. 134–150 in Gian-Carlo Rota and Fabrizio Palombi, *Indiscrete Thoughts*, Birkhauser, 1996.

[Szpiro 2003] George G. Szpiro, *Kepler's Conjecture*, John Wiley, 2003.

[Waraskewich 1937] Z. Waraskewich, "Sur les courbes planes topologiquement homogènes," *Comptes Rendus* 204 (1937), pp. 1388–1390.

[Weinberg 2001] Steven Weinberg, "Can science explain everything? Anything?" *New York Review of Books*, May 31, 2001, pp. 47–50.

[Yandell 2001] Benjamin Yandell, *The Honors Class: Hilbert's Problems and their Solvers*, A.K. Peters, 2001.

[Zeilberger 1993] Doron Zeilberger, "Theorems for a price: tomorrow's semi-rigorous mathematical culture," *Notices Amer. Math. Soc.* 40 (1993), pp. 978–981.

II

Social Constructivist Views of Mathematics

Two completely new philosophies of mathematics have been developed since 1950: structuralism and social constructivism. Structuralism is the view that mathematics is the science of structures, or patterns. That view is discussed in several of the chapters in section 3. Social constructivism has been developed primarily by mathematicians, although one can trace its origins to some discussion by philosophers such as Lakatos. Social constructivism is the view that mathematics is constructed by the community of mathematicians. In one sense, this is so obviously true that there is no need to discuss it further. Certainly, human *knowledge* of mathematics *is* developed by the community of mathematicians. However, as we discover mathematical facts, it *feels* to most of us as if there is an objective reality out there, within which these facts are either true or false. It certainly does not seem that the bunch of us can just one day decide, "the Riemann hypothesis is true," and it will be so. On the other hand, when a new mathematical *concept* is introduced and developed, things are less clear. Is there some external "natural" concept that we're grasping for? Or are we just making it up, albeit with some restrictions related to the questions we are developing it to investigate? The less extreme versions of social constructivism, represented in this volume, suggest that, once the community has developed a mathematical concept, the facts about this concept are indeed objective. However, there are philosophical issues with this viewpoint, and these are also discussed in this section.

The first two chapters of this section were written by mathematicians who have been outspoken proponents of social constructivism, and the third is by a philosopher who has been working to formulate social constructivism carefully enough for criticism by the community of philosophers of mathematics.

4

When Is a Problem Solved ?[1]

Philip J. Davis
Division of Applied Mathematics
Brown University

From the Editors

The question Philip Davis asks in his chapter, "When is a problem solved?" seems like a natural one to ask, but we have never read a discussion of this elsewhere. It is a good example of why it is important for some people who actually do mathematics to contribute to the philosophy of mathematics. There are questions of interest to mathematicians that do not occur to philosophers, who are motivated largely by the types of questions that occur in other areas of philosophy. This question might never occur to philosophers, because it is really only in mathematics that we appear *to get final answers to our questions.*

Philip Davis is a Professor Emeritus of Applied Mathematics at Brown University (www.dam.brown.edu/people/facultypage.davis.html). He came to Brown after serving as Chief for Numerical Analysis at the National Bureau of Standards in Washington, D.C. for five years. His fields of research included numerical analysis and approximation theory, in which he wrote many papers and several books, including Interpolation and Approximation *(1963),* Numerical Integration *(with Philip Rabinowitz, 1967), and* The Schwarz Function *(1974) and* Circulant Matrices *(1979). He is a prize winning expositor of mathematics, who received the Chauvenet Prize of the Mathematical Association of America in 1963 for "An Historical Profile of the Gamma Function." Professor Davis has also received the Laster Ford Award in 1982 for "Are there Considences in Mathematics?" and the George Polya Award in 1986 for "What Do I know? A study of Mathmatical Self-Awareness." In 1997, he won the Communications Award of the Joint Policy Board for the Mathematical Science. His books written jointly with Reuben*

---

[1] This article was initially submitted to this volume. However, due to the long time between initial solicitation of articles and the appearance of this volume, this article first appeared in 2006 in a collection of my articles, *Mathematics and Common Sense: A Case of Creative Tension*. It is reprinted here with permission of the publisher, AK Peters.

Hersh, The Mathematical Experience *(1980) and* Descartes' Dream *(1986), explore certain questions in the philosophy of mathematics, and the role of mathematics in society.* Mathematics and Common Sense: A Case of Creative Tension, *which appeared in 2006, contains a version of his chapter, among other philosophical articles. Readers of this volume will also be interested in his article, "When Mathematics Says No" in* No Way: The Nature of the Impossible, *which he edited with David Park (1987).*

A poem is never finished, it is only abandoned.—Paul Valéry

1 Introduction

I recently spent three days participating in MathPath, a summer math camp for very bright students aged c. 12–14 (see www.mathpath.org). One day I asked the students to pass in to me a question that was a bit conceptual or philosophical. Out of the large variety of responses, one question struck me as both profound and remarkable in that sophisticated interpretations were possible:

Elizabeth Roberts: ***How do we know when a problem is solved?***

My first reaction on reading this question—which was pencilled on a sheet of notebook paper—was "mathematical problems are never solved." Due to my limited stay at the camp, I didn't have the opportunity to ask the student what exactly she meant and so her question went unanswered at the time. I told the camp faculty—all professional mathematicians—my gut reaction. I added that my answer was ***not*** appropriate for the present age group and hoped that the faculty would take up the question after I'd left. I also told the faculty that the question inspired me to write an article. Here it is.

2 A Bit of Philosophy

Some problems are solved. A baker knows when a loaf of bread is done.[2] Yogi Berra said: "It's not over till it's over." Which implies that a baseball game gets over. But when one thinks of the problems that confront humanity—personal, medical, sociological, economic, military— problems that seem **never** to be solved, it is easy to conclude that to be truly alive is to be perpetually racked by problems.

Example: When should clinical trials for new medical procedures be terminated? This question is currently on the front pages of newspapers and is a matter of litigation and the confrontation of statisticians involved in the jurimetrics. [Finkelstein/Levin 2004]

Thus, we are concerned here with a fundamental question that can be viewed as residing at the heart of human existence itself. How can we be sure that we have solved a problem? More than this, how can we be sure we have formulated a proper question? We can't, because problems, questions and solutions are not static entities. On the contrary, the creation, formulation and solution of problems change throughout history, throughout own lifetime and throughout our

---

[2] In an amusing e-letter, Yvon Maday, a Parisian applied mathematician, pointed out to me ambiguities in the baking process.

readings and re-readings of texts. That is to say, meaning is dynamic and ongoing and there is no finality in the creation, formulation and solution to problems, despite our constant efforts to create order in the world. Our ability to create changes in meaning is great and hence our problems and our solutions change. We frequently settle for provisional, "good enough" solutions—often described as "band aid solutions." [O'Halloran 2005]

3 What Might Elizabeth Have Meant?

One might think that in the case of mathematics—that supposedly clean-cut, logical, but limited intellectual area—the situation would be otherwise. One might think that when a mathematical problem arises, then after a while (it may be a very long while) the problem gets solved. But think again; what takes place can be very complex.

The set of possible responses to the question under discussion spans the whole of mathematical methodology, history, and philosophy. Though responses are implicit everywhere in the mathematical literature, I believe that the question as framed here puts a slightly different slant on this material. I don't recall seeing it treated head on.

The question: **How do we know when a problem is solved?** can be approached at a variety of levels. The lay public tends to think that mathematics is an area where there is one and only one answer to a problem. Approached from the point of view of a school teacher, the teacher, relying on habits or traditions, and considering the age of the pupils, knows when a pupil has solved a problem. It is a matter of common sense. (I am not thinking here of multiple choice questions graded by machine.)

Approached from the point of view of the individual or the group that makes up problems either for daily work, tests, or contests, I would suppose that the act of making up the problem already implies a more or less definite notion of what the answer is. The examiner will think the problem is solved if he gets the answer he had in mind or possibly a variant that conforms to certain unconsciously maintained criteria.

One answer, appropriate to students starting algebra, might be "you plug your solution back into the equation and see if it checks." The set of possible responses that lie between this simplistic response and my seemingly dismissive "mathematical problems are never solved," spans the whole of mathematical methodology, history, and philosophy. Though responses to the question under discussion are implicit everywhere in the mathematical literature, I believe that the question as framed puts a slightly different slant on this material. I don't recall seeing it treated head on.

What did the student mean by her question? I can only guess. Perhaps she meant: "How can I tell whether my answer is correct." Well, what methods or practices of validation are available at ages 12–14? Yes, you can plug the answer back into the equation and see if it checks. But this kind of check is not available for most problems—as, for example, what and where do you plug in when asked to add a column of numbers? If you care to employ them, processes such as "casting out nines" (taught in elementary school years and years ago) or estimating the sum provide partial checks for addition.

You can "check your work" by doing the problem over again in perhaps a simpler or a more clever way and then compare. You may, in some cases, put the problem or part of it on a computer. You can ask your friend what her answer is and compare. You can look in the back of the book and see whether you get the book's answer. If the problem is a "word problem," you can ask whether your answer makes sense in the "real world." An answer of minus seven and a half dappled cows is evidence of an error somewhere.

Perhaps the student, having learned that $\sqrt{2}$ is irrational, will wonder whether or why $\sqrt{2} = 1.41421356237\ldots$ constitutes an answer. From a certain point of view, $\sqrt{2}$ can never have a completed answer. Does one have to elaborate the meaning of the three dots . . . and trot out the theory of the set of real numbers to accept this as an answer?

Iterative computations that theoretically "converge at infinity" are frequent. They must be terminated—abandoned—and an "answer" outputted. I know at least thirteen different termination criteria that are employed. It would be useful to have a full taxonomic study of such criteria, but I am not aware of such a study.[3]

Perhaps the student, having heard from the camp faculty (or from reading newspapers) that some mathematical problems have taken centuries before they were resolved, was asking me how long she should spend on a problem before abandoning it. We all abandon problems. Life calls us to other things that must get done.

4 Mathematical Argumentation as a Mixture of Materials

Here is a final conjecture as to what might have been in the student's mind in asking the question. It is a **very unlikely** conjecture, but it expresses a feeling that I occasionally have after reading through mathematical material.

What is the source of one's confidence that the informal, patched together mixture of verbal argumentation, symbol manipulation, computation and the use of visuals, whether in the published literature or of one's own devising, all click together properly as presented, and result in the confident assertion: "Yes, that certainly solves the problem!"

Let me elaborate. Consider the processes and techniques used in solving mathematical problems. The mélange of materials involved has been well described by mathematical semioticist Kay O'Halloran who studies the relationship between mathematical ideas and the symbols with which these ideas are expressed.

> "Mathematical discourse succeeds through the interwoven grammars of language, math-
> ematical symbolism and visual images, which means that shifts may be made seamlessly
> across these three resources. Each semiotic resource has a particular contribution or func-
> tion within mathematical discourse. Language is used to introduce, contextualize, and
> describe the mathematics problem. The next step is typically the visualization of the
> problem in diagrammatic form. Finally, the problem is solved using mathematical sym-
> bolism through a variety of approaches which include the recognition of patterns, the
> use of analogy, an examination of different cases, working backwards from a solution
> to arrive at the original data, establishing sub-goals for complex problems, indirect rea-
> soning in the form of proof by contradiction, mathematical induction and mathematical
> deduction using previously established results."   [O'Halloran 2005]

---

[3] Each special problem may develop its own special termination criteria. See,e.g., [Ehrich 2001].

Behind the understanding of and expertise with symbolisms, there are cognitive capacities that act to create and glue together the mathematical discourse. Lakoff & Núñez give a list required for doing simple arithmetic. They are (with these authors' elaborations omitted):

> "grouping capacity, ordering capacity, pairing capacity, memory capacity, exhaustion detection capacity, cardinal number assignment, independent-order capacity, combinatorial-grouping capacity, symbolizing capacity, metaphorizing capacity, conceptual-blending capacity." [Lakoff/Núñez 2000]

Just as logicians have wondered whether further axioms are necessary for mathematics, I wonder whether further mental capacities than those above are required to do mathematics that is more complex than simple arithmetic. I wonder whether as mathematics progresses, and as it adds new proofs and develops new theories, we are now in the possession of additional mental capacities in virtue of the work of the brilliant mathematicians of the past. I wonder also whether semantics, semiotics, and cognitive science, taken together, are adequate to explain the occurrence of the miraculous epiphany "Yes. That's it. The problem is now solved." Psychological studies and autobiographical material have not yet uncovered all the ingredients that make up the "aha" moment.

5 From a Mathematician's Perspective

I am now lead to imagine that the question *How do we know when a problem is solved?* has been put to a professional. There is no universal answer to this question. It depends on the situation at hand. The typical answers for validation just given to young math students, carry into the professional domain. Examples: product barcodes have check digits that employ modular arithmetic. When, in the first generation of computers, I computed the Gaussian weights and abscissas for approximate integration to 30 D, I plugged back to verify my output. The modes of validating a long and involved computation may involve reworking the problem with a different algorithm, with different software on a different computer and then comparing.[4]

But there is much, much more that has to be said. At the very outset, one might ask: does the problem, as stated, make sense or does it need reformulation? There are ill-posed problems, in either the technical sense or a broader sense. There are well-posed problems, weakly-well-posed problems, etc. One might also ask—but is rarely able to ask at the outset—does the problem have a solution? From the simplest problems lacking solutions, such as "express $\sqrt{2}$ as the ratio of two integers," or "find two real numbers x and y such that $x + y = 1$ and $xy = 1$ simultaneously," to the unsolvable problems implied by Gödel's Theorem, the potential solvability can be an issue that lurks in the background. We are faced with the paradoxical situation that the solution to a problem may be that there is no solution.

What kind of an answer will you accept as a solution? It is important to have in mind the purpose to which a presumptive solution will be put. (See [Uspenskii 1974], pp. 5–8, and [Wilf 1982].)

A so-called solution may be useless in certain situations and hence, not a solution at all.

---

[4] At the research level, "plugging back in" can have its own problems. See [Gautschi 1983].

Example: The expression of the determinant of an $n \times n$ matrix in terms of $n!$ monomials formed from the matrix elements is pretty useless in the world of scientific computation. One looks around for other ways and finds them.

Example: Most finite algorithm problems have a solution that involves enumerating all the possibilities and checking, but this brute force strategy is seldom a satisfactory solution and is certainly not an aesthetic solution.

Example: A differential equation may be solved by exhibiting its solution as an integral. But to a college undergraduate who has met up with integrals only in a previous semester, an integral is itself a problem and not a solution. An approximation to the solution of a differential equation may be exhibited as a table, a graph, a computer program or may be built into a chip. Is such a solution good enough in a particular situation?

Example: If the problem is to "identify" the sequence $1, 2, 9, 15, 16, \ldots$ will you accept a "closed" formula (query: what exactly do you consider as a closed formula?), a recurrence relation, an asymptotic formula, a generating function? A semi-verbal description? Do you want statistical averages or other properties? Will you try to find the sequence in *The Online Encyclopedia of Integer Sequences*? Or will you simply say that a finite sequence of numbers can be extended to an infinite sequence in an unlimited number of ways and chuck the problem out the window as ill-formulated? How would you even elaborate explicitly the verb "identify" so as not to chuck the problem?

Though a problem has been solved in one particular way, the manner of solution may suggest that it would be very nice to have an alternate solution. An interesting instance of this is the prime number theorem. Originally proved via complex variable methods, Norbert Wiener (and others) asked for a real variable proof. Since the statement of the prime number theorem involves only real numbers, the demand for such a proof was possibly a matter of mathematical aesthetics. A real variable proof was given by Paul Erdös and Atle Selberg in 1949, partly independently.

Is such and such really a solution? There are constructive solutions but, as already observed, a solution may be "constructive" in principle but in practice the construction would take too long to be of any actual use. (The dimensional effect or the n! effect.)

Then there are existential solutions in which the generic statement is "There exists a number, a function, a structure, a whatever, such that. ... " The mathematician Paul Gordan (1837–1912), when confronted with Hilbert's existential (i.e., non-constructive) proof of the existence of a finite rational integral basis for binary invariants, asked "Is this mathematics or theology?" ([Reid 1970], pp. 34–37)

Example: The Mean Value Theorem asserts that given a function $f(x)$, continuous on (a, b) and differentiable on (a, b), there exists a ξ in (a, b) such that $f(b) - f(a) = f'(\xi)(b - a)$. Some students find this statement hard to take when they first meet up with it. The ξ appears mysterious.

Example: The famous Pigeonhole Principle: Given m boxes and n objects in the boxes where n is larger than m. Then there exists at least one box that contains more than one object. Who can deny this? This may lead to an existential solution. On this basis, for example, together with some tonsorial data, one can conclude that there are two people in Manhattan that have the same number of hairs on their head. Now find them. We have been assured that they can surely be located in a platonic universe of mortals.

Example: There exist irrational numbers x and y such that x^y is rational. Proof: Set $r = \sqrt{2}^{\sqrt{2}}$. Now if r is rational, then since $\sqrt{2}$ is irrational, the selection $x = y = \sqrt{2}$ works. On the other hand, if r is irrational, then set $x = r$ and $y = \sqrt{2}$. Since $x^y = (\sqrt{2}^{\sqrt{2}})^{\sqrt{2}} = (\sqrt{2})^2 = 2$, this selection works. One may ask: is this really a solution if we can't, with our present knowledge, decide whether r is or is not rational?

There are "probabilistic solutions" as, for example, the Rabin-Miller probabilistic test for the primality of a large integer. [Rabin 1980]

Then there are the "weak solutions." In 1934, Jean Leray proved that there is a weak solution to the incompressible Navier-Stokes equations. Is there only one such? The question appears to be still open. But what, in a few sentences, is a weak solution? If there is ambiguity about the very notion of a "solution," this is equally the case for a "weak solution." Technically, if L is a differential operator, and if u = f satisfies the equation Lu = g, then f is the solution. If so, then for all "test functions" ø,

(Lf, ø) = (g, ø), (,) designating an inner product. But if only the latter is true, f is said to be a "weak solution."

Since some problems are very difficult, or even unreachable with current mathematical theory and techniques, the notion of a weak problem, possessing weak solutions, has been introduced as a framework that allows existing mathematical tools to solve them. A strong solution is a weak one but often a weak solution is not a strong one, and the relation between the two notions is still the subject of intense research.

In a numerical problem, is a weak solution really a solution if it is not computable? Despite this limitation, the knowledge that a weak solution exists can have a have considerable impact.

Apparently, the meaning of the word "solution" can be stretched quite a bit. The elastic quality of mathematical terms or definitions is remarkable, and is often achieved through context enlargement.

There are cases where a problem has been turned into its opposite. Thus, the search for the dependence of Euclid's Fifth Axiom (the parallel axiom) on the other axioms, resulted in the unanticipated knowledge of its independence. The Axiom of Choice was hopefully derivable from the other axioms of set theory. It is now known to be independent of them. An instruction to prove, disprove, or prove that neither proof nor disproof is possible, is a legitimate, though a psychologically unpleasant formulation of a problem.

There are cases where a problem was felt to be solved, and then later was felt to be open, not because an error was found, but because there was a shift in the (unconscious) interpretation of what had been given. For this, read Imre Lakatos' classic discussion *of the history of the Euler-Poincaré theorem. A very early version reads* $V - E + F = 2$ where V, E, and F are respectively the number of vertices, edges, and faces of a polyhedron. But just what kind of a 3-dimensional object is a polyhedron and what are its vertices, edges and faces? Lakatos' discussion chronicles the ensuing tug-of war—almost comic—between hypotheses and conclusions and the negotiations necessary so as to maintain a semblance of the original conclusion. This is known in philosophy as "saving the phenomenon." [Lakatos 1976].

6 When is a Proof Complete?

If the problem is to find a proof (or a disproof) of a conjecture, how does one know that that a purported proof is correct? Gallons and gallons of ink have been expended on this question as formulated generally. Are proofs stable over time? A half century after D'Alembert gave a proof of the Fundamental Theorem of Algebra, Gauss criticized it. A century after Gauss' first proof (he gave four), Alexander Ostrowski criticized it.

Is a proof legitimate if it is hundreds of pages long and would tire most of its human checkers? Is a proof by computer considered legitimate? The publicized proof by Thomas Hales of the Kepler sphere packing conjecture is said to require 250 pages of text and 3 gigabytes of programs. The mathematical community is itself split over the philosophical implications of the answers given to these and a myriad of similar questions. [Hales www]

For one criterion as to when a solution is a solution, when a proof is a proof, let's go, as bank robber Willie Sutton said he went, to where the money is. A recent answer to this question was formulated by the Clay Mathematics Institute which offers prizes of a million dollars for the solution of each of seven famous problems. The Clay criteria for determining whether a problem is solved are as follows.

(1) The solution must be published in a refereed journal.
(2) A wait of two years must ensue after which time if the solution is still "generally acceptable" to the mathematical community,
(3) the Clay Institute will appoint its own committee to verify the solution.

In short, a solution is accepted as such if a group of qualified experts in the field agree that it's a solution. This comes close to an assertion of the socially constructive nature of mathematics. The remarkable thing is the social phenomenon of (almost) universal, but not necessarily rapid, agreement , which has been cited as strengthening mathematical platonism. (See [Davis 1990], [Ernest 1998], [Rosental 2003].)

7 Applied Mathematics

In applied mathematics—and I include here both physical and social models—other answers to the basic question of this article can be put forward. Proofs may not be of importance. The formulation of adequate mathematical models and adequate computer algorithms may be all important. What may be sought is not a solution but a "good enough solution."

In introductions to applied mathematical and in philosophical texts, loops are often displayed to outline and conceptualize the process. The loops indicate a flow from

(a) the real world problem to
(b) the formulation of a mathematical model, to
(c) the theoretical consequences of the model, to
(d) the computer algorithm or code, to
(e) the computer output to
(f) the comparison between output and experiment.

Then back to any one of (b)–(f) at any stage. And even back to (a), for in the intervening time, the real world problem may have changed, may have been reconceived, or even abandoned.

In looking over these steps, it occurred to me that one additional step is missing from this standardized list. It is that (f) can lead to

(g) an action taken in the real world and to the responses of the real world to this action.

This omission might be explained as follows: at every stage of the process one must certainly simplify—but not too much, else verisimilitude will be lost. The responses of the real world are both of a physical and of a human nature, and the latter is notoriously difficult to handle via mathematical modeling. Hence there is a temptation to "put a diagrammatic wall" around (b) to (e) that emphasizes the mathematical portion as though mathematics gets done in a sanitized world of idealized concepts that does not relate to humans. Step (g) is often conflated with (f) and let go at that. Since we are living in a thoroughly mathematized world with additional mathematizations inserted by fiat every day that impact our lives in myriads of ways, it is vital to distinguish (g) and to emphasize it as a separate stage of the process.

What cannot be known in advance is how often these loops must be traversed before one says the problem has been adequately solved. Common sense, experience, the support of the larger community in terms of encouragement and funding may all be involved arriving at a judgment. And yet, one may still wonder whether steps (a)–(g) provide a sufficiently accurate description of the methodology of applied mathematics.

8 Some Historical Perspectives

One can throw historical light on the question of when a problem is solved. There are several ways of writing the history of mathematics. I'll call them the horizontal and the vertical ways. In horizontal history, one tries to tell all that was going on in, say, the period 400–300 B.C. or between 1801 and 1855. In vertical history, one selects a specific theme or mathematical seed, and shows how, from our contemporary perspective, it has blossomed over time. (See [Grattan-Guinness 2004].)

As a piece of vertical mini-history, consider the quadratic algebraic equations first met in high school. Such equations were "solved" by the Babylonians 4,000 years ago. But over the years, immense new problems came out of this equation in a variety of ways: higher order algebraic equations, the real number system as we now know it, complex numbers and algebraic geometries; group and field theory, modern number theory, numerical analysis.

Solving a polynomial algebraic equation of degree n once meant finding a positive rational solution. Today it means finding all solutions, real or complex together with their multiplicities and finding it either in closed form (rare) or by means of a convergent algorithm whose rate of convergence can be specified. But the generalizations of quadratic equations go further. Formal equations can be interpreted as a matrix or even as an operator equation in various abstract spaces. The equation $x^2 = 0$ trivially has only $x = 0$ as its solution when x is either real or complex. But this is not the case if x is interpreted as an n by n matrix: the nilpotent matrices solve this equation. And if you have the temerity to ask for all nilpotent operators in abstract spaces, you have raised a question without a foreseeable end.

A more recent example, of which there are multitudes. In 1959, Gelfand asked for the index of systems of linear elliptic differential equations on compact manifolds without boundary. The problem was solved in 1963 by Atiyah and Singer, and this opened up new ramifications with surprising features including Alain Connes' work on non-commutative geometry.

In the historical context, mathematical problems are never solved. Material, well established, is gone over and over again. New proofs, often simplified, are produced; contexts are varied, enlarged, united, and generalized. Remarkable connections are found. Repetition, reexamination are parts of the practice of mathematics.

9 A Dialogue on When is a Theory Complete

The original question as to when is a problem solved may be moved up a level to ask: when is a theory complete? Stephen Maurer, one of the MathPath faculty, provided me with a web discussion of this question he'd had with one of his most philosophical students. I present it here as Maurer sent it to me.

Andy Drucker:

"This question has been haunting me, and I know I shouldn't expect definite answers. But how do mathematicians know when a theory is more or less done? Is it when they've reached a systematic classification theorem or a computational method for the objects they were looking for? Do they typically begin with ambitions as to the capabilities they'd like to achieve? I suppose there's nuanced interaction here, for instance, in seeking theoretical comprehension of vector spaces we find that these spaces can be characterized by possibly finite 'basis' sets. Does this lead us to want to construct algorithmically these new ensembles whose existence we weren't aware of to begin with? Or, pessimistically, do the results just start petering out, either because the 'interesting' ones are exhausted or because as we push out into theorem-space it becomes too wild and wooly to reward our efforts? Are there more compelling things to discover about vector spaces in general, or do we need to start scrutinizing specific vector spaces for neat quirks—or introduce additional structure into our axioms (or definitions): dot products, angles, magnitudes, etc.?

Also, how strong or detailed is the typical mathematician's sense of the openness or settledness of the various theories? And is there an alternative hypothesis I'm missing?"

Stephen Maurer:

"This is an absolutely wonderful question—how do mathematicians know when a theory is done—and you are right that there is no definitive answer. The two answers you gave are both correct, and I can think of a third. Your two answers were 1) we know it's done when the questions people set out to answer have been answered, and 2) we know it's done when new results dry up. My third answer is 3) we don't know when it's done.

An individual probably feels done with a theory when the questions that led him/her to the subject are answered (answered in a way that he feels gives a real understanding) and he either sees no further interesting follow-up questions or can't make progress on the ones he sees. Mathematicians as a group probably feel it's done when progress peters out—the subject is no longer hot and it is easier to make a reputation in some other field that is opening up. (You called this attitude pessimistic, and I'm not so keen about it either, but it shows that math, like other subjects, is influenced by more than pure thought, and it means that mathematicians are trying to optimize results/effort.)

But finally, history shows that fields are rarely ever done. Much later a new way of looking at an old field may arise, and then it's a new ball game. Geometry is an example. The study of n-dimensions was around long before vectors and dot products (there are books of n-dimensional theorems proved by classical Euclidean methods) but the creation of these vector ideas in physics led to a new blossoming of geometry.

Another example is the field of matroids, in which I got my Ph.D. Matroids have been described as "linear algebra without the algebra." Concepts such as basis and independence make sense (and have the same theorems you have seen, such as that all bases have the same size) but there is no plus or scalar multiplication! Matroids were invented in the 1930s, for a different purpose than generalizing linear algebra, and lay fallow for some time. Then, starting in the 1960s, their general value was appreciated and they sprung to life for perhaps 30 years. We might have said that we thought linear algebra was done, but since matroids are a form of linear algebra generalization, we discovered it was not done.

Now matroids are fairly quiet again; there are still papers published in the field, but the natural questions that occurred to people when the subject was fresh have been answered or people have mostly stopped trying. It has become, like linear algebra itself, a background theory that people apply when appropriate."

Examples of revitalization abound. At the end of the 19th century, it was thought that invariant theory was finished and that Hilbert's work had killed it off. But it lives on. Where is nomography today? Its theoretical heyday seems to have been in the work of Maurice d'Ocagne [d'Ocagne 1899], but it lives on in engineering circles. See also [P. Davis 1995] for another example of revitalization in geometry.

Reading the Drucker-Maurer dialogue recalled to my mind that Felix Klein (1849–1925) and John von Neumann (1903–1957) emphasized other sources of revitalization. Felix Klein:

"It should always be required that a mathematical subject not be considered exhausted until it has become intuitively evident...." ([Kline 1972], p. 904)

By Klein's criterion, and considering contemporary proofs that require hundreds of pages or are done with a computer assist, it would appear that many mathematical subjects have a long life ahead of them before they become intuitively evident.

Von Neumann's answer contains a cautionary message which I, as an applied mathematician, appreciate. I reproduce a short portion of his article.

"As a mathematical discipline travels far from its empirical source, or still more, if it is a second and third generation only indirectly inspired from ideas coming from 'reality,' it is beset with very grave dangers. It becomes more and more purely aestheticizing, more and more purely *l'art pour l'art*. This need not be bad, if the field is surrounded by correlated subjects, which still have closer empirical connections, or if the discipline is under the influence of men with an exceptionally well-developed taste.

But there is a grave danger that the subject will develop along the line of least resistance, that the stream, so far from its source, will separate into a multitude of insignificant branches, and that the discipline will become a disorganized mass of details and complexities.

In other words, at a great distance from its empirical source, or after much 'abstract' inbreeding, a mathematical subject is in danger of degeneration. At the inception the style is usually classical; when it shows signs of becoming baroque the danger signal is up. It would be easy to give examples, to trace specific evolutions into the baroque and the very high baroque, but this would be too technical.

In any event, whenever this stage is reached, the only remedy seems to me to be the rejuvenating return to the source: the reinjection of more or less directly empirical ideas. I am convinced that this is a necessary condition to conserve the freshness and the vitality of the subject, and that this will remain so in the future." [von Neumann 1947]

10 A Possible Example of Renewal from the Outside

It may be invidious to mention a specific example of exhaustion of a field when there are people working very happily in it. But the following example and opinion is in the open literature. (See [Mumford 2000].) Classical mathematical logic, which proceeds from Aristotle through Frege, Russell & Whitehead, Tarski, and later, has lost its connection to reality and has produced mathematical monsters. The change that is suggested is to develop logics that build in theories of probability. There currently exist a number of probabilistic logics, but they are not entirely successful. Some have even said: construct logics that build in "intent" in the sense of the mathematical philosophy of Edmund Husserl.

11 Implications for Mathematical Education

What are some of the pedagogic implications of the discussions of this article?

Normally, the average student thinks of a mathematical problem as something where one arrives at a single answer as quickly as possible and then moves on to the next assigned problem. Brighter students—those who will go further with mathematics—should be encouraged to think of a problem as never really finished.

Other ways of looking at the problem may emerge and yield new insights. It is also important to examine a problem in relation to other parts of mathematics as well as to the historical and cultural flow of ideas in which it is embedded.

Discovering a sense in which a solved problem is still not completely solved but leads to new and profound challenges, is one important direction that mathematical research takes. To be fully alive in the world of mathematics is to be constantly aware of this possibility.

Finally, alluding to my MathPath experience that gave rise to this article, taking a student's question seriously can be fruitful for both the student and the professor. "Out of the mouths of babes and sucklings have I found strength."

Acknowledgements I wish to acknowledge my debt to George Rubin Thomas, the director of MathPath 2004, to its faculty and, in particular, to Elizabeth Roberts who asked the question. Thanks also to Stephen Maurer of Swarthmore College and the MathPath faculty who provided me with the dialogue that ensued when his graduate student Andy Drucker asked a higher-level question.

Thanks to the following mathematical friends who have also found the question stimulating: Bernhelm Booss-Bavnbek, Chandler Davis, Ernest S. Davis, Reuben Hersh, Yvon Maday, David Mumford, Kati Munkacsy, Kay L. O'Halloran. I have built their responses into this article. And finally, thanks to Bonnie Gold and Roger Simons for providing me with a number of textual and editorial suggestions and for including this article in their book.

The day hardly passes in which I do not receive further reponses and ramifications from additional friends. I assert firmly that I will never know when this article will really be finished. "The Song Is Ended but the Melody Lingers On"—Irving Berlin.

Bibliography

[Clay] Clay Institute criteria: http://www.claymath.org/millennium/Rules_etc/

[Davis 1990] Chandler Davis, "Criticisms of the Usual Rationale for Validity in Mathematics," pp. 343–356 in *Physicalism in Mathematics*, A.D. Irvine, ed., Kluwer, 1990.

[Davis 1987] Philip J. Davis, "When Mathematics says No," in *No Way*, Philip J. Davis and David Park, eds., W.H. Freeman, 1987.

[Davis 1995] ——, "The Rise, Fall, and Possible Transfiguration of Triangle Geometry: A Mini-history," *American Mathematical Monthly* 102 (1995), pp. 20–214.

[Ehrich 2001] Sven Ehrich, "Stopping Functionals for Gaussian Quadrature Formulas," *J. Comp. and Appl. Math.* 127 (2001), pp. 153–171.

[Ernest 1998] Paul Ernest, *Social Constructivism as a Philosophy of Mathematics,* SUNY Albany, 1998.

[Finkelstein/Levin 2004] Michael Finkelstein and Bruce Levin, "Stopping rules in clinical trials," *Chance* 17 (2004), pp. 39–42.

[Gautschi 1983] Walter Gautschi, "How and How Not to Check Gaussian Quadrature Formulae," *BIT* 23 (1983) pp. 209–216.

[Grattan-Guinness 2004] Ivor Grattan-Guinness, "History or Heritage? An Important Distinction in mathematics and in mathematics education," *Amer. Math. Monthly* 111 (2004), pp. 1–12.

[Hales www] Thomas Hales' proof: http://www.maa.org/devlin/devlin_9_98.html.

[Mumford 2000] David Mumford, "The Dawning of the Age of Stochasticity," *Rend. Mat. Acc. Lincei* 9 (2000), pp. 107–125.

[Kline 1972] Morris Kline, *Mathematical Thought from Ancient to Modern Times*, New York, Oxford University Press, 1972.

[Lakatos 1976] Imre Lakatos, *Proofs and Refutations*, Cambridge Univ. Press, 1976.

[Lakoff/Núñez 2000] George Lakoff and Rafael Núñez, *Where Mathematics Comes From: How the Embodied Mind Brings Mathematics into Being,* Basic Books, 2000.

[von Neumann 1947] John von Neumann, "The Mathematician," in *The Works of the Mind,* Robert B. Heywood, ed., Univ. Chicago Press, 1947. Reprinted in *Musings of the Masters*, Raymong G. Ayoub, ed., Mathematical Association of America, 2004, pp. 169–184.

[d'Ocagne 1899] Maurice d'Ocagne, *Traite de Nomographie,* G. Villars, 1899.

[O'Halloran 2005] Kay L.O'Halloran, e-mail correspondence. Also: *Mathematical Discourse: Language, Symbolism, and Visual Images.* Continuum, London and New York, 2005.

[Rabin 1980] M. O. Rabin, "Probabilistic Algorithm for Testing Primality," *J. Number Th.* 12 (1980), pp. 128–138.

[Reid 1970] Constance Reid, *Hilbert*, Springer Verlag, 1970.

[Rosental 2003] Claude Rosental, "Certifying Knowledge: The Sociology of a Logical Theorem in Artificial Intelligence," *American Sociological Review*, 68 (2003), pp. 623–644.

[Uspenskii 1974] V.A. Uspenskii, *Pascal's Triangle*, University of Chicago Press, 1974.

[Wilf 1982] Herbert Wilf, "What is an answer?" *Amer. Math. Monthly*, 89 (1982), pp. 289–292.

5

Mathematical Practice as a Scientific Problem

Reuben Hersh
Emeritus Professor of Mathematics and Statistics
University of New Mexico

From the Editors

Reuben Hersh is probably the best-known proponent of social constructivism as a philosophy of mathematics, which was implicit in his two books with Philip Davis and made explicit in his own What is Mathematics, Really? *His viewpoint results in his reading widely, not only philosophers of mathematics, but also sociologists, anthropologists, linguists, and others who have something to say about how mathematics develops. He tends to expand the topics generally considered part of the philosophy of mathematics. In this chapter, he explores several topics from a social constructivist viewpoint: why the existence and nature of mathematical objects are important, why it is important to study mathematical practice from a scientific perspective, and the apparent timelessness of mathematical results.*

Reuben Hersh is an Emeritus Professor of Mathematics and Statistics at the University of New Mexico (www.math.unm.edu/~rhersh/). His mathematical work has been primarily in partial differential equations and random evolutions. In addition to his research work, he has written a number of expository articles, including "Non-Cantorian set theory" (with Paul J. Cohen), Scientific American (1967), "Nonstandard analysis" (with M. Davis), Scientific American *(1972), "How to classify differential polynomials,"* American Mathematical Monthly *(1973), and "Hilbert's tenth problem" (with M. Davis),* Scientific American *(1973) (which won the Chauvenet prize). His two books with Philip Davis,* The Mathematical Experience *(1980) and* Descartes' Dream *(1986), explore certain questions in the philosophy of mathematics, and the role of mathematics in society. To explore more concretely the social constructivism that was the philosophical basis of these two books, he wrote* What is Mathematics, Really? *More recently, he edited a book,* 18 Unconventional Essays on the Nature of Mathematics *(2005) that explores some of the issues his chapter here suggests need to be studied. Among his other philosophical articles that readers of this volume are likely to find interesting are "Some Proposals for Reviving the Philosophy of*

Mathematics," Advances in Mathematics *(1979)*, *"Rhetoric and Mathematics"* (with P. J. Davis) in The Rhetoric of the Human Sciences *(1987)*, *"What is Humanistic Mathematics?"* Mathematics in College *(1990)*, *"Let's Teach Philosophy of Math!"* in The College Mathematics Journal *(1990)*, *"Mathematics Has a Front and a Back,"* reprinted in Synthese *(1991)*, *"Proof is convincing and explaining,"* Educational Studies in Mathematics *(1993)*, *"Humanistic Mathematics and the Real World"* in Essays in Humanistic Mathematics *(1993)*, *"Math Lingo vs. Plain English: Double Entendre,"* American Mathematical Monthly, *(1997)*, and *"Proof—Once More and Yet Again,"* Philosophia Mathematica *(1997)*. *His latest book with Vera John-Steiner,* Loving and Hating Mathematics: Inside Mathematical Life *will appear in 2009.*

1 Introduction

Mathematical entities do exist, they are cultural items. Mathematical experience and activity need to be studied both philosophically and empirically. Study of the nature of mathematics brings together neuroscience and cognitive science, linguistics, history, anthropology, sociology and philosophy. Phenomenological analysis can make a useful contribution: for example, in clarifying the sense in which mathematical truths are "timeless."

2 Atiyah's Pleasant Surprise

Commenting on a recent anthology [Hersh 2005], Michael Atiyah wrote: "I was pleasantly surprised to find that this book does not treat mathematics as desiccated formal logic, but as a living organism, immediately recognizable to any working mathematician."

What does it mean to say that mathematics is "a living organism"? It grows, it evolves, it interacts with its environment. It has purpose and intention. It's created and sustained by and for living human beings—with all the complexity which that fact implies.

Atiyah's comment of course conveys no disrespect for logic as a branch of mathematics. The "desiccation" refers to the philosophical reduction of all mathematics to "formal logic" (including formal set theory). That view of mathematics was plainly stated by W.V.O. Quine: "Researches in the foundations of mathematics have made it clear that all of mathematics in the above sense [i.e., all of both pure and applicable mathematics] can be got down to logic and set theory." [Quine 1966] Forty years later, that opinion still is found in the academic philosophy of mathematics. But when mathematicians wonder about the meaning and nature of our work, we look at our actual experience, creating and discovering the facts of geometry, algebra, topology, or analysis. Our reports are recognizable as mathematics by other mathematicians.

The English philosopher David Corfield urges his colleagues to get interested in mathematics as it is lived and practiced. "By far the larger part of activity in what goes by the name 'philosophy of mathematics' is dead to what mathematicians think and have thought, aside from an unbalanced interest in the 'foundational' ideas of the 1890–1930 time . . . We should be looking to inspire a new generation of philosophers to sign up to the major project of understanding how mathematics works." [Corfield 2003]

And now, in the present collection, philosophers of mathematics are addressing mathematicians! This is radical. Maybe it will start a trend. (Indeed, four or five philosophers contributed to the anthology that Professor Atiyah found "pleasantly surprising.")

Does "existence" matter?

Much current conversation among philosophers of mathematics is about "Platonism versus fictionalism." Do mathematical "things" (objects, entities, items, whatever label you like) really "exist" (whatever that means)? Or are they just "fictions" (whatever that means)?

The trouble is that mathematical items don't fall into either of Rene Descartes' two categories of existence—physical (material, ponderable, spatiotemporal) or mental (subjective, private.) The number 2, for example, is neither a physical object nor a private thought in the philosopher's head.

Perhaps out of impatience with the Platonist-fictionalist back-and-forth, some writers have even decided that existence is a matter of no concern ([Cellucci 2005], [Davies], [MacKenzie 2005], [Rota 1996]). (After all, how we calculate and prove isn't affected by philosophical existence.) And yet, *mathematical* existence is one of our frequent concerns! (As in, existence of the sporadic "Monsters" of finite group theory, or existence of a classical solution of the mixed initial-boundary value problem for the Navier-Stokes equation.)

I am only a "working mathematician," not a philosopher, so I say something "exists" if it affects us, if we need to take it into account in our actions. This violates the honorable tradition, older than Plato, in which the ephemera of daily life are mere illusion, and do not "exist" for philosophy.

If we take daily life and experience as real, the notion of existence can't be restricted only to physical existence—what can be weighed, measured, detected in the laboratory or the observatory—nor to mental existence—the private consciousness of the individual philosopher. "What exists" has to include the other things that daily life is made up of, that no one can ignore—the calendar, the schedule, the price list, the pay roll.

Laws_
  Customs_
    Family relationship_
      Nations_
        Wars_
          Special sales for Christimas_
            Et cetera, et cetera.

All that important stuff is not weighed and measured, nor is it located inside the philosopher's mind. You can call it "public," or "inter-subjective," or "cultural-historic-social." I call it "social" for short. Of course, the social is grounded in the physical and mental, in complex, fascinating ways. We must study and try to understand all that. But first of all, given the slightest degree or measure of sanity, it exists! Mathematical objects (entities, items, whatever word you like) are part of that public, social, intersubjective world. As such, they exist. They are real. They have objective properties, which we may discover, or which may elude discovery.

But someone may object, and argue, "You say something exists if it actually affects us. Yes, the speed limit and the price of gasoline are real, they do affect me. But how do the facts of mathematics directly affect anyone?" One standard answer is, "Mathematics is tied, directly and indirectly, to physics, which is embodied in the objects and processes you use every day in modern consumer society." A second answer is, "Your checks will bounce if you ignore the laws of arithmetic." A third answer is: "Mathematicians, once they enter the world of mathematics, find that they cannot do whatever they please. They must accommodate to mathematical reality."[1] (Thus it was that Andrew Wiles, working on Fermat's last conjecture, was stuck in his attic for seven years. The subtleties and complications he had to overcome were real. They had to be understood as they really are; they could not be ignored, just because he might have wished they weren't there.)

Once existence is understood to include all the stuff we have to deal with every day, the big puzzle about mathematical existence fades away. Mathematics exists, neither as a kind of physical entity, nor as a private mental experience, but socially, historically, culturally, inter-subjectively, publicly. The classical article [White 2005] by the famous anthropologist Leslie White established this simple fact once and for all. Starting from that simple observation, we should work to unravel the important properties and qualities of mathematical entities, objects, processes. White's friend, the topologist Raymond Wilder, made important contributions, with his effort to apply "culturological" analysis to mathematics. His work [Wilder 1981] deserves to be reread today.

What's missing from White's article is the special, unique character of mathematical objects—their universality, their seeming "certainty," the unanimous agreement that is forced by mathematical argumentation. *To account more deeply for these special, characteristic features of mathematics is the major open problem.*

One approach to it would be through understanding its adaptive role in the evolution of our species; another would be by investigating its basis in our nervous system and our visual and auditory centers. (See, e.g., the articles by Rav and Nunez in [Hersh 2005a].)

The trouble with Platonism is not so much that it's wrong. The trouble is, it's an easy answer, it avoids looking for scientific answers. When Newton and Leibniz believed that mathematical truths are thoughts in the mind of God, they didn't need to trouble any further about the nature of mathematical truth. Still today, a kind of Platonic faith is natural, at the moment when you're hot in pursuit of your research problem. But when you step back, and look at what you and your colleagues are doing, you can recognize another fascinating problem: to understand mathematics as a special aspect of human thought and culture.

3 For a Multi-Disciplined Study of Mathematical Practice

Once it's acknowledged that socio-cultural-historical entities, including mathematical ones, are real objects with objective properties, we can escape from the back-and-forth between Platonism and fictionalism (or logicism and formalism). We are left with an empirical or scientific question, a real phenomenon to study, to try ultimately to understand. The situation is analogous to what happened when anthropology, psychology or linguistics were recognized as autonomous disciplines, separating off from the "What is Man?", "What is Mind?", "What is Language?" wonderments. Mathematics—both mathematical practice and mathematical concepts—can be studied with every available method, as a special form of cultural life.

Of course the scientific study of special forms of cultural life is nothing new. Take economics, for instance. It started independent life as *a priori* rules of behavior of a hypothesized "economic man." But later we saw the rise of "behavioral" or "empirical" economics. A similar story can be told about linguistics. The young field of "sociolinguistics" is empirical; it may

---

[1] I am grateful to Julian Cole [Cole 2005] for taking this argument seriously enough to give it some respectable philosophical backing.

some day bring mathematical linguistics down to the ground of actual speech data, of behavior of language speakers.

4 Definition of "Mathematical Object"

If economics is the study of economic activity and behavior and linguistics is the study of language activity and behavior, then we may be ready to start a systematic study of mathematical activity and behavior. Our first step, naturally, is to ask:

What do we mean by "mathematical activity or behavior"?

Certainly it includes thinking, wondering, dreaming, learning about mathematics. Certainly it includes problem solving at all levels, from pre-kindergarten up through postdocs and Fields Prize winners. Teaching mathematics, at all levels, is also mathematical activity. (If it isn't, then we'd call it bad teaching.) Ordinary commercial calculations are too. Routine plugging of numbers into formulas by engineers and technicians is another form of mathematical behavior. So too are geometrical reasoning, and probabilistic reasoning, and combinatorial reasoning, and any formal logical reasoning.

In fact, we must expect that other, hitherto unthought-of kinds of mathematical behavior will yet arise. If that should happen, how would we identify such hitherto unseen behavior as mathematical? This question, I believe, is the crucial one that leads to a convincing, workable definition of mathematics.

Consider how it was decided in the past that some new branch of study was not just "mathematical" (containing some mathematical features) but really *mathematics*: to be included within the field of mathematics itself. Two famous examples are set theory and probability. Infinite sets were not part of mathematics before Georg Cantor explicitly based them on the notion of one-to-one correspondence. On that basis, he was able to make compelling arguments, and set theory (with some resistance) became a mathematical subject. An older example is gambling or betting. Fermat and Pascal demonstrated "rigorous" (irrefutable, compelling) conclusions about some games of chance. Therefore their work was mathematical, even though it was outside the limits of mathematics as previously understood. The subsequent work of Bernoulli, De Moivre, Laplace and Chebychev was certainly mathematics, for the same reason. Ultimately Kolmogorov axiomatized probability in the context of abstract measure theory. In doing so he was axiomatizing an already existing, ancient branch of mathematics.

It was pointed out in *The Mathematical Experience* [Davis/Hersh 1981] in 1981 that "While mathematics is a humanistic study with respect to its subject matter, it is like the sciences in its objectivity. Those results about the physical world that are reproducible—that come out the same way every time anyone asks—are called natural sciences. In the realm of ideas, of mental objects, those ideas whose properties are reproducible, that come out the same way every time anyone asks, are called mathematical objects, and the study of mental objects with reproducible properties is called mathematics." I was gratified when David Mumford quoted this approvingly. "I love this definition because it doesn't try to limit mathematics to what has been called mathematics in the past but really attempts to say why certain communications are classified as math, others as science, others as art, others as gossip. Thus reproducible properties of the physical world are science whereas reproducible mental objects are math." [Mumford 2000, p. 199]. *Reasoning about mental objects (concepts, ideas) that compels assent (on the part of everyone who understands the concepts involved) is characteristically "mathematical."* This is what is meant by "mathematical certainty." It does not imply infallibility! (On the contrary, history shows that the concepts about which we reason with such conviction have sometimes surprised us on closer acquaintance, and forced us to re-examine and improve our reasoning).

Certainly mathematics itself isn't the only place where conclusive reasoning occurs! For example, historians can use rigorous, even unimpeachable reasoning, to establish the sequence of events, or to refute anachronistic claims. Rigorous reasoning can occur anywhere—in law, in textual analysis of literature, and in ordinary daily life apart from academics. But although dates in history are subject to rigorous reasoning, they are not mathematical objects, part of mathematics. They are tied to specific places and persons. Information about them comes ultimately from reports of someone's visual or auditory perceptions.

On the other hand, when we consider abstractions, whether in law, or in theology, or in art criticism, musical criticism, or literary criticism, we certainly do find argument and reasoning, but it is not usually conclusive. It is usually subject to continuing unresolved dispute and disagreement. And if in a part of some field of abstract thought, such as linguistics for example, some concepts arise which lend themselves to conclusive and decisive reasoning, then that field is characterized as "mathematical," and we have mathematical linguistics.

So mathematics itself isn't the only place where conclusive reasoning occurs. But its objects of study are *all those abstractions that lend themselves to conclusive, irresistible reasoning— to "proof:"* "proof," not in any formal or formalized sense, but in the sense mathematicians talk about proof—conclusive demonstrations that compel agreement by all who understand the concepts involved. *Abstract concepts subject to such conclusive reasoning or proof are called mathematical concepts.*

Beginning with Aristotle, formal logic has helped to clarify mathematical reasoning, and rigorous argument in general. But most mathematical argument is done on the basis of the *content* of mathematical statements, not on their logical form. It is done, not only without reference to the rules of formal logic, but often even without awareness of them. And now formal logic itself is well-established as another *part of* mathematics! As such, it is subject to conclusive reasoning that generally is *informal*, as in the rest of mathematics. That is to say, logicians reason *informally* in proving theorems about *formal* logic. (This remark, made forcefully by Imre Lakatos 30 years ago [Lakatos 1976], is now a commonplace).

Donald MacKenzie [2005] uses the term "rigorous proof" in contrast to "formal proof," to mean "all those arguments that are accepted by mathematicians (or other relevant specialists) as constituting mathematical proofs, but that are not formal proofs." I could amend him, and say "rigorous proof" is any argument that compels assent from everyone who understands the concepts involved. Then my definition of mathematics could be shortened: any set of ideas is mathematical, to the extent that it is subject to rigorous proof.

Saunders MacLane [1986], among others, has written that mathematics is characterized by "precision." But what is meant by "precision"? He could not have meant numerical precision, for a huge part of modern mathematics, including his own contribution, is geometrical or syntactical, not numerical. By "precise" did he mean "formally explicit"—expressed in a formal symbolism? But it has been said, by several famous mathematicians, that you don't really understand a mathematical concept until you can explain it to the first person you meet in the street. And of course there are many beautiful examples in mathematics of conclusive *visual* reasoning, which are accepted and recognized as *mathematical proofs* independent of any *post hoc* formalization and formal proof.

Probably the correct interpretation of "precise" should be simply, "subject to conclusive, irrefutable reasoning." So my argument here amounts to accepting the familiar claim that "Mathematics is characterized above all by precision" and simply "unpacking" what is meant by "precise."

Lakoff and Nunez [2000] have shown that mathematical proof often can be understood as based on "embodied metaphors." That explanation of proof cannot be formalized. In fact, mathematical proof is too varied to be pinned down in a single precise, universal description

I will repeat: Mathematics can be defined as the subject that consists *primarily, characteristically,* of conclusive, irresistible reasoning about abstract concepts. Such reasoning is what we call "proof:" "proof," not in the formal or formalized sense, but in the sense mathematicians say "proof"—a conclusive demonstration that compels agreement by all who understand the concepts involved.

And mathematical objects (or entities or items, or whatever word you like) are simply *those concepts which are subject to such conclusive reasoning or proof.*

Wouldn't this include chess problems? Yes, of course. "A chess problem is simply an exercise in pure mathematics... Chess problems are the hymn-tunes of mathematics." [Hardy 1992, p. 87]

5 The Basic Problem

The basic problem is the same for us as it was for Kant:
"How is pure mathematics possible?" [Kant 1950]
In the context of the philosophical discourse of his time, Kant answered, "intuition."
In the context of empirical science, we can ask:
"How is it possible for people to create reasoning which is indisputable?"

This is a major intellectual challenge, which will take decades to unfold: to use the methodologies of history, sociology, anthropology, psychology, cognitive and neuroscience, and no doubt still others yet to be invented, to develop a coherent, empirically-based, overall understanding of the nature of mathematical practice and knowledge, as a major part of our larger understanding of what it is to be human.

This is much more than saying "We need a sociology of mathematics." Mathematics is a near-universal, almost all-pervasive aspect of humanity, of being human, and should be studied from all points of views, by all available methods, not as distinct, disconnected academic departments, but as related ways of focusing on the same mysterious phenomenon.

Empirical studies of mathematics are already taking place, in considerable variety. The wonderful book *The Number Sense* by Stanislas Dehaene [1997], reports on neurological, linguistic and educational studies, all pointing to a bodily (neurological or biological) foundation for arithmetic. (See also [Butterworth 1999] and [Campbell 2005].)

Bettina Heintz [2000] carried out an extended ethnological study of ongoing mathematical research at the Max Planck Institute in Bonn.

Where Mathematics Comes From by George Lakoff and Rafael Núñez [2000], and *The Math Gene* by Keith Devlin [2001], are path-breaking attempts to connect mathematical thinking with the language ability. Núñez [2005] is carrying out a precise quantitative study of gesture in mathematical communication, as a clue to the connection between mathematical abstractions and embodied metaphor .

Anthropologists have long studied mathematical understanding and language in many different cultures and recently, under the label of "ethnomathematics," they have been joined by mathematicians Ubiratan D'Ambrosio [1985] and Marcia Ascher [1991]. Among sociologists, the writings of David Bloor [1976] in Edinburgh stirred up considerable resistance and controversy. The sociologists Andrew Pickering and Donald Mackenzie contributed to my recent collection, *18 Unconventional Essays on the Nature of Mathematics* [Hersh 2005a]. The long-standing work of developmental and educational psychologists, especially Jean Piaget [1958, 1960, 1065, 1965a, 1967, 1969, 1970, 1975] and Lev Vygotsky [1978, 1986] and their followers, has yielded a great deal of knowledge on how children learn mathematics.

The oldest specialty in "mathematics studies" is, of course, the history of mathematics. Traditional history concentrated on describing mathematical findings and results as embedded or embalmed in print, but today many historians are broadening their focus, to see mathematics as the product of *individuals in communities*—the professional community of fellow-mathematicians, and also the larger political-economic-ideological community, whose support makes them possible.

It would be redundant to try to summarize the work of these different authors.

Here are some questions to which we do not have adequate answers:

The dialectic between discrete and continuous, between arithmetic and geometry, between logical and visual, is a fundamental pervading theme throughout mathematics. It was manifested to the Pythagoreans, when they discovered that no fraction can measure the diagonal of the unit square. It manifested in the acceptance, rejection and acceptance of infinitesimals, from Archimedes, Newton and Leibniz to Abraham Robinson [1996]. It surfaced again in Andre Weil's conjectures [Weil 1979], linking the number of solutions of Diophantine equations to the cohomology of differentiable manifolds.

If we trace back to our animal origin, it is clear that both human and pre-human primate hunters and gatherers had to think about direction and distance. This kind of thinking is intimately connected with seeing, with the visual function of the brain. It is also clear that language is primarily auditory. Words and language, first spoken, later written, are the soil from which spring logic and counting—number and arithmetic. Are there two different brain activities or potentialities, one for visual-geometric and another for logical-arithmetical thinking, associated with the two different brain centers—visual and auditory? Is there a neurological basis for these two contrasting worlds of mathematical thought—at once incompatible and inseparable? This question can be investigated even today, using current methods of brain localization on subjects carrying on mathematical work.

Another fundamental dialectic is between "existence as construction" and "existence as logical possibility." How are we able to think and draw conclusions about things that we do not know how to find? This has played out as the dispute of intuitionism and constructivism (Brouwer, Bishop) against mainstream mathematics, whether formalist or Platonist (Hilbert, Gödel). (See [Bishop 1967], [Brouwer 1975], [Gödel 1995], [Hilbert 1964].)

Unconscious mathematizing is a huge mystery. Mathematical ideas can come into consciousness by surprise, as if by a gift from nowhere—evidently from some subconscious process. This is attested to by many anecdotes. The most famous is Poincare's discovery of the theta-fuchsian functions, a mystery calling for explanation. [Poincaré 1923]

The disconnect between verbal and mathematical abilities is another unexplained common observation. So is the prominent correlation between ability in mathematics and ability to perform music.

If the nature of mathematics can be identified as a scientific problem, where does that leave philosophy of mathematics? Of course philosophers of mathematics will still carry on. Many will continue their traditional ongoing conversation, without serious attention to actual mathematical practice. But some others are already trying to study mathematical behavior, activity and experience ([Asprey/Kitcher 1988], [Cellucci 2005], [Corfield 2003]). In that effort, phenomenological analysis deriving from Edmund Husserl and his followers is helpful. (See [Hauser], [Livadas preprint, 2005, 2006], [Rota 1974, 1996], [Tieszen 2005, 2006a, 2006b], [Tragesser 1984], and my own [Hersh 2005b].) Here I offer a phenomenological analysis of an aspect of mathematical practice.

6 Timely or Timeless?

One puzzle about mathematics is its timeliness or timelessness. People say mathematical truths are timeless, even eternal—"Always were true, always will be." The squares on the sides of a right triangle add up to the square on the hypotenuse—presumably they "did so" before the notion of right triangle or hypotenuse had crossed anyone's head. This way of thinking seems to force one to allow the "existence" of all right triangles with their hypotenuses, even before the famous Big Bang that gave birth to our Cosmos. And of course, the same thinking applies to all complex determinants, say, of order 29,146,298,979, or to all Grothendieck toposes. They all existed, "somewhere, somehow."

Once, at a math department colloquium in a certain state university, I called for a vote for or against the following proposition: "Resolved, that the spectral representation of self-adjoint operators in Hilbert space was true before the Big Bang that created the Cosmos." The vote was 3 to 1 in favor.

I would like to suggest a different way of looking at these matters. Once a new, well-founded mathematical question is asked, the answer usually is already determined, but still unknown. In that sense, the answer now exists. But the answer didn't exist in advance of the question. Once people conceived of right triangles and of the area of a square, the question about the squares on the sides of the triangle was meaningful, and we would say that the Pythagorean theorem was true, even before it was stated. But before there were triangles and sums of squares, there was no object to which the Pythagorean theorem could refer, so it is senseless to say it was true already at that time.

Similarly, theorems on "faithful functors" or "Abelian sheaves" were neither true nor false before the mathematical concepts they describe had been conceived or formulated. They simply didn't refer, they had no content. Once those objects came into being, as mathematical objects, that is as thoughts and objects of discourse, then we could say that certain theorems were already true—waiting to be discovered, so to speak.

In this respect mathematical facts are different from physical ones. The Earth rotated on its axis before there were people to care about night and day. But mathematical concepts or entities or objects, whichever you prefer to call them, are called into being by our questions. Only then can answers to such questions be true or false.

It is tempting for mathematicians and philosophers of mathematics to look for mathematical precision in thinking about mathematics. But it will not do to expect a definite yes-or-no, either-or answer to such questions as exactly when some mathematical fact became true.

Mathematics is not a fixed, static, eternal piece of abstract hardware. It is an evolving, growing, developing world of ideas, problems, algorithms, conjectures, proofs, analogies—a cultural world, a world existing first of all in the thoughts of people, and in the activities in their brains that correspond to these thoughts, and only subsequently in the records on paper and on microchips where they have recorded their thoughts.

A new mathematical concept may arise in the course of a mathematical conversation, or in the course of an individual's thinking or writing about a mathematical topic. In its first state of coming into being, it is transient, evanescent, subject to quick disappearance and oblivion. It may be held onto, communicated to others, developed, it may become a topic of conversation among several people. If it is written down and preserved somehow, or if it becomes a widespread topic of conversation over a considerable period of time, we recognize it as a stable part of the cultural world of mathematics, as a new topic or concept in mathematics. But there would be no single moment when it was created or invented.

If it is the predetermined solution of a definitely stated problem, there is a sense in which it existed latently, *in potentio,* from the time that problem was stated. And if that problem was a natural consequence, predictable as part of an already existing theory, we could say that the object in question had a potential existence, even before the explicit statement of the problem to which it is a solution. This is not a particularly mysterious situation. In exactly the same sense, one could say that a leaf on an oak tree was potentially present in the DNA of the acorn from which the tree grew.

The creation of novelty in mathematics, as in life, is precisely the passage from potentiality to actuality.

The intermediate region between the potential and the actual is the area of active growth at any particular time. It is the social counterpart of what the psychologist Lev Vygotsky called "the zone of proximal development." Colloquially, we talk about new ideas being "in the air." We call on this notion to account for a very common event: multiple near-simultaneous inventions or discoveries.

On the other hand, when we are doing mathematics rather than reflecting on ourselves as doers of mathematics, we set aside, disregard the temporality of a mathematical object (or "item" or "entity," if you like.) For the purpose of doing mathematics, its temporality is irrelevant, so we set it aside, put it out of consideration. (In Husserl-style vocabulary, we "bracket" it.) It is in that sense, and for that reason, that mathematics can be said to be timeless. We make it so, because its temporality is not to our purpose most of the time.

An analogy can be made to watching a movie. When we watch a movie in a movie theater, we are looking at a flat screen on which are projected patterns of color. In order to watch it *as a movie* (that's our purpose in being there) we intentionally put aside our awareness that all we are seeing is patterns of color on a flat screen. That "putting aside" is what enables us to "see" the story, the movie.

Putting aside the temporality and concrete historicity of mathematical objects is necessary for us to enter the mathematical dream world and live there, to make discoveries and creations there. On the other hand, when we are not doing mathematics, but talking about what mathematics is, we can look at it from the outside and see that it is a temporal part of human culture (just as movie-goers have no trouble, before or after the movie, in seeing the blank screen where the images will be projected).

This may be what some people mean by calling mathematics "a fiction." Call it "a fiction" if you want—but a "necessary fiction," not an arbitrary fiction—a "fiction" with laws of its own that must be obeyed, if it is to be entered and lived in successfully. Because it does have its own laws, which compel us to obey, I call it a reality.

7 Educational Implications

Identifying the nature of mathematics as a form of human activity has important implications for education. Mathematics educators easily accept this identification. They often seem to take it for granted. A forthcoming article [Umland/Hersh] develops some of the interactions between the philosophical and the educational issues.

Arithmetical skills that used to be essential for cashiers have long vanished from the checkout counter. Pocket calculators are everywhere, so cheap they're almost free. So it's no surprise when a high school teacher tells me her students can't multiply or divide!

Yet the industry which has made universal skill in arithmetic obsolete, the computer industry, is totally dependent on mathematical thinking! This is a paradox. There is a simultaneous degradation and glorification of mathematics in our culture and our economy. These two opposing trends create a tremendous stress and dislocation for mathematics education.

Numerical and geometric thinking are rooted in our brains, our genes and our culture. If this becomes better understood, it might help move mathematics education away from fruitless preoccupation with obsolete skills, and toward a liberating part of human enlightenment. If it were well understood that rigorous thinking is as much a part of our nature as playing ball or making music, then mathematics might come to be seen as part of everyone's educational birthright.

8 Conclusion

The existence of mathematical items or entities does matter. Their seeming timelessness is an artifact of our practice, a necessary way of framing them so that we can "enter into" their world. The study of the nature of mathematics and mathematical practice is an interdisciplinary task. It is too big for philosophy alone, or even for half a dozen separate, isolated academic specialties. It is a central problem in the ongoing study of humankind.

Acknowledgments Conversations with Vera John-Steiner resulted in significant improvements in this work. Repeated criticisms of earlier versions of it, by the editors of this volume, were painful and helpful.

References

[Archimedes] Archimedes, *Works*, ed. R. Netz, Cambridge University Press, 2004; ed. T. L. Heath, Dover Publications, 2002.

[Ascher 1991] M. Ascher, *Ethnomathematics. A Multicultural View of Mathematical Ideas*, San Francisco: Brooks/Cole, 1991.

[Asprey/Kitcher 1988] W. Asprey and Philip Kitcher, eds., *History and Philosophy of Modern Mathematics*, University of Minnesota Press, 1988.

[Bishop 1967] E. Bishop, *Foundations of Constuctive Analysis*, New York: McGraw-Hill, 1967.

[Bloor 1976] D. Bloor, *Knowledge and Social Imagery*, Boston: Routledge and Kegan Paul, 1976.

[Brouwer 1975] L. E. J. Brouwer *Collected Works*. Amsterdam: North-Holland, 1975.

[Butterworth 1999] B. Butterworth, *What Counts,* New York: Free Press, 1999.

[Campbell 2005] J. I. D. Campbell, ed., *Handbook of Mathematical Cognition*, New York: Psychology Press, 2005.

[Cellucci 2005] C. Cellucci, "Introduction," in [Hersh 2005a].

[Corfield 2003] D. Corfield, *Towards a Philosophy of Real Mathematics*, Cambridge University Press, 2003.

[Cole 2005] J. Cole, in the present volume; also, "Practice-Dependent Realism and Mathematics," Dissertation, Ohio State University, 2005.

[D'Ambrosio 1985] U. D'Ambrosio, "Ethnomathematics and Its Place in the History and Pedagogy of Mathematics," F*or the Learning of Mathematics* 5(1) (1985), pp. 4–48.

[Davies] E. B. Davies, "A Defence of Mathematical Pluralism," Preprint.

[Davis/Hersh 1981] P. J. Davis and R. Hersh, *The Mathematical Experience*, Birkhauser, Boston, 1981.

[Dehaene 1997] S. Dehaene, *The Number Sense,* New York: Oxford University Press,1997.

[Devlin 2001] K. Devlin, *The Math Gene,* Basic Books, 2001.

[Gödel 1995] K. Gödel, *Collected Works*, Oxford University Press, New York, 1995.

[Hardy 1992] G. H. Hardy, *A Mathematician's Apology*, Cambridge University Press, 1992.

[Hauser] K. Hauser, "Godel's program revisited", *Husserl Studies*, forthcoming.

[Heintz 2000] B. Heintz, *Die Innenwelt der Mathematik*, Heidelberg, Springer, 2000.

[Hersh 1997] R. Hersh, *What is Mathematics, Really?* Oxford University Press, New York, 1997.

[Hersh 2005a] ——, ed. *18 Unconventional Essays on the Nature of Mathematics*, Springer-Verlag, New York, 2005.

[Hersh 2005b] ——, "Wings, not foundations!" pp. 155–164 in *Essays on the Foundations of Mathematics and Logic*, Milan, Polimetrica, 2005.

[Hilbert 1964] D. Hilbert, "On the Infinite," in P. Benacerraf and H. Putnam, *Philosophy of Mathematics*, Prentice-Hall, New York, 1964.

[Kant 1950] I. Kant, *Prolegomena to any future metaphysics*, Indianapolis, Bobbs Merrill, 1950.

[Lakatos 1976] I. Lakatos, *Proofs and Refutations*, Cambridge, Cambridge University Press, 1976.

[Livadas preprint] S. Livadas, "Impredicativity of continuum in phenomenology and in non-Cantorian theories", preprint.

[Livadas 2005] ——, "The phenomenological roots of nonstandard mathematics," *Romanian Journal of Information Science and Technology*, 8,2 (2005), pp. 115-136.

[Livadas 2006] ——, "A phenomenological analysis of observation," preprint, June 21, 2006.

[Lakoff/Núñez 2000] G. Lakoff and R. E. Núñez, *Where Mathematics Comes From*, New York, Basic Books, 2000.

[Mac Lane 1986] S. Mac Lane, *Mathematics: Form and Function,* New York: Springer-Verlag, 1986.

[MacKenzie 2005] D. MacKenzie, "Computers and the Sociology of Mathematical Proof," in [Hersh 2005a].

[Mumford 2000] D. Mumford, "The dawning of the age of stochasticity," in V. Arnold, M. Atiyah, P. Lax, B. Mazur, eds., *Mathematics: Frontiers and Perspectives*, American Mathematical Society, 2000.

[Núñez 2005] R. Núñez, "Do Real Numbers Really Move?" in [Hersh 2005a].

[Piaget 1958] J. Piaget, Growth of Logical Thinking from Childhood to Adolescence. New York: Basic Books, 1958.

[Piaget 1960] ——, *The Child's Conception of Geometry*, with B. Inhelder and A. Szeminka, New York: Basic Books, 1960.

[Piaget 1965] ——, *The Child's Conception of Number*, New York: Norton, 1965.

[Piaget 1965a] ——, *The Child's Conception of Physical Causality*, Totowa N. J. Littlefield, 1965.

[Piaget 1967] ——, *The Child's Conception of Space*, New York: Norton, 1967.

[Piaget 1969] ——, Early Growth of Logic in the Child: Classification and Seriation, New York: Norton, 1969.

[Piaget 1970] ——, *The Child's Conception of Movement and Speed*, New York: Basic Books, 1970.

[Piaget 1975] ——, *Origin of the Idea of Chance in Children*, with B. Inhelder, New York: Norton, 1975.

[Poincaré 1923] H. Poincaré, *Science and Method*, New York: Dover, 1923.

[Quine 1966] W. V. O. Quine, "The Scope and Language of Science," in *The Ways of Paradox*, Cambridge, Harvard University Press, 1966.

[Robinson 1996] A. Robinson, *Non-standard Analysis,* Princeton University Press. 1996.

[Rav 2005] Y. Rav, "Philosophical Problems of Mathematics in the Light of Evolutionary Epistemology," in [Hersh 2005a].

[Rota 1974] G.-C. Rota, *The End of Objectivity*, MIT lecture notes, 1974.

[Rota 1996] ——, *Indiscrete Thoughts*, Birkhauser, 1996.

[Tieszen 2005] R. Tieszen, *Phenomenology, Logic, and the Philosophy of Mathematics*, Cambridge: Cambridge University Press, 2005.

[Tieszen 2006a] ——, "Review of Edmund Husserl, *Philosophy of Arithmetic*," *Philosophia Mathematica* 14, 1 (2006), pp. 112–130.

[Tieszen 2006b] ——, "After Godel: Mechanism, Reason and Realism in the Philosophy of Mathematics," *Philosophia Mathematica* 14, 2 (2006), pp. 229–254. [TI2006b]

[Tragesser 1984] R. S. Tragesser, *Husserl and Realism in Logic and Mathematics*, Cambridge: Cambridge University Press, 1984.

[Umland/Hersh] K. Umland and R. Hersh, "Mathematical Discourse: the link from pre-mathematical to fully mathematical thinking," *Philosophy of Mathematics Education Journal 19 (2006)*.

[Vygotsky 1978] L. Vygotsky, *Mind in Society: The Development of Higher Psychological Processes* (M. Cole, V. John-Steiner, & E. Souberman, Eds.), Cambridge, MA: Harvard University Press, 1978.

[Vygotsky 1986] ——, *Thought and Language* (A. Kozulin, Ed. and Trans.), Cambridge, MA: MIT Press, 1986.

[Weil 1979] A. Weil, *Collected Papers*, Springer-Verlag, 1979.

[White 2005] L. White, "The Locus of Mathematical Reality," in [Hersh 2005a]

[Wilder 1981] R. Wilder, *Mathematics as a Cultural System*, Pergamon Press, 1981.

6

Mathematical Domains: Social Constructs?

Julian Cole
Department of Philosophy and Humanities
Buffalo State College

From the Editors

Social constructivism is probably the philosophy of mathematics that has seen the greatest growth in support among mathematicians in the last twenty-five years. However, until now, because of assorted difficulties that this view appears to imply, it has not received serious attention from philosophers of mathematics. For example, see Balaguer's quick dismissal of it in his chapter (section 2.2 and elsewhere). Julian Cole is one of the first philosophers to seriously attempt to deal with these problems. This chapter gives you an introduction to the philosophical issues and how he is attempting to deal with them. His view is still being developed. After reading this chapter, you may want to follow his future work (and that of those who respond to it). In particular, his upcoming article, "Creativity, Freedom, and Authority: A New Perspective on the Metaphysics of Mathematics" seems likely to be of interest.

Julian Cole is an Assistant Professor of Philosophy in the Department of Philosophy and Humanities at Buffalo State College. His interests are in logic and the philosophy of mathematics. He recently finished a doctoral dissertation on social constructivism as a philosophy of mathematics, under the direction of Stewart Shapiro at The Ohio State University. Prior to his work on social constructivism he wrote a doctoral dissertation in multifractal geometry at the University of St. Andrews under the direction of Lars Olsen.

1 Introduction

There can be little doubt that mathematics is a social activity. Among other things, mathematicians often work together in groups, they frequently choose to work on problems because other mathematicians deem them important or difficult or worthy, they rely on other mathematicians to verify the correctness of their work, they present their work in public forums, more than one mathematician (or group of mathematicians) can work on the same problem, and mathematicians compete with each other for sparse funding. That mathematics is social in all of these senses—and several others—is uncontentious. In the last ten years, however, two books[1] have been published that advocate that mathematics is social in a much deeper—and correspondingly more controversial—sense than any of these. The authors of these books—respectively, Reuben Hersh, a professional mathematician, and Paul Ernest, a specialist in mathematics education—suggest that the *subject matter* of mathematics is social. More precisely, they advocate the thesis that the subject matter of mathematics—mathematical domains or structures—is constructed by or created by—quite literally brought into existence by—the social activities of mathematicians. This is a contentious thesis if ever there was one, as, on a standard interpretation, it implies there were, for example, no numbers until mathematicians invented them.

The details of Ernest's and Hersh's accounts of how the subject matter of mathematics is socially constructed[2] are quite different. Further, anybody who has read their books will be aware that both Ernest and Hersh have a much broader agenda[3] than the mere advocacy of the thesis that the subject matter of mathematics is socially constructed. Indeed, it would probably be accurate to say that the articulation and defense of this thesis are in many ways secondary to both authors' primary goals. Nonetheless, both Ernest and Hersh do promote this thesis.

I am fascinated by the suggestion that mathematical domains (structures)[4] are socially constructed. What I would like to do in this chapter is to explore this suggestion with the aim of making one version of it reasonably precise and evaluating its merits and weaknesses. I shall undertake the latter task by comparing it with Platonism, its best-known rival. Ideally, I would also compare this version of social constructivism with the two popular versions of Nominalism, Fictionalism and Modal Nominalism. Unfortunately, space does not allow. For similar reasons, I shall be unable to discuss all of the details of this account of mathematics.

2 Ernest's and Hersh's View of Mathematics

Let us begin by considering what Ernest and Hersh say about the social construction of **mathematical ontology**—the mathematical items that exist. The following quotes indicate Ernest's general position about mathematical objects:

> According to the social constructivist view the discourse of mathematics creates a cultural domain within which the objects of mathematics are constituted by mathematical signs in use.   ([Ernest 1998], p. 193)

---

[1] Reuben Hersh's *What is Mathematics, Really?* [Hersh 1997] and Paul Ernest's *Social Constructivism as a Philosophy of Mathematics* [Ernest 1998].

[2] I shall provide an extensive discussion of social construction in Section 3.

[3] This broader agenda is, as a matter of fact, quite similar for both authors. They are both interested in exploring philosophical issues concerning mathematics using a much broader range of mathematical examples than is typical in the contemporary (analytic) philosophy of mathematics literature. They are both interested in discussing informal mathematical activities, e.g., the nature of mathematical intuition and how mathematical theories are generated and refined before they are formalized. (Discussions of this type are almost completely lacking in contemporary analytic philosophy of mathematics.) Additionally, they are both interested in combating what they believe—I think mistakenly—is a dominant conception of mathematical knowledge as certain and infallible.

[4] From this point onwards I shall simply talk about mathematical domains. I leave it open whether some or all mathematical domains are structures.

... signifiers have ontological priority over the signified—especially in mathematics, for the signifiers can be inscribed and produced, or at least instantiated, whereas the signified can be indicated only indirectly, mediated through signifiers.

([Ernest 1998], p. 196)

... the ontology of mathematics is given by the discursive realm of mathematics, which is populated by cultural objects, which have real existence in that domain ... mathematical discourse as a living cultural entity creates the ontology of mathematics.

([Ernest 1998], p. 202)

While there is much in these quotes that the reader is likely to find perplexing and in need of further explanation, two points can be gleaned from them. First, Ernest's belief that mathematical objects are constructed by or created by—made real by—the activities of mathematicians. This is the **basic thesis of social constructivism** (about mathematics). Second, Ernest takes the discursive elements of mathematics to be central to the construction of mathematical ontology. Indeed, the first and second quotes indicate that Ernest believes that the constructive work of mathematical practices[5] is done, at least primarily, by the presence of mathematical signs and signifiers in the discursive elements of those practices.

The most natural interpretation of "mathematical signs" and "signifiers" in Ernest's quotes is one according to which they are **lexical items**—such items as the marks written down by mathematicians and the sounds uttered by mathematicians. Yet, under this interpretation, Ernest's suggestion is problematic. Mathematical discursive practices only contain a finite number of such signs and signifiers. Thus, if such signs and signifiers are responsible for the existence of all mathematical entities, then some of them must be responsible for the existence of collections of mathematical entities with infinite—indeed, extremely large infinite—cardinalities. *How* can they be so responsible? At least to my knowledge, Ernest has not provided an answer to this question.[6] So, Ernest's discussion of the social construction of mathematical ontology is unhelpful with respect to a key aspect of that construction. We shall explore this issue further in Section 4.

Let us investigate whether Hersh can provide us with a more helpful account of mathematics. Here are some quotes from his book:

> Fact 1: Mathematical objects are created by humans. Not arbitrarily, but from activity with existing mathematical objects, and from the needs of science and daily life.
>
> Fact 2: Once created, mathematical objects can have properties that are difficult for us to discover.   ([Hersh 1997], p. 16)
>
> 4. Mathematical objects are a distinct variety of social-historical objects. They're a special part of culture.   ([Hersh 1997], p. 22)

---

[5] A **practice** is a collection of activities governed by standards of correctness and incorrectness. A practice is **discursive** if the activities in question are ones that center about **assertoric content**, i.e., the thing that we can assert, assume, consider, etc. Many mathematical activities are discursive practices because they involve assertions, proofs, etc. For a detailed discussion of assertoric content, I refer the reader to Crispin Wright's discussion in Chapters 1 and 2 of [Wright 1992].

[6] Interestingly, non-mathematicians tend to find the criticism I level at Ernest in this paragraph obvious, while mathematicians sometimes have difficulties understanding the problem. I suspect that this is because mathematicians are so used to representing infinite collections of entities with a finite number of symbols that they intuitively fill in an answer to my question. My point is simply this, while there most certainly is an answer that can be provided, Ernest has failed to provide it in his book.

In Fact 1, Hersh expresses the basic social constructivist thesis with a minor twist: he recognizes the need to account for why human beings created mathematical domains[7] and hints at such an account. In Fact 2, Hersh indicates his sensitivity to a certain type of independence that mathematical domains have from mathematical practices—let us call it **epistemic[8] independence**, for it relates to our knowledge of mathematical domains. Just below Fact 2, he tells us

> Once created and communicated, mathematical objects are *there*. They detach from their originator and become part of human culture. We learn of them as external objects, with known and unknown properties. Of the unknown properties, there are some that we are able to discover. Some we can't discover, even though they are our own creations.
>
> ([Hersh 1997], p. 16)

The second part of this quote reinforces Hersh's sensitivity to the epistemic independence of mathematical domains from mathematical practices. The first part of this quote goes further than this, however. It indicates that mathematical domains detach—in some sense—from their specific creator. We shall return to this point in Section 4.

Perhaps Hersh's most interesting claim, however, is that "mathematical objects are . . . social-historical objects" ([Hersh 1997], p. 22). What are we to make of this claim? I believe that the following quote is helpful:

> Frege showed that mathematical objects are neither physical nor mental. He labeled them "abstract objects." What did he tell us about abstract objects? Only this: They're neither physical nor mental.
>
> Are there other things besides numbers that aren't mental or physical?
>
> Yes! Sonatas. Prices. Eviction notices. Declarations of war.
>
> Not mental or physical, but not abstract either!
>
> The U.S. Supreme Court exists. It can condemn you to death!
>
> Is the court physical? If the Court building were blown up and the justices moved to the Pentagon, the Court would go on. Is it mental? If all nine justices expired in a suicide cult, they'd be replaced. The court would go on.
>
> The Court isn't the stones of its building, nor is it anyone's minds and bodies. Physical and mental embodiment are necessary to it, but they're not *it*. *It's a social institution*. Mental and physical categories are insufficient to understand it. It's comprehensible only in the context of American society.
>
> What matters to people nowadays?
>
> Marriage, divorce, child care.
>
> Advertising and shopping.

---

[7] The observant reader will have noticed that both Ernest and Hersh talk about the construction of mathematical objects while I talk about the construction of mathematical domains. There are two reasons for this. First, it seems to me that (at least most) mathematical objects are the objects they are in virtue of their relationships to the other objects in the domain of which they are a member. So, in order to construct a particular mathematical object, one really needs to construct all of the objects in the domain of which that object is a member. Second, in constructing some aspect of mathematical reality, one is presumably not only constructing the objects in that aspect of mathematical reality, but also the properties of those objects and the relationships between those objects. A domain, at least as I am using this notion, is a collection of objects that have properties and stand in relations to one another.

[8] **Epistemology** is the branch of philosophy that investigates the nature of knowledge and justification.

Jobs, salaries, money.

The news, and other television entertainment.

War and peace.

All these entities have mental and physical aspects, but none is a mental or a physical entity.
Every one is a social[-historical] entity.   ([Hersh 1997], pp. 13–14)

In this passage, Hersh mentions a wide variety of social-historical entities, some legal (e.g., eviction notices and the U.S. Supreme Court), some political (e.g., declarations of war and peace), some financial (e.g., money and salaries), and others recreational (e.g., sonatas and television programs). All of these items exist, and their existence has very real consequences. Yet they owe their existence to the power of certain types of acts, decisions or practices undertaken by human beings to make certain items real simply by happening or being undertaken. In suggesting that mathematical entities are social-historical entities, Hersh is suggesting that the same is true of mathematical domains. That is, mathematical domains exist and they owe their existence to the power of certain mathematical activities undertaken by human (and other rational) beings to make them real simply by being undertaken. Let us call this the **social-institutional understanding of the nature of mathematics**.

In what follows, when I talk about mathematical domains as social constructs, I shall have in mind Hersh's social-institutional understanding of the nature of mathematics. I believe that it is more promising than Ernest's signifier-signified understanding of mathematics. While many of the practices that constitute social-historical entities involve signs and signifiers, the presence of these signs and signifiers is not, in general, central to these practices' constructive power. The above discussion of Ernest's account of mathematics certainly suggests that, if the basic social constructivist insight is correct, then the same is true in the mathematical case.

3 Social Construction and Dependence

In this section, I provide one framework for how objects come to be socially constructed. I do this so that I can locate the social construction of mathematical domains within this framework. In "Ontology and Social Construction" ([Haslanger 1995]), Sally Haslanger gives expression to a variety of ways in which social acts, decisions, or practices might be involved in social construction. The most basic distinction she makes is that between "causal social construction" and "constitutive social construction." These are two ways of constructing existent items. Haslanger offers the following characterizations of these two varieties of social construction:

Causal social construction: Something is causally socially constructed if [9] social factors play a *causal* role in bringing it into existence or, to some substantial extent, in its being the way that it is.

Constitutive social construction: Something is constitutively socially constructed if a *correct definition* or account of what it is for something to be an item of the type in question *must make reference to social factors*.[10]

---

[9] Throughout I follow the mathematical convention of leaving 'only if' out of definitions.

[10] These definitions are taken from page 98 of [Haslanger 1995], though I have slightly modified the second.

Consider first such items as cars, scissors, alarm clocks, and telephones. These are spatio-temporal entities that have been manufactured for some particular purpose. Let us call such items **artifacts**. Clearly, artifacts would not exist if there were no social acts, decisions, or practices. So, artifacts are dependent on[11] certain social acts, decisions, or practices. Further, the primary mechanism of artifacts' dependence on social acts, decisions, or practices is well understood. Artifacts are *causally dependent* on the social acts, decisions, or practices that bring them into existence. Thus, artifacts are **causal social constructs**—the products of causal social construction.

Next, consider the examples that Hersh mentions in the long quote in Section 2, and such items as legal borders between pieces of property (land), political borders between countries, property itself, countries themselves, laws (in the sense of statutes),[12] and games like baseball and tennis. It should be uncontroversial that all of these items exist. Further, a moment's reflection should make it clear that if various types of legal, political, financial, cultural and recreational practices had not developed on Earth, then none of these items would exist. Thus, all these items are dependent on social practices. Yet the *mechanism* of these items' dependence on social practices is different from that of artifacts' dependence on social acts, decisions, or practices. Social practices need not causally manipulate previously existing spatio-temporal items in order to bring legal and political borders, countries, laws, etc. into existence. Rather, these items simply owe their existence to certain social acts. It is this type of a dependence of an item on a social act that is characteristic of constitutive social construction. This type of dependence ensures that social factors have to be talked about in a correct definition or account of what the item is. Thus, these items are **constitutive social constructs**.[13]

Constitutive social constructs *can* have influence over the spatio-temporal world and the spatio-temporal world can have influence over which items we construct constitutively. One only need reflect on the impact of declarations of war to recognize this. What our contrast emphasizes is that the *means* by which an item becomes a *constitutive* social construct is not causal in the strict sense characteristic of *causal* social construction.

While my exposition so far might suggest that constitutive and causal social construction are mutually exclusive, this is not the case. Many cases of social construction involve both elements, though one or the other might be dominant in any particular case. An excellent example of this is a "regulation baseball" for Major League play. Two distinct types of considerations are involved in something's being a regulation baseball. First, the ball in question must have certain physical characteristics, e.g., it must be a certain size, shape, color, etc. Regulation baseballs are manufactured to have these characteristics. Thus, regulation baseballs are causal social constructs. The second consideration is that the ball has to have been deemed regulation by an individual

---

[11] Let us say that an item X is **dependent on an act, decision, or practice** Y if X would not exist if Y did not occur or exist. Additionally, let us say that an item X is **independent of an act, decision, or practice** Y if X would exist even if Y did not occur or exist.

[12] Whenever I talk about laws in this chapter, I shall be talking about laws in the sense of statutes rather than laws in the sense of laws of nature or the laws of probability.

[13] I have to confess a certain level of dissatisfaction with Haslanger's definition of constitutive social construction. This dissatisfaction is rooted in the fact that her definition obscures the importance of the particular mechanism of dependence of an item on social factors that my examples serve to illustrate. This mechanism of dependence is central to my own thought about constitutive social construction.

acting on behalf of the League and be signed by the League's commissioner. This consideration makes regulation baseballs constitutive social constructs.

Many acts of constitutive social construction are accompanied by acts of causal social construction, or provide already existing objects with additional features. For example, in composing a sonata, a composer will usually write a score. When declaring war, a country will usually produce a written proclamation of war. In legally dividing a single piece of land into two pieces of land, the owners of the two properties will usually either construct a barrier of some description to mark the division or divide the land using a natural barrier. Some acts of constitutive social construction *require* an accompanying object or act of causal construction. For example, a representative of the Major League can only deem a baseball regulation if it has certain physical characteristics. You don't have an eviction *notice*—as opposed to an eviction *order*—without the piece of paper on which the eviction order is written. Other acts of constitutive social construction do not require any kind of associated object. For example, in legally dividing a single property into two smaller properties, there is no need to place a barrier between the two properties, and in declaring war, there is no need to write a proclamation.

Let us call constitutive social constructs that do not require any kind of associated object **pure constitutive social constructs**, and those that do **impure constitutive social constructs**. Pure constitutive social constructs exist wholly in virtue of the undertaking of certain acts, decisions, or practices of social significance. Legal statutes are pure constitutive social constructs: roughly speaking,[14] a collection of statements has the property of being a legal statute wholly in virtue of its having appropriately proceeded through the process of approval and having been passed by a legitimate legislative authority.[15] Political borders are also pure constitutive social constructs. Roughly speaking, a certain line's marking a political border is wholly a matter of certain decisions made by relevant political groups; there is no need for such a border to be marked in any particular way.

With the above conceptual tools in place, let us refine the basic thesis of social constructivism (about mathematics) into the **central thesis of social constructivism** (about mathematics): *mathematical domains (and the items of which they are composed) are pure constitutive social constructs constituted by mathematical practices.* That is, particular mathematical domains (and the items of which they are composed) exist wholly in virtue of the undertaking of mathematical practices of a specific type. In the next section, we shall consider what specific type of mathematical practice is required. In the remainder of this chapter, when I talk about social constructivism, I shall be talking about this thesis, not the wider agenda of most social constructivists (see Footnote 3).

Social constructivism's advocacy of the dependence of mathematical domains on mathematical practices is what distinguishes it from all forms of Platonism. For our purposes, **Platonism** is the conjunction of three theses about mathematical domains: a) some exist, b) they (and the items

---

[14] There are other considerations involved. For example, a statute must not be declared unconstitutional and it must not be overridden by later legislative activities. None of these further considerations undermine the claim that legal statutes are pure constitutive social constructs.

[15] It is part of the procedure for passing a federal statute that various written versions of it are produced, including the version signed by the President. Yet, after signing, should all these required written versions of the statue be destroyed—perhaps by a nuclear attack in the D.C. area—it would remain law without them. Consequently, the statue itself has a certain type of independence from all of its written versions; they are not required for its continued existence.

of which they are composed) are paradigm cases of abstract entities[16]—so, for example, they are acasual, non-spatio-temporal, eternal, and changeless, and c) they (and the items of which they are composed) are independent of all social acts, decisions, and practices—they would exist even if there were no social acts, decisions, or practices.

4 Logic and Ontological Structure

We now have a basic understanding of the account of mathematical domains offered by social constructivists—mathematical domains are socially constituted by mathematical practices. Yet we still lack an answer to one important question: *according to social constructivists, how, exactly, do mathematical practices manage to socially constitute mathematical domains?*

As a preliminary to answering this question, it will be useful to ask, "Why, according to social constructivists, is the purported construction that takes place within mathematics social in nature rather than individual in nature?" After all, it would appear that many mathematical domains are introduced by *individual* mathematicians rather than by groups of mathematicians. For example, it would appear that William Hamilton introduced the domain of quaternions and Georg Cantor introduced the domain of transfinite numbers.

In order to answer this question, we first need to be clear about what is meant by 'social in nature' rather than 'individual in nature'. I mean to be asking "Why are mathematical constructs *sharable*?" That is, why can both you and I—and any reasonably sophisticated human being— theorize about the *same* mathematical construct rather than each of us theorizing about, and thus constructing, a different mathematical domain. For example, it could be that you construct *your* domain of natural numbers and I construct *my* domain of natural numbers, where these two constructs are *different* entities.[17]

Consider for a moment another class of constitutive social constructs, sonatas. In general, one individual is responsible for composing any given sonata, yet this does not undermine the social—sharable—nature of sonatas. An individual's musical creation can be shared by many, because that individual uses socially recognized tools in its construction. For example, sonatas are composed using the twelve-tone scale, a social convention standardized around "middle C" having the frequency of 440Hz, and sonatas are composed for standard—socially recognized— musical instruments. It is precisely because shared musical tools of this type are used in the construction of sonatas that they are constructs of a social nature rather than constructs of an individual nature.

---

[16] I shall provide a somewhat more detailed discussion of abstract entities in Section 5.

[17] There are some passages in Arend Heyting's work that suggest that he took mathematical entities to be individual mental entities rather than sharable entities in the sense that I am concerned with here (see, e.g., [Heyting 1931]). My worries about the sharability of mathematical constructs are a direct response to Gottlob Frege's criticisms of psychologism (see [Frege 1884]). My interest in this notion of "social", i.e., sharability, distinguishes me from Ernest and Hersh. Reflection on constitutive social construction will reveal that it is frequently achieved by providing certain individuals or groups of individuals with certain rights, responsibilities, authorities, etc. Consequently, it involves complex social dynamics. At least as I read Ernest and Hersh, when they claim that mathematical constructs are social in nature, they are acknowledging the importance of these social dynamics. I certainly do not want to deny the importance of these social dynamics. Likewise, I presume that Ernest and Hersh would not want to deny that mathematical constructs are sharable. We are merely emphasizing different things with our respective uses of the word 'social'.

According to social constructivists, a similar situation arises in mathematics. Frequently, one mathematician is responsible for the mathematical community taking an interest in a particular mathematical domain.[18] Consequently, from the perspective of a social constructivist, one individual is responsible for introducing the mathematical practice that constitutes that domain. Yet mathematical domains are sharable because mathematicians use shared logical tools (e.g., first and higher-order quantification) to characterize and constitute those domains.[19] These shared logical tools allow mathematicians to characterize the domains they seek to theorize about, specifically, how they are structured into objects, properties and relations. For example, characterizing the structure of the domain of natural numbers involves characterizing an ω-sequence.

Characterizing the structure of a mathematical domain is precisely what we take categorical axiom systems to do. For example, Hilbert's axioms characterize the structure of a two-dimensional Euclidean plane. The production of a categorical axiom system within a mathematical practice is, usually, the formal culmination of a long process. From the early stages of their development, mathematical practices that concern a single domain incorporate features that informally characterize the structure of the domain they concern. Perhaps the most important such features are the informal proofs and counterexamples given and accepted within the practice in question. Close consideration of which such proofs are judged legitimate, which illegitimate, and which purported counterexamples are taken to be actual and which not provides extensive information about the structure of the domain the mathematical practice in question is about. These and other features of the early development of these types of mathematical practices contribute to those practices determining how their subject matters are structured into objects, properties, and relations.

Let us now return to the question asked at the outset of this section, i.e., according to social constructivists, how, exactly, do mathematical practices manage to socially constitute mathematical domains? The optimal answer to this question—or at least part of that optimal answer—is that it is their ability to provide a (coherent) characterization of a particular structure.[20]

A social constructivist should maintain that all that there is to a particular mathematical domain existing is the undertaking of a mathematical practice that centers about a (coherent) characterization of the structure of the domain in question.[21] So, for example, when William

---

[18] There are, of course, cases where two mathematicians are independently responsible for introducing a particular mathematical domain. I am not aware of any analogous cases in the musical world. This difference is best explained by the very specific purposes for which mathematical domains are introduced. Further, this difference in no way undermines the point I am making in this discussion.

[19] A second difference between the mathematical and musical cases relates to community involvement in the characterization of a construct. While occasionally close friends of a composer do make suggestions for change prior to the completion of a composition, typically, other musicians are extremely uncomfortable making any changes to another's (finalized) work. By contrast, it is common for members of the mathematical community to seek more conspicuous characterizations of newly introduced mathematical domains. Ernest and Hersh both emphasize this type of social negotiation as important to the nature of mathematics. I agree, but wish to note that this aspect of mathematics' social nature is independent of, and secondary to, the type of sociality that I am discussing. In order for a mathematician to offer a different characterization of a newly introduced domain, he or she must already be sharing the domain in question with the individual who introduced it. Thus, that domain must be sharable.

[20] There is, in fact, a lot more involved in providing an optimal answer to this question. Some further details can be found in Chapter 2 of my Ph. D. dissertation [Cole 2005].

[21] Early characterizations of new domains are often less than ideal. Frequently, later development shows them to be ambiguous. Difficult questions need to be asked about when the practices surrounding these characterizations actually

Hamilton first started to discuss entities with a noncommutative algebra to help represent and reason about 3-dimensional vectors, he introduced the practice responsible for the existence of quaternions.

With this social constructivist conception of mathematics in place, we should note the following features of it. First, it vindicates Hersh's claim that mathematical domains "detach from their originator" ([Hersh 1997], p. 16). They do this in a similar way to the way that a piece of music detaches from its composer or composers. Both types of detachment are made possible by the use of sharable tools in the social construction/constitution of the respective items.

Second, mathematical domains have objective features. Sonatas have objective features because of the objective features of the sharable tools that are used in their composition. Similarly, the objective nature of the logical tools used in the characterization and constitution of mathematical domains provides them with objective features. Roughly speaking, mathematical domains inherit the objectivity of logical consequence because they are constituted using logical tools.[22]

Third, the detachment of mathematical domains from mathematical practices allows for the epistemic independence of mathematical domains from mathematical practices highlighted in Section 2. Our imperfect knowledge of mathematical domains can be accounted for in the following way: mathematical domains are constituted using logical tools, yet human beings do not immediately perceive all of the logical consequences of a given characterization of a domain.

Fourth, the above social constructivist conception of mathematics at least points in the direction of an account of how finite mathematical practices have the ability to socially constitute mathematical domains with extremely large cardinalities. Mathematical practices do so simply by (coherently) characterizing those domains. There is, of course, an interesting question that one might ask about how mathematical practices manage to so characterize extremely large domains. Yet it is clear that mathematicians do take themselves to do this all of the time. Thus, *any* philosophy of mathematics will have to face this question concerning characterization (and offer an answer to it) unless it wants to claim that mathematical practices are riddled with massive amounts of error.

Fifth, according to the above account of the nature of mathematical domains, they are socially constituted by mathematical activities that concern particular mathematical domains (e.g., arithmetic, early Euclidean geometry, real analysis, complex analysis, and set theory). Those aspects of mathematics—such as group theory, ring theory, etc.—that do not concern particular domains (but rather all domains that share some structural features) do not, at least in general,[23] contribute to the social constitution of mathematical objects.

---

become responsible for the existence of the domains they characterize. We need not address these difficult questions for our purposes in this chapter.

[22] Issues concerning the objectivity of logic and, consequently, the inherited objectivity of mathematics are complex from the perspective of a social constructivist. Unfortunately, I cannot hope to treat them adequately in this chapter.

[23] It seems to me that there are (probably) historical exceptions. Algebraic theories are only considered of interest if there are particular domains that have the structural features they center about. Consequently, a mathematician working on an algebraic theory will generally produce examples that have the structural features her theory centers about. Nowadays, set theory provides such examples. But, before this rich collection of structures was constituted by set-theorists, those working with algebraic theories produced their own examples. I suspect that occasionally this resulted in them characterizing new particular domains.

5 Abstract Entities

There is one final piece of metaphysics[24] that is worth exploring before we turn to the evaluation of social constructivism as an account of mathematical domains. This is the issue of whether or not mathematical domains and the items of which they are composed are abstract entities. Traditional—by which I mean Platonistic accounts of mathematical entities take them to be abstract entities—indeed, paradigm cases of abstract entities.[25] You might recall, however, that Hersh denies that mathematical entities (and social-historical entities in general) are abstract entities. Yet his argument for this thesis is peculiar. First, all he tells us about abstract entities is that they are neither mental nor physical. Second, he maintains that social-historical entities are neither mental nor physical. Why, then, does Hersh deny that social-historical entities are abstract? The reason, I suspect, is that Hersh's concept of an abstract entity is—unnecessarily—restricted to the concept of a paradigm case of an abstract entity.

A concept F is said to be a **cluster concept** if the application of F is determined by several features, known as "the cluster constitutive of F." If F is a cluster concept, then an item x is F if and only if x has a sufficiently large number of the features in the cluster constitutive of F. An item that has *all* of the features in the cluster constitutive of F is said to be a **paradigm case of F**.[26] I contend that 'abstract' is a cluster concept. It is difficult to specify all members of the cluster constitutive of '**abstract**', but the following are the most important members: **acausality**—the item neither exerts a (strict) causal influence over other items nor does any other item influence it in a (strict) causal way,[27] **non-spatio-temporality**—the item does not stand in spatio-temporal relations to other items, **eternality**—the item exists timelessly, and **changelessness**—none of the item's (intrinsic[28]) properties change. I conjecture that, for Hersh, for something to be an abstract entity, it must have all of these features and the others in the cluster constitutive of 'abstract'.

Hersh's restricted use of abstract is quite understandable and his claim that social-historical entities are not abstract is reasonable. Many social-historical entities fail to have some of the features constitutive of 'abstract'. For example, the U.S. Constitution has a causal impact on people, was constituted at a certain time and so is not eternal, and—perhaps[29]—goes through revisions of its intrinsic properties: amendments to it have been, and probably will continue to

---

[24] **Metaphysics** is the branch of philosophy that investigates the nature of reality. A metaphysical account of some subject matter is a theory about the nature of that subject matter. The following theses are popular parts of *Platonistic* metaphysical accounts of mathematics: mathematical entities exist, mathematical entities would exist even if there weren't any human beings or other types of beings, mathematical entities are not spatio-temporal entities, mathematical entities do not causally influence other entities, the properties of mathematical entities do not change over time, etc. I hope that these theses give the reader some understanding of what it is to provide a metaphysical theory (or interpretation) of mathematics.

[25] I shall provide an account of what an abstract entity is and what a paradigm case of an abstract entity is shortly.

[26] This notion of a cluster concept is prefigured in a number of places in the philosophy literature. Perhaps the most useful discussion is Hilary Putnam's (see [Putnam 1962]).

[27] The relevant sense of strict is the one I identified while I was discussing causal social construction.

[28] The intrinsic properties of an item are those that it has independently of its relationships to other items. This modifier is needed, because it is clear that the extrinsic properties of all things change. For example, the extrinsic properties of the number 7 would change were I to decide that it is no longer my favorite (natural) number.

[29] There is a very tricky issue here about whether such amendments result in a new Constitution or a modified version of the original Constitution.

be, made. It is thus quite reasonable that Hersh should take mathematical entities to be like other social-historical entities in this regard.

However, I don't see any convincing reason why a social constructivist *has* to deny that mathematical domains and the items of which they are composed are acausal, non-spatio-temporal, eternal (or at least timeless), and changeless. I have sketched an argument elsewhere that this suggestion is intelligible (see [Cole 2005], Section 2.1). In fact, in [Cole 2005], I actually endorse it. Unfortunately, I do not have space here to provide a full argument for my endorsement of this suggestion. At the heart of this argument is a recognition of the universal representational function that mathematical domains serve. In essence, the argument is that the universal representational function of mathematics would be undermined by our taking mathematical domains to be causal, spatio-temporal, of limited duration, or changeable.

For clarity, let me briefly illustrate what I mean by the universal representational function of mathematics. The natural numbers can aid us in representing *all* subject matters—including past, future, spatio-temporal, abstract, and counterfactual subject matters. For example, I can claim that the number of people on planet Earth was smaller one hundred years ago than it is today and than it is likely to be in one hundred years time. Mathematics' ability to help represent all subject matters is what is meant by the claim that mathematics' representational function is universal.

A further reason a social constructivist should maintain that mathematical domains are abstract entities is the abundance of tenseless forms of representation in mathematical practices. Another is the fact that this contention allows for the vindication of the intuition that $2 + 2 = 4$ has always been true, as have all well-established mathematical truths.[30]

In addition, maintaining that mathematical domains and the items of which they are composed are (at least close to) paradigm cases of abstract entities would allow a social constructivist to sidestep some tricky issues. For example, it is well-known that Newton's and Leibniz's early developments of calculus were riddled with inconsistencies, yet practiced users of Newton's and Leibniz's tools were able to avoid these inconsistencies. Does the presence of this stable mathematical practice *force* a social constructivist to acknowledge the existence of a domain of infinitesimals with inconsistent properties constituted by this practice? On the present proposal, the answer is no. She *could*[31] take Newton and Leibniz to have been making a range of false assumptions about the real numbers as constituted by our contemporary practice of real analysis— presuming, of course, that our practice of real analysis does constitute the domain of real numbers. Further, the contention that mathematical domains and the items of which they are composed are (at least close to) paradigm cases of abstract entities would allow a social constructivist to account for mathematical practices progressing toward optimal characterizations of mathematical domains. It would also provide for a sense in which a social constructivist could account for early participants in a mathematical practice—individuals like Newton and Leibniz—getting things wrong about the domain the practice in question concerns. Both the claim that mathematical practices progress toward optimal characterizations of mathematical domains and the claim that

---

[30] Some social constructivists (e.g. Ernest and Hersh) would deny or criticize this intuition. I do not share their views on this matter.

[31] She is not, however, forced to offer this answer. A careful investigation of the early practices surrounding the calculus might warrant her accepting the constitution of a domain having inconsistent properties.

early participants in mathematical practices get things wrong about the domain the practice in question concerns find widespread acceptance in our everyday thought about mathematics.[32]

It is for the types of reasons mentioned above that I take the optimal variety of social constructivism to be one that takes mathematical domains and the items of which they are composed to be constituted as (at least close to) paradigm cases of abstract objects. For convenience, let us call this variety of social constructivism **practice-dependent realism (PDR)**—"realism" because it maintains that many mathematical domains genuinely exist and have objective features, and "practice-dependent" because their existence is dependent on the existence of the mathematical practices that constitute them. Officially, practice-dependent realism—like Platonism—is the conjunction of three theses about mathematical domains: a) some exist, b) they and the items of which they are composed are (at least close to) paradigm cases of abstract objects, and c) they and the items of which they are composed are dependent on mathematical practices—in fact, they are pure constitutive social constructs constituted by mathematical practices.

6 *Why Accept Practice-Dependent Realism?*

So far, I have done little more than explicate social constructivism in general and PDR in particular. I have given little or no reason to accept PDR—or any variety of social constructivism—as an account of the nature of mathematical domains. All that I have done is show that PDR is compatible with various aspects of mathematical practices. So, why might one endorse PDR? Ernest pays little attention to this aspect of his social constructivist proposal. I find no argument in his book for social constructivism—the thesis that mathematical domains are socially constituted by mathematical practices. Hersh, on the other hand, provides two arguments in favor of social constructivism.

Hersh's first argument is an extended historical discussion of Platonism, social constructivism's best-known rival. This discussion shows why Platonism has been the historically dominant account of mathematical domains. It also demonstrates why the historical factors that have made Platonism dominant do not provide it with genuine support.

It is all very well to show that Platonism has been accepted for dubious reasons: despite this, it might be true. As Hersh notes in connection with mathematical discovery, it doesn't matter how you come to believe a thesis, what matters is whether that thesis is true. What is needed is not an argument that historical arguments for Platonism have been flawed, but an argument that Platonism is false or at least that social constructivism is preferable to Platonism as an account of mathematical domains. Even better would be a positive argument for the conclusion that social constructivism is true.

Hersh's second argument takes a very different approach from his first. It cites historical evidence from mathematical practices concerning the creative nature of mathematicians' activities. Specifically, Hersh argues that introducing new mathematical theories is a creative endeavor, i.e., involves genuine creativity. No doubt Hersh is correct; the introduction of new mathematical

---

[32] The arguments of this paragraph rely on the assumption that a mathematical discursive practice is able to pick out a mathematical domain as the one it is about, even if it does not characterize that domain perfectly. This is a controversial assumption. Yet my overall argument on behalf of social constructivism can be provided without the support of the arguments made in this paragraph.

theories is a creative endeavor. Yet this fact does not establish the truth of social constructivism. The introduction of new theories about the spatio-temporal world is also a creative endeavor. It doesn't follow that the spatio-temporal world is a pure constitutive social construct. So, Hersh's second argument is no more successful than his first.

Yet Hersh's second argument does point in the direction of a better argument for social constructivism, or at least the thesis that social constructivism is preferable to Platonism as an account of the nature of mathematical domains. In order to make this better argument, one needs to provide historical evidence for more than the thesis that the introduction of new mathematical theories is a creative endeavor. One needs to provide historical evidence that it is a creative endeavor that is not—in fact, cannot be—influenced by **Platonistically construed** mathematical domains—domains that are abstract and independent of all social practices. To start with, such evidence would distinguish the mathematical case from the spatio-temporal case; clearly, theories about the spatio-temporal world are generated under the influence of that (independent) world. In addition, however, it would go some way toward establishing that Platonistically construed mathematical domains are a kind of metaphysical extravagancy that we can—and therefore *should*—do without.[33]

In fact, Hersh even makes observations about mathematical practices that support the conclusion that new mathematical theories are introduced by mathematicians without influence from Platonistically construed mathematical domains. Recall, for example, the second half of Hersh's Fact 1, where he tells us that the creation of mathematical domains is "[not arbitrary], but from activity with existing mathematical objects, and from the needs of science and daily life" ([Hersh 1997], p. 16). Here Hersh observes that new mathematical theories are, on the whole, introduced for two reasons. Most (particularly contemporary) mathematical theories are introduced in response to needs internal to mathematics, such as answering questions raised within already existing mathematical practices. Consider, for example, Hamilton's introduction of the quaternions as a tool for representing and reasoning about three dimensional vectors. The other important reason why new mathematical theories are—or at least were—introduced is in response to a need from science or everyday life, frequently the kind of representational need mentioned in Section 5.

It is important to recognize that new mathematical theories are not introduced simply to describe Platonistically construed mathematical domains that, to use Gödel's famous phrase, "force themselves upon" us. Yes, once mathematical domains have been characterized by some individual or group of individuals, discovering their properties can—and does—feel like discovering the properties of something external to the individual.[34] But this kind of feeling comes after the introduction of a new mathematical theory; it does not motivate that introduction. The introduction of mathematical theories occurs for a variety of (other) reasons, primarily the two mentioned in the last paragraph. These reasons can force a mathematician to include certain features in her new theory, but these external constraints are not constraints from a Platonistically construed mathematical domain, but rather from the problem that she is introducing her theory to solve. Further, mathematical domains can—and, a social constructivist will argue, do—perform

---

[33] Further details of how this argument is meant to go will be provided in Section 8.

[34] Indeed, according to social constructivists, these properties are external to individual mathematicians in the sense that they are determined, at least to a large extent, by the objective logical tools used to characterize mathematical domains.

the roles demanded by these reasons without them needing to exist independently of mathematical practices.

In an ideal world, I would provide further empirical evidence showing that new mathematical theories *are* introduced without influence from Platonistically construed mathematical domains. But space is limited. So, instead, I shall offer a philosophical argument that new mathematical theories *must be* introduced without influence from these domains. In order to make this argument clear, it will be useful to relate it to well known epistemological worries about Platonism. So, let us consider those.

7 *Platonism and Epistemology*

In his 1973 paper "Mathematical Truth" [Benacerraf 1973], Paul Benacerraf made explicit an epistemological concern about Platonism that has inspired much discussion. It is now generally agreed that Benacerraf's original formulation of the challenge is not damaging to Platonism because it rests on a false assumption. Benacerraf's original formulation assumes that there needs to be a causal relationship between a knower and any domain of which she has knowledge. Yet the influence of his challenge remains, as it has been reformulated without mention of this false premise. Perhaps the most forceful such reformulation is Hartry Field's (see [Field 1989]).

According to Platonists, there are two distinct realms that are connected in a specific way: first, a mathematical realm consisting of Platonistically construed mathematical domains, and second, a collection of beliefs, shared by many mathematicians (and others), about this mathematical realm. Further, according to Platonists, the mathematical domains that make up the mathematical realm in question are those things that make the mathematical beliefs in question true or false.[35] Thus, the connection that Platonists claim holds between these two realms is that the first makes many of the second true. Given mathematicians' (and non-mathematicians') causal isolation from any Platonistically construed mathematical realm, there is a need for an explanation of this connection existing. Field challenges Platonists to provide such an explanation.

Let us call an explanation **non-mysterious** if it does not appeal to any mechanisms that would be found illegitimate by a reasonable individual engaged in a natural scientific investigation of the world. The specific form of Field's challenge to Platonists is to provide a non-mysterious, even if only rough, explanation of the *systematic* truth of mathematicians' (and non-mathematicians') pure mathematical beliefs. In other words, Field challenges Platonists to identify some collection of mechanisms that are scientifically investigable and which, in principle, could be the basis of an explanation of mathematicians (and others) having systematically true beliefs about a Platonistically construed mathematical realm.

Field's challenge is legitimate because we share a belief that non-mysterious explanations are, in principle, available for many types of relationships, including our knowledge and beliefs about the world. It is therefore unacceptable to provide an account of the nature of mathematical reality that rules out the possibility of there being a non-mysterious explanation of our having systematically true beliefs about that reality. Field challenges Platonists to show that they have not made this unacceptable move.

---

[35] More precisely, the mathematical realm is that in virtue of which the mathematical beliefs are true or false.

It is our lack of causal connection with any Platonistically construed mathematical realm that motivates Field's challenge. Yet Field's challenge is stronger than can be recognized simply by noting our lack of causal connection with such a realm. 'Abstract' is, in fact, defined in opposition to 'spatio-temporal'. Thus, the abstract nature of any Platonistically construed mathematical realm makes it likely that *all* explanations grounded in features of the spatio-temporal world are unavailable to a Platonist in answering Field's challenge. In other words, it is likely that there are no scientifically investigable mechanisms that could be the basis of an explanation of a Platonistically construed mathematical realm influencing mathematicians and their practices.

Logical deduction is likely to occur almost immediately to the reader as a different kind of potential tool for responding to Field's challenge. Yet noting the role of deduction in mathematics does not provide a full answer to Field's challenge, because beliefs established by means of deduction are only systematically true if the basic beliefs from which they are deduced are systematically true. So, for example, the many arithmetical truths that one can establish by deduction from the Peano Axioms are only systematically true if the Peano Axioms are. A Platonist must thus account for the systematic truth of the basic truths about mathematical domains. Most mathematicians will be tempted to suggest that the basic truths about mathematical domains are true in virtue of something like stipulation. But *why* can we simply stipulate these basic truths? The independence of Platonistically construed mathematical domains from mathematical practices seems to ensure that there might be no mathematical domain that answers to the stipulations in question.

A similar challenge has no force against a PDRist, because, according to her, mathematical *practices* are responsible for mathematicians' (and non-mathematicians') basic pure mathematical beliefs being true. The Peano Axioms are true because they have been accepted as an optimal characterization of a collection of objects, i.e., the natural numbers, appropriate for mathematicians' purposes. More generally, because mathematicians get to decide which mathematical objects should be constituted to serve their purposes, and get to decide which basic claims best characterize such objects, roughly speaking, mathematicians do indeed stipulate the basic truths about mathematical domains. Further, mathematical practices, as spatio-temporally instantiated activities, can causally influence human beings to become (at least minimally) competent participants in them. Consequently, mathematical practices can causally influence human beings to have systematically true pure mathematical beliefs. Think, for example, of how school teachers influence their pupils to become minimally competent participants in mathematical practices. This influence begins with such rudimentary lessons as how to add together two natural numbers, includes an introduction to axiomatic characterization and deduction from axioms, usually in the form of Euclidean Geometry, and will, in the mathematically sophisticated classroom, incorporate discussions of how to characterize the continuity of real valued functions using epsilons and deltas.

There are those who have not been persuaded by Field's challenge. While there have been many responses, the only promising one has been of the following type.[36] Field's challenge— and other challenges inspired by Benacerraf—rests on a false assumption: that there is some need for the mathematical realm to *influence* human beings in order for human beings to have mostly true beliefs about that realm. As Mark Balaguer (see [Balaguer 1998]) and Stewart

---

[36] A defense of this claim can be mounted along the lines found in Chapter 2 of [Balaguer 1998].

Shapiro (see [Shapiro 1997]) have realized, such influence is not required. All that is needed is that no matter which axiom systems mathematicians choose to believe, provided only that they are coherent,[37] they will be true of that realm. Balaguer and Shapiro suggest that the mathematical realm is so "large" that every coherent axiom system (characterization) will be true of some collection of objects, properties, and relations in that realm. So, their response to Field is simple: *all* mathematical statements true in a coherent mathematical theory are true because the mathematical realm is so large that it has enough objects, properties, and relations to accommodate the existential commitments of the theory in question. This suggestion provides a partial[38] solution—indeed, the only known and ever likely to be produced partial solution—to the epistemological challenge to Platonists.[39]

For our purposes, it is important to note that this partial solution does not rest on any assumption that a Platonistically construed mathematical realm influences mathematicians or their practices. Indeed, according to both Balaguer and Shapiro, mathematical practices float free of any influence from the mathematical realm that they countenance. This mathematical realm is not involved in the best explanation of why mathematical practices are the way that they are. Nor should it be invoked as the basis of internal justifications of mathematicians' beliefs and choices.

Now, why have I gone to the trouble of discussing these epistemological worries about Platonism? First, one of the theses that is central to Field's challenge is that there is no respectable sense in which a Platonistically construed mathematical domain can be taken to influence mathematicians and their practices. My argument for the preferability of PDR to Platonism relies on this thesis. Second, some readers might be aware of the Balaguer-Shapiro response to the epistemological worries about Platonism. For this reason, I wanted to be clear that my argument is distinct from the argument to which Balaguer and Shapiro respond. Third, by giving the details of Field's challenge and the Shapiro-Balaguer response to it, I can make it clear that the Shapiro-Balaguer response does not undermine the thesis on which my argument relies.

8 *Platonism vs. Practice-Dependent Realism*

Let us turn to my argument for the preferability of PDR to Platonism as an account of mathematical domains. It is very simple: *Platonistically construed mathematical domains are explanatorily and justificationally superfluous. Consequently, we should not accept their existence.* Let me make some observations about this argument. First, the conclusion follows from the premise by means of an application of **Occam's razor**—don't multiply types of entities without necessity. The idea is that if Platonistically construed mathematical domains are explanatorily and justificationally superfluous, we can do without them.

---

[37] The notion of coherence in play here is a technical one developed by Shapiro (see [Shapiro 1997]). It is closely related to, though not identical with either, deductive consistency and set-theoretic satisfiability.

[38] I describe this solution as merely partial, because it leaves unjustified Platonists' metaphysical claims about the mathematical realm. A full solution to the epistemological worries about Platonism should have the resources to justify these claims.

[39] This claim is controversial. The most significant challenge to it concerns worries about our ability to refer to items in a realm consisting of Platonistically construed mathematical domains (see, e.g., [Azzouni 2000]). It falls outside the scope of this chapter to respond to this worry. Yet if this worry were to be well-founded, it would only strengthen the case against Platonism.

Those familiar with mathematical practices might be wary of applying Occam's razor to mathematical domains. Mathematics is not governed by Occam's razor. Rather, it is an underlying methodological feature of many mathematical practices that mathematicians should seek maximal generality. This feature of those practices can, particularly in foundational areas such as set theory and category theory, result in the characterization of ever larger mathematical domains. There is no problem here, however, because my application of Occam's razor is not internal to some mathematical practice, but rather takes place within the practice of naturalistic metaphysics (i.e., metaphysics guided by the methodological practices of natural science). Occam's razor is a legitimate tool within this practice, because it is a legitimate tool within the non-mathematical aspects of natural science.

Further, I take it to be a benefit of PDR that it predicts this methodological difference between the mathematical and non-mathematical aspects of natural science. If mathematical domains are pure constitutive social constructs, then Occam's razor governs mathematics if and only if it governs the practices that constitute pure constitutive social constructs. Does it? No! Consider for a moment the collection of legal statutes of the United States of America. Without doubt, the system of law embodied in this collection could be represented in a simpler and theoretically more elegant way by a collection of statutes with fewer members than there are in the actual collection. Despite this, we claim that the number of legal statutes is exactly the number in the actual collection. That number is, at least roughly speaking, the number felt necessary in order for them to serve the social functions for which they are constituted. So, the proposal that mathematical domains are pure constitutive social constructs should bring with it two predictions: first, that Occam's razor does not govern mathematical practices, and second, that the number of mathematical domains that in fact exist is linked with the purposes for which mathematical domains are constituted. Both predictions are accurate.

The discussion in the preceding paragraph points toward the following: entities that are *dependent* on social practices are not the kind of entities whose existence should be denied on the grounds of Occam's razor, while entities that are *independent* of social practices are the kind of entities whose existence can be—and, in some cases, should be—denied on the grounds of Occam's razor.

Let us now consider the premise of my argument, viz., Platonistically construed mathematical domains are explanatorily and justificationally superfluous. The thesis that there is no respectable sense in which these domains can be thought to influence mathematicians and their practices is central to the justification of this premise. Yet my premise requires further justification, for it might be possible for these domains to play some kind of explanatory or justificatory role without influencing mathematicians or their practices. Indeed, this belief has been embedded in a number of arguments for Platonism. For example, the existence of Platonistically construed mathematical domains has been argued to be required in order for mathematical statements to have the truth-value ascribed to them by mathematicians. Also, their existence has been considered necessary for providing mathematics with a semantics that resembles the semantics of everyday discourses sufficiently closely to account for the way in which these two types of discourses are intermingled.[40]

---

[40] See [Benacerraf 1973] for arguments of both types.

Taking PDR seriously undermines both of these reasons for postulating Platonistically construed mathematical domains. First, PDR takes mathematical statements to have the truth-value ascribed to them by mathematicians. Second, since pure constitutive social constructs are among the entities talked about using everyday discourses, an adequate semantics for everyday discourses must be able to accommodate them.

Perhaps there are other explanatory or justificatory benefits that Platonistically construed mathematical domains might yield without influencing mathematicians or their practices. The most natural suggestion would be that they are indispensable to an account of the objectivity of mathematics. Yet—as I have indicated above—I believe that a PDRist has the resources to provide such an account. If such an account can be fleshed out, then Platonistically construed mathematical domains are not required for this purpose.

In fact, it is difficult to see what work Platonistically construed mathematical domains can do that the mathematical domains countenanced by PDR cannot do. And, unless some such work can be found—indeed, a fairly significant amount of such work can be found—we should not countenance these domains, for to do so would be to multiply types of (independent) entities without necessity.

9 Conclusion

Obviously, there is still much work that could be done in defending the premise of my argument that PDR is preferable to Platonism as an account of mathematical domains. Most importantly, I need to give the details of a PDRist's account of the objectivity of mathematics. There is also a need for arguments that PDR is preferable to the other accounts of mathematical domains found within the philosophy of mathematics literature. Such arguments require PDRists to show that they have the resources to account for the other traditional features of mathematics (e.g., its apriority and necessity). I don't have the space to explore these topics in this chapter. What I hope I have achieved in this chapter is to have given you a clearer understanding of what social constructivism about mathematics is and to have given you an idea of why you might want to be a social constructivist about mathematics.

Acknowledgements I wish to thank Stewart Shapiro for his assistance in writing this chapter and in developing Practice-Dependent Realism. I also wish to thank Bonnie Gold, Roger Simons, and Barbara Olsafsky for their many comments on previous drafts of this chapter.

References

[Azzouni 2000] Azzouni, J., "Stipulation, logic, and ontological independence," *Philosophia Mathematica*, 8(3) (2000), pp. 225–243.

[Balaguer 1998] Balaguer, M., *Platonism and Anti-Platonism in Mathematics*, Oxford University Press, New York, NY, 1998.

[Benacerraf 1973] Benacerraf, P., "Mathematical truth," *Journal of Philosophy*, 70 (1973), pp. 661–679.

[Benacerraf/Putnam 1983] Benacerraf, P. and Putnam, H., eds., *Philosophy of Mathematics: Selected Readings*, 2nd ed., Cambridge University Press, New York, NY, 1983.

[Cole 2005] Cole, J., *Practice-Dependent Realism and Mathematics*, Ph.D. dissertation, The Ohio State University, Columbus, OH, 2005.

[Ernest 1998] Ernest, P., *Social Constructivism as a Philosophy of Mathematics*, State University of New York Press, Albany, NY, 1998.

[Field 1989] Field, H., *Realism, Mathematics, and Modality*, Basil Blackwell, Oxford, England, 1989.

[Frege 1884] Frege, G., *Die Grundlagen der Arithmetik*. Translated by Austin, J. (1960). *The Foundations of Arithmetic*, 2nd ed., Harper, New York, NY, 1960.

[Haslanger 1995] Haslanger, S., "Ontology and social construction," *Philosophical Topics*, 23(2) (1995), pp. 95–125.

[Hersh 1997] Hersh, R., *What Is Mathematics, Really?* Oxford University Press, New York, NY, 1997.

[Heyting 1931] Heyting, A. "The intuitionist foundations of mathematics," *Erkenntnis*, (1931), pp. 91–121. Reprinted in [Benacerraf/Putnam 1983], pp. 52–61.

[Putnam 1962] Putnam, H., "The analytic and the synthetic," pp. 358–397 in Feigl, H. and Maxwell, G., eds., *Scientific Explanation, Space, and Time*, volume 3 of Minnesota Studies in the Philosophy of Science, University of Minnesota Press, Minneapolis, MN, 1962. Reprinted in [Putnam 1975], pp. 33–69.

[Putnam 1975] ——, *Mind, Language and Reality*, Cambridge University Press, New York, NY, 1975.

[Shapiro 1997] Shapiro, S., *Philosophy of Mathematics: Structure and Ontology*, Oxford University Press, New York, NY, 1997.

[Wright 1992] Wright, C., *Truth and Objectivity*, Harvard University Press, Cambridge, MA, 1992.

III

The Nature of Mathematical Objects and Mathematical Knowledge

In the last forty years, philosophers of mathematics who are not working strictly in foundations have concentrated on questions about the nature of mathematical objects and how we come to have mathematical knowledge. Because this work has resulted in hundreds of papers and dozens of books, we have four chapters summarizing it. They were written by philosophers with very different perspectives. While there is a common set of questions running through these chapters, each has chosen different aspects in his summary, because of the difference of perspective. Of the philosophers whose chapters are in this section, Chihara has spent his career working on various versions of nominalism, the view that there are no mathematical objects. Shapiro has leaned toward the realist side, currently in a version called structuralism, which has origins in Bourbaki's mother-structures and the view of mathematics as the science of patterns. Balaguer has most recently suggested that there may be no testable distinction between the most appealing versions of platonism (or realism) and nominalism. Linnebo, the youngest of the authors in this section, appears to be working on developing a very minimal version of platonism (that is, a commitment to mathematical objects that involves a minimal "ontological" commitment). Each of these chapters sets forth the general argument overall and then gives the individual author's perspective on where the delicate points are. We end the section with a chapter by a mathematician, offering a very different approach to the question of mathematical objects via category theory.

7

The Existence of Mathematical Objects

Charles Chihara
Emeritus Professor of Philosophy
University of California, Berkeley

From the Editors

Charles Chihara is the most senior of the philosophers contributing to this book. He appears to be genuinely interested in having his philosophy of mathematics be one that is acceptable to mathematicians. One would think that this is requisite; what is the point of a philosophy of X that people who work in X view as absurd? But there has not been much interaction between the two communities (mathematicians and philosophers of mathematics) in the last half century. As he notes in his chapter, Charles Chihara started out as a mathematician and has both a brother and a niece who are mathematicians. He thus has a better feel for what will make sense to a mathematician than do many philosophers of mathematics. His writings are normally quite accessible to mathematicians, and this one is especially so.

His chapter walks a rather delicate line. Since Chihara is a nominalist, he is not willing to commit to the existence of any mathematical objects, including structures. Yet it is important from his viewpoint that we do *have mathematical knowledge. Chihara's solution is a sort of structuralism, but without a commitment to the existence of structures. It is a rather delicate balance, but it is certainly a thoughtful one.*

Charles Chihara is an Emeritus Professor of Philosophy at the University of California, Berkeley (sophos.berkeley.edu/chihara/). Chihara has published nearly fifty articles in his principal areas of interest: philosophy of mathematics and philosophy of logic. He has also published widely in the philosophy of science and confirmation theory, as well as on the philosophies of Wittgenstein, Russell, Quine, Goodman and Davidson. He is the author of Ontology and the Vicious Circle Principle *(1973),* Constructibility and Mathematical Existence *(1990),* The Worlds of Possibility: Model Realism and the Semantics of Modal Logic *(1998), and* A Structural Account of Mathematics *(2004). Among his articles likely to be of interest to readers of this volume are "On Alleged Refutations of Mechanism Using Gödel's Incompleteness Results,"* The Journal of

Philosophy *(1972); "A Gödelian Thesis Regarding Mathematical Objects: Do They Exist? And Can We Perceive Them?"* The Philosophical Review *(1982); "Burgess's 'Scientific' Argument for the Existence of Mathematical Objects,"* Philosophia Mathematica *(2006); and "The Burgess-Rosen Critique of Nominalistic Reconstructions,"* Philosophia Mathematica *(2007). He continues to work on various problems in the philosophy of mathematics.*

1 Introduction

Many mathematicians believe in the existence of mathematical objects of various sorts, and they think that mathematics is the study of these objects. It is the contention of this paper that such beliefs are fundamentally mistaken and that mathematics can more appropriately be regarded as a particular kind of study of structures: one that does not imply the existence of special mathematical objects. But before explaining in detail in what way and why I believe mathematics should be so regarded, some misconceptions about the nature of philosophy need to be cleared away.

Shortly after I began my teaching career at Berkeley, I had lunch with two of my former teachers—mathematicians who were in the city to attend a conference. Intrigued by the fact that I had switched fields from mathematics to philosophy, one of these professors rather pointedly said to me: "So, you are now a *philosopher* of mathematics! So far as I can see, philosophy of mathematics is either logic or mysticism. Which do *you* do?"

Needless to say, I do not regard my work in the philosophy of mathematics as being just logic, even though logical work and logical reasoning clearly have an important place in my view of mathematics. Nor can I find anything in my philosophical views that is in the least way mystical. What follows is a detailed description, from my perspective, of the sort of undertaking philosophy is and of what I, as a philosopher, am trying to accomplish. It will then be clear to the reader why what I do in philosophy is neither logic nor mysticism.

2 What is Philosophy?

The philosopher seeks an understanding of the world. But the sort of understanding sought might be called "Big Picture understanding." What one seeks in philosophy is the really "Big Picture": what, in general and in broad outlines, is the universe like? What, in general and in broad outlines, is our (i.e. humanity's) place in the universe? How, in general and in broad outlines, do we (humans) gain an understanding of the universe? This "Big Picture" goal explains a striking feature of philosophy: the fact that, for practically any heavily studied area X of serious intellectual work, there is a philosophy of X. There is philosophy of biology, philosophy of physics, philosophy of language, philosophy of religion, philosophy of art, philosophy of history, and so on. For each X, one seeks to fit X into this Big Picture.

In this search for the Big Picture, *coherence* is an essential ingredient. We seek an understanding of X that is consistent with our other beliefs about the universe and us. Take the philosophy of language, for example. Here, we seek an understanding of the nature of language and our mastery of language that is consistent with both our common sense beliefs and also our scientific views about the universe we inhabit and also about us as organisms with the features attributed to us by science. Any account of the nature of language that conflicted with the prevailing scientific accounts of how we learn a language would be considered by most philosophers of language to be in serious trouble. We seek a coherent and comprehensive Big Picture, where all the different *X*s fit together. Thus, one would expect a contemporary philosopher's account of mathematics to be consistent with our generally accepted views of science and scientific knowledge.

One can see, then, why in philosophy there is great attention to uncovering and solving paradoxes. A paradox is an argument that starts with premises that seem to be incontestable, that proceeds according to rules of inference that are apparently incontrovertible, but that ends in a conclusion that appears to be obviously false. In many cases, a paradox ends in an outright absurdity or even a self-contradiction. Thus a paradox evidently shows us that either one or other of our "incontestable" premises are not true or that some "apparently incontrovertible" rule of inference that we used in our reasoning is not valid. Ultimately, it seems to show that the totality of our beliefs do not form a coherent whole and hence that there is a need to repair our beliefs—which is one reason philosophers of mathematics continue to work on the paradoxes of mathematics and set theory discovered in the late Nineteenth and early Twentieth Centuries.[1] It is all part of the philosopher's ongoing project of refashioning our beliefs into a coherent whole.[2]

3 The Platonic (Realistic) Conception of Mathematics

Given the above conception of philosophy, one can see why some philosophers are dissatisfied with one of the most widely held philosophical views of mathematics: the view known as "Mathematical Realism" ("Realism" for short) or "Platonism." Realism is frequently characterized to be the philosophical doctrine that *mathematical objects exist*.[3] Anyone who held that such things as numbers, sets, functions, vectors, matrices, spaces, etc., in fact exist could then be said to be a Realist.

What about the mathematician who accepts the theorem that there are prime numbers greater than five? Is such an acceptance tantamount to an acceptance of the existence of prime numbers? And should such a mathematician be classified as a Realist? Well, suppose that the mathematician's acceptance amounts to no more than the belief that among the theorems of arithmetic is one that can be expressed by the sentence 'There are prime numbers greater than five'. Strictly speaking, a mathematician who believes that 'there are prime numbers greater than five' is a theorem of arithmetic need not conclude that *there are such things as prime numbers*: such a conclusion requires an additional acceptance of a substantial philosophical thesis (to be discussed in detail later).

In any case, most Realists maintain a much more robust view of mathematics than is expressed by the above existential characterization. Kurt Gödel expressed a Realist view of mathematics that is more typical than the bare bones existential doctrine indicated above. In discussing *Cantor's Continuum Hypothesis* (henceforth 'CH'), Gödel rejected the belief of a number of researchers that, if CH were proven to be independent of the standard axioms of set theory, the question of its truth or falsity would simply lose its meaning, just as the question of the truth or falsity of the Fifth Postulate of Euclidean geometry was thought to have lost its meaning with the discovery of

---

[1] See [Chihara 1973, Chapter 1] for a discussion of the paradoxes and of Russell's attempt to solve them.

[2] The reader can find a fuller account of my view of the nature of philosophy in the Introduction to [Chihara 2004].

[3] For more details on this topic, see [Chihara 2004, Chapter 5], and [Maddy 1990].

its independence from the other postulates.[4] Gödel was convinced that CH had a truth value that was independent of whether or not it was formally decidable from the axioms of standard versions of set theory. This conviction was tied to his belief that sets truly exist.[5] Thus, he argued in his paper [Gödel 1964b] that such an independence result in set theory would render the question of the truth or falsity of CH meaningless only if set theory were regarded as a hypothetico-deductive system in which the meanings of the primitives of set theory were left undetermined. But, Gödel argued, set theory is not that sort of system. According to Gödel:

(1) the objects of set theory "exist independently of our constructions;"

(2) we have "an intuition of them individually" (the term "intuition" here being used by Gödel to refer to something like a "perception" of individual sets[6]);

and

(3) the general mathematical concepts we employ in set theory are "sufficiently clear for us to be able to recognize their soundness and the truth of the axioms concerning" these objects.

He concluded that "the set-theoretical concepts and theorems describe some well-determined reality, in which Cantor's conjecture must either be true or false," even if the conjecture is independent of the other axioms.[7]

Gödel's views about CH illustrate some of the principal features of most robust forms of Mathematical Realism. These Mathematical Realists maintain that mathematical objects truly exist and that the mathematician is attempting to provide us with information about these objects. In the case of set theory, these Realists believe that the theorems are true statements that tell us what sets in fact exist and how these mathematical objects are related to one another by the membership relationship. One implication of such a view of set theory is that the standard axioms of set theory, such as those of Zermelo-Fraenkel set theory (or **ZF** for short), are literally true statements—the statements correctly describe objects that in fact exist and that in fact are related (by the membership relation) in the way implied by the axioms. According to such Realists, the axioms are not statements that the set theorist merely postulates or arbitrarily lays down. They are supposed to be truths that the mathematician has, in some way, *discovered*.[8] Since mathematical entities are not supposed to be things that can be seen, touched, heard, smelled, tasted, or even

---

[4] Such a position is suggested by the following quote: "Probably we shall have in the future essentially different intuitive notions of sets just as we have different notions of space, and will base our discussions of sets on axioms which correspond to the kind of sets we wish to study ... everything in the recent work on foundations of set theory points toward the situation which I just described." [Mostowski 1967, p. 94].

[5] It should be noted that Gödel did not believe that the question of the truth or falsity of CH rested solely upon the belief in the existence of sets. He argued that the "mere psychological fact of the existence of an intuition which is sufficiently clear to produce the axioms of set theory and an open series of extensions of them suffices to give meaning to the question of the truth or falsity of propositions like Cantor's continuum hypothesis." [Gödel 1964b, p. 272].

[6] [Gödel 1964b, p. 271].

[7] [Gödel 1964b, p. 262]. It is now known, as a result of Paul Cohen's proof (see [Cohen 1966]), that CH is indeed independent of the axioms of standard versions of set theory.

[8] Cf. G. H. Hardy's Platonic assertion: "I believe that mathematical reality lies outside us, and that our function is to discover or *observe* it, and that the theorems which we prove, and which we describe grandiloquently as our 'creations' are simply our notes of our observations." [Hardy 1941, pp. 63-4] Cf. also [Hardy 2002, p. 182]. A recent work defending a robust Realist view similar to Gödel's is [Brown 1999].

detected by our most advanced scientific instruments, we seem to have, according to the picture of mathematics advocated by these thinkers, two causally isolated worlds. There is the mathematical world of sets, numbers, functions, etc., from which we are excluded, and the physical world of which we humans are members—with apparently no causal links between any member of one of these worlds and any member of the other.[9] Anti-Realists find it hard (if not impossible) to reconcile the above Realistic position with generally accepted scientific views of how humans are able to obtain knowledge of things "outside" their minds. The idea that set theorists, just sitting in their offices, are somehow able to discover the truths that mathematicians enshrine as "the axioms of set theory"—this about entities that are supposed to be completely undetectable by us and yet independent of our thoughts and intentions—strikes anti-Realists as bizarre or even unintelligible.[10]

Realists have, of course, tried to deal with the difficulty of accounting for the mathematician's supposed knowledge of the existence and properties of mathematical entities. For example, Gödel postulated that we have something like a perception of the objects of set theory. Then he argued:

> I don't see any reason why we should have less confidence in this kind of perception, i.e. in mathematical intuition, than in sense perception, which induces us to build up physical theories and to expect that future sense perceptions will agree with them and, moreover, to believe that a question not decidable now has meaning and may be decided in the future.[11]

The idea here seems to be that mathematical intuition plays a role in mathematics analogous to the role that sense perception plays in the empirical sciences. In both cases, we are pictured as constructing theories that have implications about future "perceptions," so that, in favorable instances, the theory is confirmed by "perceptions."[12] Such a view of set theory suggested to Gödel that:

> There might exist axioms so abundant in their verifiable consequences, shedding so much light upon a whole field, and yielding such powerful methods for solving problems... that, no matter whether or not they are intrinsically necessary, they would have to be accepted at least in the same sense as any well-established physical theory.[13]

Anti-Realists (more commonly called "nominalists"[14]) have found Gödel's epistemology of set theory to be paradox-ridden, fantastic, or simply unscientific.[15]

---

[9] It should be noted that Gödel explicitly asserted that the objects of transfinite set theory "clearly do not belong to the physical world." [Gödel 1964b, p. 271].

[10] The reader should be warned that not all Realists accept all the doctrines being attributed here to Gödel. There are many different brands of Mathematical Realism; see the article by Balaguer in this volume for a discussion of them.

[11] [Gödel 1964b, p. 271] Of course, not all versions of Realism, even "robust" forms, appeal to a kind of "perception" of mathematical objects being characterized here by Gödel.

[12] For additional insights into Gödel's philosophical views about the nature of mathematics, the reader should study [Gödel 1964a]. For another Realist's view of how mathematicians obtain knowledge of mathematical objects, see [Brown 1999, Chapter 3]. I give criticisms of Brown's view in [Chihara 2004, Chapter 10, Section 2].

[13] [Gödel 1964b, p. 271] In one of his Alfred Tarski lectures given in April, 2001, Ronald Jensen described the view expressed in the above quote as being the most influential (in the community of set theorists) of all of Gödel's philosophical views.

[14] See [Chihara 2005] for a detailed defense of the nominalist's view of mathematics. For a critical assessment of nominalism, see [Burgess/Rosen 1997].

[15] For a detailed criticism of Gödel's views about his postulation of a kind of perception of the objects of transfinite set theory ("mathematical intuition"), see [Chihara 1982] and [Chihara 1990, Chapter 1, Section 3].

Also, anti-Realists find it hard to make sense of the Realist's picture of the set theorist somehow picking out and referring to *specific* mathematical entities (such as the empty set). If we accept the Realist's assumption that sets are entities that are completely undetectable by us or our scientific instruments, then how can this picking out and referring take place? Well, can't we pick out and refer to specific things that we have never seen, touched, or experienced in any way? No scientists have ever perceived a dinosaur, but no one doubts that they are able to pick out and to refer to specific ones. How is this done? By way of the traces dinosaurs have left. For example, we have fossil remains of specific dinosaurs by means of which we can refer to, say, "that specific Tyrannosaur whose fossilized bones are in the Smithsonian museum." But suppose that dinosaurs left no causal traces. Suppose that they were things in a completely separate world from which we were totally isolated. How could we then pick out and refer to specific dinosaurs? That is the sort of situation we are in vis-á-vis the sets postulated by Realists.

Still, the Realist might argue that we can pick out and refer to a particular set by saying, for example, that "it is that set that has no members." Of course, this strategy assumes that the meaning of the word 'set' somehow actually distinguishes those mathematical entities that are sets from those that are not (since, even if one had adequately defined the membership relation, there is no reason to believe that the null set is the only *mathematical entity* that has no members). It also assumes that we succeeded in specifying what the membership relation is in a sufficiently definite way so that only one mathematical object could be the null set. But we haven't. To see why, imagine that God has informed you that 'A' and 'B' name two sets. Now try to answer the question: what would have to be the case for A to be a member of B? Can one answer this question, giving necessary and sufficient conditions, in a way that is not merely another way saying that A is a member of B (such as "A belongs to B," "B contains A as a member" or "A is in B")? In fact, we cannot specify what must be the case for A to be a member of B in a way that would enable one to pick out one particular mathematical object as the empty set.

It has been suggested to me that many mathematicians believe that there a "standard interpretation" of 'membership' which could be used to pick out that unique mathematical object that is the null set. Supposedly, this is because there is only one set, under the standard interpretation of membership, that has no element whatsoever. Talk of a "standard interpretation" is usually about an interpretation of some axiom system such as **PA** (see p. 144). Take the "standard interpretation" (or "standard model") of **PA**: it can be said that *in this model*, there is only one element that is the successor of 0. But "in this model" is not the same as "in the universe of mathematical objects." Thus, *in a group structure*, there is only one thing that is the right identity element. No realist would be led to believe by such talk that, in the vast realm of mathematical objects, there is only *one thing* that is the right identity element of a group.

The idea that the "standard interpretation" of membership can enable us to pick out a single mathematical object as the null set may have arisen from the fact that when we were first introduced to set theory, we were given such examples of sets as the set of all dogs or the set of all natural numbers (or more generally the set of all Fs, where F is some condition). One can indeed give necessary and sufficient conditions for a thing x to be a member of the set of dogs: one can say that a thing x is a member of the set of dogs iff x is a dog. One may be led to think that this shows that one really knows what the membership relation is, when in fact, all one knows is that, when one is informed that some set C is the set of all Fs, then one can state that x is a member of C iff x is an F. We can state necessary and sufficient conditions in this case only because we were informed that a thing x is a member of C iff x is an F. But without this extra information, we would still be in the dark about what must be the case for something to be a member of a set, as the above thought experiment illustrates.

Typically, what is done in text books on set theory is to take the axiomatic approach: one simply lists the axioms that are to govern the universe of sets and the relation of membership that the sets may bear to one another. Obviously, such a listing will not enable us to pick out a unique mathematical object as the one thing that is the null set. After all, the axioms of standard formalized set theories such as **ZF** fail to pick out a unique totality of objects as the universe of sets: **ZF** has many different models.

Realists may contend that set theorists have a kind of "intuition" or "mind's eye" by means of which they are able to distinguish the empty set from all the other mathematical objects that are supposed to exist. But any such postulation of special mental powers would be met by much skepticism among contemporary philosophers. Clearly, the Realist is faced with a daunting task if she is to convince anti-Realists of her ability to single out and to refer to the empty set.[16]

The mystery is deepened by the fact that questions of reference of the above sort never seem to arise in actual mathematical practice.[17] As Jodi Azzouni has noted:

> The current philosophical concern with how mathematical terms pick out what they refer to is an oddity from the point of view of mathematical practice, which, in broad respects, is simply not concerned with reference. Any view that fails to explain why this is the case has not explained something crucial about mathematics.
>
> [Azzouni 1994, p. 31]

Surprisingly, the Realists have, for the most part, simply ignored the problem of how one is able to pick out and refer to specific mathematical objects. It is a problem that they should grapple with since it seriously stains their portrait of mathematics.

There is another implication of the Realist's view of mathematics that anti-Realists have difficulty making sense of. It will be agreed by everyone that the empirical scientist makes heavy use of mathematics. Evidently, the scientist needs to know a significant amount of mathematics in order to understand the workings of the physical world. But these apparent facts about science, interpreted in the metaphysical way the Realist advocates, imply that the empirical scientist needs to know a great deal about how various non-physical entities are related to one another in a world from which we are forever isolated—this in order to understand the workings of the physical entities in the world we do inhabit. That this should be the case strikes the anti-Realist as, if not totally implausible, at least very counterintuitive.

The above problem is closely related to a puzzle that Azzouni labels "the epistemic role puzzle." [Azzouni 1994, p. 58] This puzzle arises when we ask: what role do mathematical objects play in our attempts to gain knowledge of the entities in our world? It is clear that mathematical objects are not involved in the causal processes that occur around us, since they are not supposed to interact causally with any physical entities. It is hard to see what possible function

---

[16] There are, of course, other attempts to answer the anti-Realist's skeptical doubts, and many responses to responses. See [Chihara 2004, pp. 15–16], for references to works discussing the topic.

[17] Some readers may ask: Why should anyone be interested in such questions of reference, if they never seem to arise in actual mathematical practice? One answer is: because such questions of reference naturally arise when one accepts the Realist's view of mathematics. The fact that such questions never seem to arise in actual mathematical practice is, as I see it, a reason for questioning the Realist's view of mathematics.

the existence of these mathematical objects can have in the scientist's theoretical understanding of the physical world.[18] But if so, then why should we postulate such things?

Despite these problems, many Realists continue to hold fast to their metaphysical picture of mathematics. Of course, it is open to Realists to search for a resolution of their difficulties by, say, questioning the scientific or epistemological views that conflict with the implications of their own views or that raise difficulties for their ontological position.[19] But any serious philosopher of mathematics should, I would think, at least consider the possibility that such conflicts are the products of the Platonic doctrines themselves.

4 Reasons for Accepting the Realist's View

There is no doubt that many mathematicians and philosophers continue to find the Realist's view irresistibly attractive. What makes Platonism so plausible? I suggest that it is ultimately a thesis I shall call "the central assumption of philosophy of mathematics" or "the Central Assumption" for short. This is the proposition that *the theorems of classical mathematics are truths*. Of course, no one denies that there could be errors in classical mathematics, whereby what are taken to be genuine theorems are in fact not truths at all. But most mathematicians and philosophers believe that, apart from a few exceptions, what mathematicians call "theorems" are in fact truths.

Every Mathematical Realist I know of has accepted the Central Assumption in some form or other.[20] For example, John Burgess and Gideon Rosen signal their acceptance of the Central Assumption in [Burgess/Rosen 1997] when they advance a version of a Realistic view of mathematics which they call "minimal anti-nominalism" (a view they clearly advocate). "Having studied Euclid's Theorem," they write, "we are prepared to say that there exist infinitely many prime numbers." Furthermore, they understand Euclid's theorem in the standard straightforward way that logicians tend to understand such theorems: "Moreover, when we say that there exist infinitely many prime numbers, we say so without conscious mental reservations or purpose of evasion. . . . " [Burgess/Rosen 1997, pp. 10–11] Thus, they are willing to accept the implications of their acceptance of the truth of the theorem by also asserting that there are prime numbers (and hence that mathematical objects exist).

4.1 Why the Central Assumption Leads to Mathematical Realism

Using the above reasoning of the minimal anti-nominalists as a model, one can develop an explicit argument for the existence of mathematical objects. (In the following discussion, I shall use the term 'classical mathematics' to include set theory, number theory, and the theory of functions of real and complex numbers).

Argument T:

(Central Assumption) The theorems of classical mathematics are true.

---

[18] See [Azzouni 1994, Section 7] for a fuller discussion of this puzzle.

[19] For an example of an attempt to defend the Realist position against such doubts, see [Brown 1999, pp. 15–18]. For criticisms of Brown's defenses, see [Chihara 2004, pp. 13–15].

[20] It should be noted, however, that not all philosophers of mathematics have accepted the central assumption. Hartry Field, for example, argued in [Field 1980] that most of the theorems of classical mathematics are false.

(E) Some of the theorems of classical mathematics assert the existence of mathematical objects.

(Conclusion) There are mathematical objects.

4.2 *Reasons for Believing the Central Assumption*

What reasons do we have for accepting the Central Assumption? Here are some of the principal ones.[21] Certainly classical mathematics is constantly being fruitfully used by all sorts of people in a variety of situations, both practical and theoretical, in ways that suggest that the assertions of mathematics are truths. In particular, mathematics is *applied* in science and engineering to draw conclusions upon which even our most brilliant thinkers rely in dealings with the world and other people. If the mathematical theorems were not true, how could we rationally place so much confidence in the conclusions that we infer using mathematics? Would we not be basing our inferences upon falsehoods and shouldn't we accumulate false beliefs about the world? If these theorems were not true, shouldn't we expect the bridges we build using mathematics to collapse, and shouldn't the rockets we program using mathematics go off course, etc.? Perhaps one can see why Michael Resnik would include the following as a premise in an argument he once produced:

(R-1) *We are justified in drawing conclusions from and within science only if we are justified in taking the mathematics used in science to be true.*[22]

Resnik does not provide a convincing justification for accepting (R-1), perhaps because he believes that the claim is obviously true.[23]

Another related reason for accepting the Central Assumption is based upon the fact that many mathematical theorems seem to be directly verified to be true. Consider, for example, the *Fundamental Theorem of Calculus*. This theorem not only has been successfully applied in countless ways for hundreds of years, apparently it has also been confirmed in a variety of empirical ways by graphing specific continuous functions and empirically calculating the areas under the graph, say on engineering paper. For example, it would seem that what the theorem says regarding $\int_3^7 (2x + 7)\,dx$ can be checked, empirically, by graphing the function in the interval [3, 7] and counting the relevant squares. Of course, this particular example is so simple as to appear trite. But obviously one can produce examples in which the values of far more complicated and sophisticated definite integrals are estimated by making the same sort of empirical estimations of areas under curves, yielding the type of "confirmation" of the fundamental theorem discussed above.

A closely related reason for accepting the Central Assumption is based on the undeniable fact that mathematical theorems frequently convey information to researchers. It is not just that mathematics is continually being applied in science and engineering (the central fact underlying the first reason given above), but it also seems to provide scientists and engineers with *information*.

---

[21] Other reasons are given in [Chihara 2004], e.g. see "Maddy's mystery" (pp. 229–30, 291–2).

[22] [Resnik 1997, p. 48]

[23] Actually, Resnik does put forward a kind of argument for (R-1) in [Resnik 1998, p. 233], but there he uses the premise that "science assumes . . . the truth of much mathematics," which is itself not justified by Resnik.

Since declarative sentences that convey information are, for the most part, true, we seem to have additional evidence that theorems of mathematics are true. Indeed, if the theorems were not true, how could they provide us with information?

The above reasons, however, do not convince me of the correctness of the Central Assumption and later I shall explain why. But first, I should emphasize that I am not suggesting here that mathematical theorems are false but only that we should not simply assume that they are true— that a serious investigation is called for. Let me say straight off that I am one of those anti-Realists who have come to reject the kind of metaphysical view of reality that the Realists have adopted. Despite the fact that some renowned mathematicians have espoused a Realist philosophy of mathematics, I am convinced that a reasonable view of mathematics can only be achieved by abandoning the Platonic view.[24] Thus, I have set about trying to fashion a more accurate view.

5 The Hilbert-Frege Dispute

I believe that some helpful insights into the sort of view of mathematics I shall be sketching in this paper can be obtained by examining a dispute between two outstanding mathematicians that was carried on by letter between 1899–1903. During this period in which David Hilbert was conducting his pioneering research on geometry—a period which gave rise to, among other things, the publication of his groundbreaking *Foundations of Geometry*—it was widely accepted that the axioms of geometry were truths. Hilbert himself seems to have accepted this widely held view during this period. For example, in 1898, in an introduction to a course he was giving in mechanics, Hilbert characterized geometry as a sort of empirical science, asserting:

> [G]eometry [like mechanics] emerges from the observation of nature, from experience. To this extent, it is an *experimental science* . . . But its experimental foundations are so irrefutably and so *generally acknowledged*, they have been confirmed to such a degree, that no further proof of them is deemed necessary. Moreover, all that is needed is to derive these foundations from a minimal set of *independent axioms* and thus to construct the whole building of geometry by *purely logical means*.[25]

In the introduction to his *Festschrift* on geometry, Hilbert had written: "Geometry requires . . . for its consequential construction only a few simple facts. These basic facts are called axioms of geometry."[26] Notice that, in this quotation, Hilbert is claiming that the "few simple facts" that geometry requires are called axioms of geometry—what are called "axioms" are said by Hilbert to be facts. In his *Foundations of Geometry*, he tells his readers that the axioms express "certain related facts basic to our intuition." [Hilbert 1971, p. 23] Thus, it can be seen that Hilbert, at least some of the time, thought of the axioms of geometry as expressing true statements.

However, in section 3 of his *Foundations of Geometry*, Hilbert asserts that the axioms there "define the concept 'between'," and he goes on to say, in section 6, that the "axioms of this group define the concept of congruence or of motion." These characterizations elicited the following

---

[24] Of course, I am not alone in rejecting the Realist's view of mathematics. As Paul Cohen as noted, "probably most of the famous mathematicians who have expressed themselves on the question have in one form or another rejected the Realist position." [Cohen 1971, p. 13] Cf. also Saunders Mac Lane's view that "save for mythology, all the variants of Platonism shatter on the actual practice of mathematics." [Mac Lane 1986, p. 449]

[25] This quotation is taken from the introduction to a course on mechanics Hilbert taught in the winter semester of 1898. See [Corry 1999, p. 152] (italics in the text).

[26] This is quoted in [Frege 1971, p. 25] and then criticized.

critical question from Gottlob Frege: "How can axioms [that express facts basic to our intuition] define something?" [Frege 1971, p. 25] If the axioms are definitions, then for Frege they are laid down by fiat or stipulation. On the other hand, if the axioms are facts or express facts, then they are truths and they are truths whether or not we take them to be truths: not something that is stipulated to be so. So how can the axioms be both true statements that express facts and also definitions?[27]

Evidently, Hilbert had adopted a significantly new approach to geometry in his book, but, unfortunately, the old traditional approach remained in the background. It still exerted an influence upon his thoughts, thus producing the conflicting characterizations of his axioms described above. To obtain a more perspicuous grasp of Hilbert's new point of view, let us imagine that Hilbert's geometrical theory had been developed as a formal theory of first-order logic in such a way that the *undefined* terms 'is a point', 'is a line', and 'is a plane' of Hilbert's book are given as non-logical constants in the vocabulary of this formalized theory. Within such a logical setting, let us investigate why Hilbert might regard his axioms as "definitions."

We get some idea of how Hilbert regarded his axioms from his letters to Frege. In response to Frege's claim that, from the truth of one's axioms, it follows that the axioms do not contradict one another, Hilbert wrote back that, as long as he had been thinking about these matters, he had been saying "just the reverse":

> If the arbitrarily given axioms do not contradict one another, then they are true and the things defined by the axioms exist. [Hilbert 1980, p. 40]

By this reasoning, Hilbert thought that one can prove that the set of axioms of the real numbers do define something and the things defined do exist. In other words, Hilbert thought that one can simply lay down axioms about some new mathematical entity—axioms stating how these new entities are interrelated—and by proving that the set of axioms is consistent, one would prove that the axioms legitimately define something that can be said to "exist."

We can make sense of these ideas of Hilbert within the setting of first-order logic. Imagine that a mathematician lays down a set of sentences as axioms, and the set is found to be consistent. This would imply that the set of axioms succeeds in singling out a class of models. Any first-order structure satisfying the axioms would have to be such that the individual constants refer to specific "parts" of the structure—the "parts" being related to each other in definite ways. In this way, the axioms can be regarded as "implicitly defining" the non-logical constants themselves. The axioms implicitly tell us what the individual constants refer to (and hence "mean") in the various models of the theory. Thus, it would be natural for Hilbert to claim, as he did in a letter to Frege, that "to try to give a definition of a point in three lines is to my mind an impossibility, for the whole structure of axioms yields a complete definition." [Hilbert 1980, p. 40] It is, of course, the whole set of axioms—and not just a single axiom—that determines what properties each model of the theory must have.

There are passages in Hilbert's letters to Frege that suggest that Hilbert was definitely thinking in terms of models of the axioms when he called his axioms 'definitions'. He wrote, for example:

> [I]t is surely obvious that every theory is only a scaffolding or schema of concepts together with their necessary relations to one another, and that the basic elements can

---

[27] I am here focusing on only one aspect of the dispute between Frege and Hilbert. For a more detailed discussion of this dispute, see [Chihara 2004, Chapter 2, Section 1].

be thought of in any way one likes. If in speaking of my points I think of some system of things, e.g. the system: love, law, chimney sweep . . . and then assume all my axioms as [specifying] relations between these things, then my propositions, e.g. Pythagoras' theorem, are also valid for these things. In other words: any theory can always be applied to infinitely many systems of basic elements.   [Hilbert 1980, p. 40]

Paul Bernays, writing in *The Encyclopedia of Philosophy*, describes Hilbert's axiom system "not as a system of statements about a subject matter but as a system of conditions for what might be called a relational structure." [Bernays 1967, p. 497] Similarly, Ian Mueller describes the content of Hilbert's geometrical axioms as "structural" and characterizes the Hilbertian geometry as "the study of structure." [Mueller 1981, p. 9]

It should be noted that, when we view geometry in the structural way Hilbert did, then the axioms of geometry cannot be said to be literally true. Such sentences could be taken to be "true" only in the technical sense of being true *in a structure* or true *under an interpretation*. Thus, Hans Freudenthal describes this revolutionary aspect of Hilbert's geometry with the words: "[T]he bond with reality is cut. Geometry has become pure mathematics. . . . Axioms are not evident truths. They are not truths at all in the usual sense." [Freudenthal 1962, p. 618]

6 Mathematics Regarded as a Theory About Structures

Might we not view practically all mathematical theories in the way Hilbert regarded geometry? In the case of axiomatized mathematical theories, we can take the axioms of the theory as, in effect, characterizing (or implicitly defining) a type of structure (not necessarily first-order). We could then regard the theorems of the theory as being about this type of structure without worrying about what the assertions really say about mathematical objects or about "reality"?

6.1 First-Order Logic

To make the above ideas both more specific and also more precise, I shall restrict my discussion of mathematical theories, for now,[28] to just axiomatized theories expressed in a first-order logical language whose non-logical constants include only individual constants and predicates.[29] I shall assume that the readers of this work are familiar with the fundamentals of first-order logic. The semantics of first-order logic yields a definition of 'true under an interpretation' (or 'true in a structure') for the sentences of the formal language. An interpretation for such a theory consists of: (1) a first-order structure, and (2) an assignment of appropriate "parts" of the structure to the non-logical constants of the language.[30] A first-order structure consists of: a non-empty set of elements called "the domain" of the structure, and n-ary relations among the elements of the domain ($n = 1, 2, 3, \ldots$). Thus, each individual constant is assigned an element of the domain, and each n-ary predicate is assigned an n-ary relation among the elements of the domain.[31]

---

[28] Later in this paper, I shall consider the cases of mathematical theories expressed in natural languages and even those that are not axiomatized.

[29] I do not include operation symbols for the sake of simplicity of exposition: operation symbols can be defined as special kinds of predicates.

[30] See [Mates 1972, Chapter 4, Section 3] for an elementary account of these definitions.

[31] See [Mates 1972, Chapter 4, Section 1] for additional details and examples.

There is, however, another sort of "interpretation" of formal logical languages that is important for my analysis: they are what I call "natural language interpretations" (or "NL Interpretations" for short). These are the sort of "interpretations" of first-order languages that *philosophically trained* logicians are apt to consider when "translations" of the logical language into some natural language are seriously contemplated. These "interpretations" do more than what the first-order interpretations described above do. They not only single out a first-order structure and assign the relevant sort of objects and relations to the non-logical constants of the formal language in question —they also supply meanings or senses to the constants and quantifiers. "Interpretations" of this sort specify the domain of the structure, using a specific English name or description of the domain. They assign to each individual constants the sense or meaning of some English name or definite description. For example, it may assign to the individual constant '*b*' the meaning of the English definite description 'the governor of Florida', so that '*b*' can then be understood to mean the very same thing as the phrase 'the governor of Florida'. They also provide each predicate of the language with the sense or meaning of an *English predicate*, where English predicates are obtained from English declarative sentences by replacing occurrences of names or definite descriptions with occurrences of circled numerals.[32] For example, if the NL interpretation $\mathscr{I}$ specifies that the domain is to be the set of living human beings, and it assigns to the binary predicate 'R^2' the sense of the English predicate '① is younger than ②', and it assigns to the individual constants '*a*' and '*b*' the senses of 'The father of the President of the United States' and 'Hillary Clinton' respectively, then the sentence 'R^2ab' will express the statement 'the father of the President of the United States is younger than Hillary Clinton' and '$(\exists x)R^2xa$' will express the statement that there is a living human who is younger than the father of the President of the United States. In this interpretation, the former sentence will be false and the latter will be true.

There is an important respect, then, in which an NL interpretation differs from a standard first-order interpretation. A sentence of a formal language that has been given an NL interpretation will express a statement. However, a sentence of a language that has been given only a first-order interpretation cannot reasonably be said to express a statement, since the non-logical constants occurring in it have not been given meaning or sense. It is only in the former case that a sentence of the language can be said to be true or false (and not merely true under an interpretation). I shall say that a logical language equipped with the meanings or senses provided by an NL interpretation is an *NL interpreted language*.[33]

One reason philosophical logicians have made use of NL interpretations is because, from the time when Logical Positivism[34] was the dominant school in the area of philosophy of science, a question that has received much attention from philosophers is whether genuine scientific theories can be (and ought to be) formalized as axiomatized deductive theories of first-order logic. Of

---

[32] See [Mates 1972, p. 77] for a fuller explanation of what English predicates are.

[33] See [Chihara 2004, chapters 5, 6, and 7] for more detailed discussions of NL interpretations and NL interpreted languages.

[34] Logical Positivism arose out of the Vienna Circle and became a dominant force in the philosophy of science in the Twentieth Century, especially during the thirties and forties, and even into the fifties (so much so that the Positivist's view of scientific theories has frequently been called "the Received View"). For a detailed discussion of the development of the philosophy of science of the Logical Positivists, see [Suppe 1974, Sections I, II, and III]. Suppe notes that the Positivists "construed scientific theories as axiomatic theories formulated in a mathematical logic" meeting a number of specific conditions (p. 16).

course, if such a formal theory is to serve as a genuine scientific theory, its assertions must express statements that can be said to be true or false—hence the need for NL interpretations.

6.2 The Basic Idea of the Structural Account of Mathematics

Let us now reconsider the idea of treating all axiomatized mathematical theories in the way Hilbert regarded his geometry. We can regard the assertions of any axiomatized first-order mathematical theory as characterizing its first-order models. Notice, that we can ignore NL interpretations of the language for this purpose, since we can treat the non-logical constants occurring in mathematical sentences in the way Hilbert treated the non-logical constants in his geometry: as being parameters. One advantage of understanding mathematical theories in this way is that we can avoid having to justify any analysis of what the assertions of the mathematical theories truly mean. Trying to understand what mathematical assertions mean usually comes down to trying to figure out what actual individual mathematicians, scientists, engineers, and everyday ordinary working people have in mind when they utter mathematical sentences—a none too easy task.

7 The Structural Content of Theorems of Mathematics

Let us start with the precise case in which a theorem ϕ is derived in an axiomatized formal theory of first-order logic, say Peano Arithmetic (or what I shall call '**PA**' for short).[35] It follows that:

> Any model of **PA** would have to be a model of ϕ.

The above displayed sentence gives what I call "the structural content" of ϕ. More generally, one can give the structural content of any theorem δ of a logically formalized mathematical theory (not necessarily first-order) with axioms Γ as follows:

> Any structure that would satisfy Γ would also have to satisfy δ.

The above displayed sentences are what philosophers characterize as "modal sentences" because they crucially involve the concept of necessity.[36] It is not just that any model of **PA** happens to be a model of ϕ; any model of **PA** *would have to be* a model of ϕ. That is, *necessarily*, any model of **PA** is a model of ϕ.

Notice that I have *not* been claiming that the structural content of any theorem ϕ of **PA** gives the meaning of ϕ or tells us what ϕ asserts. Had I done so, I would have been advancing a view about mathematics that philosophers call "if-thenism." If-thenism is the type of view Hilary Putnam once proposed in claiming that "pure mathematics consists of assertions to the effect that *if* anything is a model for a certain system of axioms, *then* it has certain properties." [Putnam 1967, p. 294] In the above passage, Putnam was putting forward a thesis about the *meaning* or semantic form of all the theorems of pure mathematics. I, on the other hand, do not advocate any such thesis. I certainly do not claim that '$2 + 2 = 4$' asserts that if anything is a model of, say, Peano's axioms, then this model has certain properties. Nor do I want to be committed to the

---

[35] See [Chihara 2004, p. 241] for a specific version of **PA**.

[36] A modal sentence asserts *how* a proposition is true (or false). Thus, the following is a modal sentence: "That there is life on Mars is possibly true."

position that all the theorems of pure mathematics have such an if-then form. I see no plausible way of justifying any such semantic position, and in fact I see a number of serious problems that such a view implies.[37] Instead, what I have been suggesting is that, regardless of what a theorem of a first-order theory may actually mean or assert, a proof of the theorem can be seen to give us the above sort of if-then information.

Some readers may find the differences I have been noting here to be slight, if not trivial. Some indication of the philosophical importance of these differences will become evident from the paragraph following the next, in which Newton and Leibniz are discussed, and also from Section 10 below. The previous paragraph also distinguishes my position from that of Mathematical Structuralists, who claim that mathematical theorems are always assertions about structures. Such a view is expressed by the Structuralist Michael Resnik:

> In mathematics, I claim, we do not have objects with an "internal" composition arranged in structures, we have only structures. The objects of mathematics, that is, the entities which our mathematical constants and quantifiers denote, are structureless points or positions in structures. As positions in structures, they have no identity or features outside a structure. [Resnik 1981, p. 530]

Charles Parsons writes that such a view is a familiar one with "a long history, going back to the late nineteenth century." It is the view that "reference to mathematical objects is always in the context of some structure, and that the objects involved have no more to them than can be expressed in terms of the basic relations of the structure." [Parsons 1995, p. 74] Thus, the Structuralist claims to be providing an account of "what mathematics is about."[38] My position, by contrast, is in no way an account of what mathematical theorems actually assert. "The structural content of a theorem" is not supposed to tell one what the terms of the theorem refer to and what the range of the quantifiers is. Hence, it would be misleading to identify my account of mathematics with that of the Mathematical Structuralist.

There are a few other points of clarification of my position that should be made. I do not restrict the structural information provided by mathematical theorems to only cases in which a theorem is proved in an axiomatized formal theory such as **PA** or **ZF**. Newton and Leibniz, for example, proved theorems even though they were reasoning in unformalized and unaxiomatized systems of mathematics. They were, as I see it, theorizing about a kind of structure, and hence their proofs were providing mathematicians with structural information.[39] Thus, a proof they gave of some theorem ϕ would provide the following sort of information:

> Any structure of the kind about which the mathematician(s) producing the proof is (are) theorizing would have to be a model of ϕ

---

[37] See [Chihara 2004 Chapter 10, Section 4] for a detailed discussion of some of the problems that "if-thenism" faces. For a fuller exposition of Putnam's "if-thenism" and how my structural account differs from his account, see [Chihara 2004, p. 245–49].

[38] This is explicitly claimed by the Structuralist Stewart Shapiro in his [Shapiro 1997, p. 8]. For Shapiro, "mathematics is the science of structure." [Shapiro 1997, p. 75] Shapiro also claims that "the subject matter of, say, arithmetic is a single abstract structure, the natural-number structure." [Shapiro 1997, p. 9] For additional material (as well as references) on Structuralism, the reader should consult the two articles on Structuralism (by G. Hellman and F. MacBride respectively) in *The Oxford Handbook of Philosophy of Mathematics and Logic* edited by Shapiro.

[39] However, just what specific kind of structure these pioneers had in mind is not easy to specify with confidence.

Thus, given such a structure **S**, each quantifier in ϕ would have as its range the domain of **S** and each individual constant in ϕ would denote an appropriate element of the domain of **S** and each *n*-ary predicate would denote an appropriate *n*-ary relation among the elements of the domain of **S**. And ϕ, so interpreted, would be true in the structure. The situation is, in important respects, similar to the case of Hilbert's geometrical theory axiomatized in his *Foundations of Geometry*. The axioms are expressed in an ordinary natural language (say, English)—not a formal logical language. Still, it can be seen that the theorems had a structural content of the sort described above.

For cases like the Newton-Leibniz one, there is admittedly a great deal of vagueness and unclarity as to the kind of structures intended. Hence, for such cases, one should not expect the precision and definiteness of first-order logic. In general, the structural content of any theorem ϕ of mathematics can be given in the above way, but obviously the earlier ones expressed in terms of logically formalized axiomatized theories should be used wherever possible. The advantage of dealing with formalized axiomatic theories is that it allows much more clarity and precision.

Finally, I should emphasize that by "a proof of a theorem," I do not intend to restrict what are to count as proofs to only formal proofs ("derivations") or to proofs that only establish what *logically* follows from a set of axioms. The non-formal proofs that practicing mathematicians regularly produce can be, according to my view, perfectly rigorous and convincing.[40] In this respect, too, my position on proofs differs from what Putnam advocated at the time he advocated if-thenism, since he specifically asserted:

> [I]n pure mathematics, the business of the mathematician is not in discovering *materially* true propositions of the form 'If *M* is a model for *T* then so-and-so', but in discovering *logically true* propositions of that form. Even if a proposition of the form in question is true, if it is only 'true by accident' ..., then it will not be provable by purely formal means, and hence will not be asserted by the mathematician.   [Putnam 1967, p. 291]

My account accepts that Newton, Leibniz, and Euler gave genuine proofs of mathematical theorems, even when their proofs could not be translated into a formal proof or could not be said to yield the sort of logical knowledge that, say, a proof in first-order logic establishes.

8 *A Structural Account of Applications of Mathematics*

Recall Resnik's claim that we are justified in drawing conclusions from and within science only if we are justified in taking the mathematics used in science to be true. Let's see how we can make use of a mathematical theorem to draw some conclusion in science even though we are not justified in taking the theorem used to be true.

---

[40] Cf. what William Thurston has written about formal proofs: "[W]e should recognize that the humanly understandable and humanly checkable proofs that we actually do [produce] are what is most important to us, and that they are quite different from formal proofs. For the present, formal proofs are out of reach and mostly irrelevant: we have good human processes for checking mathematical validity." [Thurston 1994, p. 171] Cf. also Hardy's comments about formal and informal proofs in his [Hardy 2002].

8.1 A Theory of Light Rays[41]

Definition 1: The term 'point' is used to refer to points in Euclidean 3-space, i.e. the space of ordered triples of real numbers, the standard metric being used, and the points being ordered in the standard way.

Definitions and Axioms Assumed: In the following, all the usual axioms and definitions of the geometry of Euclidean 3-space will be assumed, as well as the usual properties of the real numbers—in particular, such terms as 'angle', 'normal', 'plane', 'half space bounded by a plane', 'lies on', and 'continuous curve' will be assumed to have been defined in standard ways.

Definition 2: Any continuous curve from point a to point b will be called a *path from* a *to* b. If **f** is any path from a to b, then a may be called a 'source' and b may called an 'observer'.

Definition 3: A *path of a light ray* is a path from any source to any observer, which is the shortest path from the source to the observer.

Definition 4: A *light reflection instance* consists of a path **f**, a plane **p**, and points a, b, and c such that:

> c lies on **p**; a and b are not on **p**, but are both in the same half space bounded by **p**; and **f** is the union of a path of a light ray from a to c and a path of a light ray from c to b, all the parts of **f** being in the same half space bounded by **p**.

Such an **f** will be called a 'path of reflection (relative to **p**) from a to b'; c will be called a 'point of reflection' of **f**; and the half-space bounded by **p** in which **f** is present will be called the 'space of reflection' of **f**, whereas the half-space bounded by **p** in which **f** is absent will be called the 'complement space of reflection' of **f**.

This theory has one *principal axiom* (in addition to all of the axioms assumed above to have been given):

Axiom: *If* **p**, **f**, *a*, *b*, *and c constitute a light reflection instance*—**f** *being a path of reflection (relative to* **p***) from a to b*—*then if b* *is that point in the complement space of reflection of* **f** *that is symmetric[42] to b with respect to* **p***, point c must be such that the path consisting of the union of that part of* **f** *that goes from a to c and the path of a light ray that goes from c to b* *is the shortest path from a to b**.

It can be shown that:

Theorem: *Let* **p**, **f**, *a*, *b*, *and c constitute a light reflection instance*, **f** *being a path of reflection (relative to* **p***) from a to b. If n is the normal to* **p** *at c directed into the space of reflection, then the angle between that part of* **f** *that is a path from a to c and n is equal to the angle between n and that part of* **f** *that is a path from c to b.*

This theory is a purely mathematical theory, despite the fact that such expressions as 'observer', 'source' and 'path of a light ray' may suggest otherwise. For example, by examining

---

[41] This theory is based upon a theory given in Sec. 2.3.3 of [Maki/Thompson 1973].

[42] The sense of 'symmetric' here is that **p** is a perpendicular bisector of the path of a light ray from b^* to b.

the definitions above, it can be seen that the first two of the above terms are only suggestive of how the theory is to be applied and in fact are said to refer only to points. Also, seeing that not all mathematical structures satisfy the principal axiom, one can conclude that the term 'path of reflection' functions as a parameter that can refer to different "entities" in different structures. Thus, sentences of the theory involving this term are not true (or false), but only true (or false) in certain structures. **In particular, the above theorem is not true (but only true *in certain structures*)**.[43]

However, this theory can in fact be applied to actual physical situations involving the behavior of light, by regarding paths of *actual* light rays to be approximately that of paths of light rays from a source to an observer (as characterized above). In so applying the theory, we need to make certain idealizations about the physical space in which we operate (e.g. that it is Euclidean and three dimensional). Also, we need to restrict the scope of the theory to light traveling in "homogeneous media"—say, reasonably clear (fogless, smokeless) air—which is reflected by a flat object, such as a mirror, approximating a plane in certain geometrically relevant ways. The structural content of the theorem mentioned earlier tells us that, in such conditions, *the angle of incidence of the light is equal to its angle of reflection*. Needless to say, knowing such a result can be used to make practical predictions in a variety of situations.[44]

The above reasoning undercuts the Resnik premise (discussed earlier) that we are justified in drawing conclusions from and within science only if we are justified in taking the mathematics used in science to be true, since it is clear from the above example that a mathematical theorem can be legitimately applied in science even when it is not a true assertion (but only true in certain structures).

8.2 *The Fundamental Theorem of Calculus Reexamined*

The above example shows us how to understand the kind of "empirical verification" of the Fundamental Theorem of Calculus discussed earlier. We have on the one hand a purely mathematical theory (analysis) in which the Fundamental Theorem is proved. On the other hand, we have an empirical theory (say, about the printed squares of engineering paper) according to which certain kinds of entities are definitely related in ways that approximate, more or less accurately, the ways that various elements of structures are asserted to be interrelated by the structural contents of the axioms of analysis. It is natural to speak, in this context, of a sort of "empirical model" that is structurally identical, given certain idealizations, to the models (or sub-models) of the mathematical theory.[45] Then, the fact that the theorems of the mathematical theory are true of all the models of the theory explains why the theorem would hold of the "empirical model" used to carry out the "empirical verification." Thus, it is not the actual truth of the mathematical theorem that is "empirically verified" but rather only its structural content. It can be seen that this kind of "verification" no more presupposes the literal truth of the Fundamental Theorem itself than does an "empirical verification" (made by observations that verify the empirical law of reflections of light rays) presuppose the truth of the theorem of the previously discussed theory of light. In

---

[43] Hence, there is no need to argue in this argument that such mathematical terms as 'point' and 'continuous curve' do not refer to particular entities but only places in structures.

[44] For an indication of how the above theory of light rays can be developed into a more complex and versatile theory, see [Maki/Thompson 1973, Sec. 2.3.4].

[45] Cf. [Chihara 2004, Ch. 9, Sec. 12, especially fn. 59].

both cases, all that is really presupposed in the verification is that the structural contents of the respective mathematical theorems in questions hold.[46]

Of course the above discussion is not meant to provide a rigorous argument to undermine the convictions of a staunch supporter of the Central Assumption or a devoted Realist. It is only meant to provide the reader with the central ideas of how the structural content of mathematical theorems can supply the information needed to apply mathematics in an empirical science or, more substantially, to show how one can have a nominalist philosophy that accords with our picture of mathematics. For a more detailed argumentation that supports my rejection of the thesis that the applicability of mathematics in science presupposes its truth, the reader should study [Chihara 2004, Chapter 9, Sections 8-12].

9 Fermat's Last Theorem

Since Andrew Wiles' 1993 proof of Fermat's Last Theorem depended upon proving the Taniyama-Shimura conjecture, and involved reasoning about modular forms and elliptic curves, one can see that his proof involves theorizing about the field of complex numbers and about a four-dimensional space called hyperbolic space.[47] The above facts prompt the following questions: why were investigations into the nature of such conceptually complicated mathematical concepts as modular forms in hyperbolic space needed to solve a problem about the natural numbers? One might think that a question about the natural numbers would be best answered by reasoning directly about the natural number structure. *Why was it necessary for mathematicians to theorize about positions in the much more complicated structure of hyperbolic space in order to prove the theorem?*

To see how something like this can come about, let's look at a somewhat simpler case of the use of a larger structure to prove theorems about a smaller one. Note that most mathematicians learn to prove the *Fundamental Theorem of Algebra* in their studies of the theory of functions of complex variables.[48] Then, from the Fundamental Theorem, it is proved as a corollary that:

> Every polynomial of degree n, with only real coefficients, can be factored into a product of real linear and real irreducible quadratic polynomials.[49]

The latter is a theorem about the structure of the real numbers. Yet, the above sort of proof of the corollary involves theorizing about the field of complex numbers.[50] This can be understood to be a case in which the mathematician obtains important information about the type of structure that forms the subject matter of the algebraic study of the field of real numbers, and she does this by

---

[46] For a more detailed explanation of the underlying point being made here, see [Chihara 2004, Ch. 9, Sec. 12]. From the perspective of the structural account, one can also rebut the third reason given earlier for accepting the central assumption. One can explain how a mathematical theorem can provide researchers with genuine information, without having to assume that the theorem is true. For the theorem does not have to be true to provide researchers with valuable information about what must hold in all structures of a certain sort.

[47] Ken Ribet had earlier proved that the Taniyama-Shimura conjecture implied Fermat's last theorem. For a highly readable account of Wiles' proof, see [Singh 1997]. See also [Laubenbacher/Pengelley 1999, Chapter 4], for a nice historical discussion of Fermat's last theorem, giving helpful references about the details of Wiles' proof.

[48] As, for example, in [Knopp 1945, pp. 113–114].

[49] See [Birkhoff/Mac Lane 1953, p. 110].

[50] I am not suggesting, however, that one cannot prove the corollary without theorizing about the realm of complex numbers. See, for example, [Fine/Rosenberger 1997].

investigating a more complicated structure in which is embedded the type of structure in question. One can see how reasoning about the larger, encompassing structure can yield information about the embedded structure, since the embedded structure will be related mathematically to the larger structure in countless ways.

Still, it may be wondered how it can be that a mathematician, faced with a problem about a given structure, may find a solution to the problem only by reasoning about a more complex structure in which the given structure is embedded. I am sure there are many explanations that can be given in answer to such wonderment. One thing stands out as evident: theorizing about features of the more complex structure may make apparent to the "mind's eye" a number of mathematically significant relationships, features, and regularities involving "entities" in an embedded structure that are difficult, if not practically impossible, to "visualize" when one is thinking only in terms of the simpler structure.[51] The reason may be that certain sorts of relationships, features and regularities concerning the "entities" of the simpler structure can only be readily comprehended when these "entities" are seen in their relationships with the "entities" of the more complex structure. This is a bit like the case in which certain traits of character of a certain member of a family can only be easily noticed in contexts in which this member interacts with a more diverse group of people than those only in the immediate family, as say when the whole family gets together with all their friends and relations at a wedding.

A mathematical example that illustrates the above idea may be helpful here. Consider the well-known "diagrammatic proof" of the Pythagorean Theorem (see the figure).

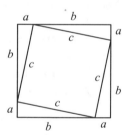

The reasoning would proceed as follows:

[1] The area of the outer square $= (a + b)^2$.
[2] The area of the outer square is also $= c^2 + 4$ x (area of triangle abc).
[3] From the above and by simplification, we get: $a^2 + b^2 = c^2$.

The role of the diagram is clear. This diagram represents the "structure" (pattern) of the triangle abc as part of a larger, more complex encompassing "structure" (pattern), which then

---

[51] In speaking of "the mind's eye", I am not making any sort of appeal to the sort of mental powers that Gödel postulated to when he wrote about "mathematical intuition." I am not suggesting, for example, that humans have the ability to "perceive" or "intuit" mathematical objects that do not exist in the physical world. When I spoke above of what may be made apparent to the "mind's eye", I only intended to be speaking metaphorically about what normal humans can be made to grasp, to comprehend, or to understand.

makes apparent a number of geometric and algebraic relationships, involving the sides of the triangle *abc*—relationships that are essential to the above reasoning but that would not be apparent in the absence of the more complex "structure."[52]

Let's return to the discussion of Wiles' use of very high-powered theorems of analysis to prove Fermat's theorem. This can be seen to be a case in which reasoning about very complicated and sophisticated structures is used to draw conclusions about the much simpler embedded structure of the natural numbers. This is similar to what we saw in the above examples involving the Fundamental Theorem of Algebra and the Pythagorean Theorem.

All of the above is meant to give some additional insights into how mathematicians are able to provide physicists with, as Arthur Jaffe and Frank Quinn put it, "reliable new information about the structures they study." Physicists frequently formulate their theories in terms of mathematical structures. Mathematician may study these structures directly, or they may investigate generalized versions of these structures, perhaps via more abstract structures possessing substructures of which the physicist's structures are instances. The above examples indicates how the mathematician's researches into features of the structures they single out for study can yield theoretically useful (and even essential) information for the physicist. We can now also see how mathematicians are able to accomplish such feats without requiring of the theorems of mathematics they prove that they be true assertions. It is required only that the structural contents of the theorems be true.[53]

Notice that to regard mathematical theories in this way is to regard mathematicians as reasoning about structures. That mathematicians reason about structures is not controversial. That they even see themselves as reasoning about structures is evident. William Thurston, for example, describes mathematical progress with the words: "As our thinking becomes more sophisticated, we generate new mathematical concepts and new mathematical structures: the subject matter of mathematics changes to reflect how we think." [Thurston 1994, p. 162] Jaffe and Quinn describe the stages of mathematical discovery in the following way:

> Typically, information about mathematical structures is achieved in two stages. First, intuitive insights are developed, conjectures are made, and speculative outlines of justifications are suggested. Then the conjectures and speculations are corrected; they are made reliable by proving them.   [Jaffe/Quinn 1993, p. 1]

A crucial role that mathematics plays in physics is specified when they write: "It is now mathematicians who provide [physicists] with reliable new information about the structures they study." [Jaffe/Quinn 1993, p. 3]

---

[52] Taking diagrams to be structures is something that Resnik has advocated with his pattern-theory of structures. See in this regard his [Resnik 1997, p. 206]. In opposition to the widely held view that diagrams and pictures are only heuristic devices that have no place in a mathematical proof, Jon Barwise and John Etchemendy write (in connection with the above proof of the Pythagorean Theorem): "Once you have been given the relevant diagram, the rest of the proof is not difficult to figure out. It seems odd to forswear nonlinguistic representation and so be forced to mutilate this elegant proof by constructing an analogous linguistic proof, one no one would ever discover or remember without the use of diagrams." [Barwise/Etchemendy 1991, p. 12].

[53] To see how I treat applications of arithmetic, see [Chihara 2004], especially pp. 244–5.

Of course, the classic statement of the importance of structure for mathematics was given by Bourbaki when he wrote:

> [M]athematics appears thus as a storehouse of abstract forms—the mathematical structures; and it so happens—without our knowing why—that certain aspects of empirical reality fit themselves into these forms, as if through a kind of preadaptation.... The unity which it [i.e. structure] gives to mathematics is not the armor of formal logic, the unity of a lifeless skeleton; it is the nutritive fluid of an organism at the height of its development, the supple and fertile research instrument to which all mathematical thinkers since Gauss have contributed.[54]

10 The Big Picture

From discussions I have had with mathematicians, I have gotten the impression that many mathematicians think that philosophy is not a serious topic of study and that, in philosophy, one can believe and say practically anything without being thought a fool. Such a view of philosophy may have obtained currency among mathematicians because a superficial perusal of the history of philosophy can convince a sober mind that the most absurd and unbelievable doctrines have been espoused by some highly regarded philosophers. Certainly, some celebrated philosophers have espoused such remarkable theses as "the real things that we know by experience last for a very short time, one tenth or half a second . . . the things we call real, like tables and chairs, are . . . logical fictions." [Russell 1956, p. 274] But such an apparently absurd doctrine is taken seriously by other philosophers because it can be seen to be an essential part of an ingenious "Big Picture"[55]—a "Big Picture" that is not obviously inconsistent or incoherent. Indeed, such a doctrine would be especially attractive to some philosophers if it provided an effective way of obviating or overcoming certain serious theoretical difficulties that had long troubled philosophers.

Can the structural account I have sketched here be fitted into such a "Big Picture"? There is no way I could convincingly defend a positive answer to this question in the limited space I have left, but I can give some indications of how my structural account is consistent with more

---

[54] [Bourbaki 1950, p. 231] Some may argue that what mathematicians mean by 'structure' is not what philosophers or physicists mean by it, in which case the above quotes do not provide any support for the view I have put forward. In response, I would argue that, although there are some minor differences in how the three groups use the term, there is a core meaning to the terms 'structure' and 'model' which are common to all these groups, and that the above quotes all use this core meaning. Thus, in the article from which the above quotation was taken, Boubaki undertakes to make clear what is to be understood by 'mathematical structure': the term, we are told, "can be applied to sets of elements whose nature has not been specified: to define a structure, one takes as given one or several relations, into which these elements enter . . . ; then one postulates that the given relation, or relations, satisfy certain conditions (which are explicitly stated and which are the axioms of the structure under consideration)." This is basically a characterization of what philosophers take structures to be. Consider what Mac Lane writes, after giving the Peano axioms: "This is a typical description of a structure by axioms" [Mac Lane 1986, p. 44]—which with relatively minor differences fits my view beautifully. He also gives the "general notion" of an algebraic structure {note: there are also **non**-algebraic structures!} as follows: "A set X with nullary, unary, binary, ternary . . . operations satisfying as axioms a variety of identities between composite operations." [Mac Lane 1986, p. 26] See also [Mac Lane 1986, p. 33]. Mac Lane's characterization of structure is essentially what philosophers of mathematics have in mind when they speak of "structures." See [Chihara 2004, Chapter 3, Section 1]. Cf. [Barbut 1970]. See also [Suppes 1967a], where Suppes argues that "the meaning of the concept model is the same in mathematics and the empirical sciences." (p. 289) Cf. also [Suppes 1967b, pp. 57–59].

[55] Recall the discussion of philosophy in Section 2 and the philosopher's goal of achieving a kind of understanding I called "Big Picture" understanding.

than just our beliefs about mathematics, by showing how the structural account does not conflict with standard views about knowledge and reference in the way the Realist's account did. In this way, I can illustrate how the structural account affords us an effective way not only of avoiding some of the theoretical problems that troubled the Realist's view of mathematics, but also of accommodating standard views about knowledge and reference, thus indicating how the account fits into a larger picture of the world than can be obtained by just focusing on mathematics.[56]

Recall, from section 3, that one difficulty with the Realist's account of mathematics was the problem of understanding how set theorists, just sitting in their offices and, apparently merely thinking, are able somehow to obtain knowledge of the existence and properties of sets— mathematical entities from which humans are completely cut off physically and causally. This prompted some outstanding Realists to postulate some kind of faculty of mathematical intuition by means of which set theorists are able to "perceive" sets (in a way analogous to the way we can perceive physical objects). Recall also that another problem the Realist ran into was of comprehending how mathematicians are able to pick out and refer to specific mathematical entities (such as the empty set). Additionally, there was the "oddity" noted by Azzouni that mathematical practice "is simply not concerned with reference." And recall Azzouni's follow up assertion that any "view that fails to explain why this is the case has not explained something crucial about mathematics."

Before beginning my sketch of how, within the framework of my structural account, mathematics is to be fitted into a "bigger picture," I should like to review the reasons I am reluctant to call my view of mathematics "Structuralism" (I prefer instead to call it a "structural account of mathematics"). Certainly, my view of mathematics is in harmony with many of the views of Mathematical Structuralists. For example, I am willing to grant the Structuralist that mathematical theorems can be viewed as being about structures. But this is because I hold that the theorems have a structural content that is about structures. My account differs importantly from the Structuralist's in so far as: (1) I do not require the theorems of mathematics to be genuine statements about structures; and (2) I do not require the theorems to be literally true assertions.

These differences figure in my "Big Picture" investigations. My structural account takes no stand either on what the theorems assert, if anything, or on whether they are true or false. Applications of mathematics are not explained in terms of the truth and meaning of the theorems used in the application. One simply makes use of the fact that the *structural content* of each theorem is true: that is, that the theorem is true in every model of the theory. This eliminates the task of explaining how the set theorist could know the truth of the axioms and theorems of set theory. It thus eliminates the need to attribute special faculties or powers of "perception" to set theorists.

---

[56] In this paper, I have sought to avoid getting bogged down working out confusing details and messy logical reasoning. My aim has been only to produce an overview of the structural account, stressing essential ideas and general principles, rather than to provide a full account of my positions. This is because of constraints of space and because the reader can find a much more complete and in-depth account in my book [Chihara 2004].

For similar reasons, I do not respond to various objections that may occur to the alert reader. For example, a critic might observe that my structural account seems to make reference to, and to assume the existence of, "structures"— entities which are widely regarded as mathematical entities (because they apparently have just the features attributed to mathematical entities that raise serious epistemological difficulties for the Realist). Thus, an objector might conclude that my structural account has failed to avoid the many theoretical problems that bedevil the Realist's metaphysical account of mathematics. For my response to this objection, see [Chihara 2004], especially Chapters 7 and 8.

For similar reasons, the structural account eliminates the problem of explaining how the mathematician is able to pick out and refer to mathematical entities —a problem that arises out the Realist's metaphysical view of mathematics. The structural account eliminates this problem by eliminating the need for there to be mathematical entities to be picked out. Thus, this account is consistent with that feature of mathematical practice noted by Azzouni: *mathematical practice is simply not concerned with reference.* From the perspective of the structural account of mathematics, individual mathematicians do well *not* to concern themselves with questions of how they succeed in referring to particular mathematical objects.

Acknowledgements My thanks go to my niece Laura Chihara and my wife Carol Chihara for their useful comments about earlier versions of this paper. I would especially like to thank the editors of this collection for their many insightful corrections and suggestions: the time, energy and thought these mathematicians have put into their editing of this paper was truly impressive.

References

[Azzouni 1994] Azzouni, J., *Metaphysical Myths, Mathematical Practice*, Cambridge: Cambridge University Press, 1994.

[Barbut 1970] Barbut, M., "On the Meaning of the Word 'Structure' in Mathematics," in M. Lane (Ed.), *Introduction to Structuralism* (pp. 367–388), New York: Basic Books, Inc., Publishers, 1970.

[Barwise/Etchemendy 1991] Barwise, J. and Etchemendy, J., "Visual information and valid reasoning," in *Visualization in teaching and learning mathematics*, W. Zimmermann & S. Cunningham (Eds.), Washington, DC: Mathematical Association of America, 1991.

[Bernays 1967] Bernays, P., "Hilbert, David," in P. Edwards (Ed.),*The Encyclopedia of Philosophy* (Vol. 3, pp. 496–504), New York: MacMillan Publishing & The Free Press, 1967.

[Birkhoff/Mac Lane 1953] Birkhoff, G., and Mac Lane, S., *A Survey of Modern Algebra* (Revised ed.), New York: MacMillan, 1953.

[Bourbaki 1950] Bourbaki, N. (1950) "The Architecture of Mathematics," *American Mathematical Monthly* 57 (1950), pp. 221–232.

[Brown 1999] Brown, J. R., *Philosophy of Mathematics: an Introduction to the World of Proofs and Pictures*, London: Routledge, 1999.

[Burgess/Rosen 1997] Burgess, J., & Rosen, G., *A Subject With No Object: Strategies for Nominalistic Interpretations of Mathematics*, Oxford: Oxford University Press, 1997.

[Chihara 1973] Chihara, C. S., *Ontology and the Vicious-Circle Principle*, Ithaca: Cornell University Press, 1973.

[Chihara 1982] Chihara, C. S., "A Gödelian Thesis Regarding Mathematical Objects: Do They Exists? And Can We Perceive Them?" *Philosophical Review* 91 (1982), pp. 211–227.

[Chihara 1990] Chihara, C. S., *Constructibility and Mathematical Existence*, Oxford: Oxford University Press, 1990.

[Chihara 2004] Chihara, C. S., *A Structural Account of Mathematics*, Oxford: Oxford University Press, 2004.

[Chihara 2005] Chihara, C. S, "Nominalism," pp. 483–414 in S. Shapiro (Ed.), *The Oxford Handbook of Philosophy of Mathematics and Logic*, Oxford: Oxford University Press, 2005.

[Cohen 1966] Cohen, P., *Set Theory and the Continuum Hypothesis*, New York: W. A. Benjamin, 1966.

[Cohen 1971] Cohen, P., "Comments on the Foundations of Set Theory," pp. 9–15 in D. Scott (Ed.), *Axiomatic Set Theory,* Providence, Rhode Island: American Mathematical Society, 1971.

[Corry 1999] Corry, L., "Hilbert and Physics (1900–1915)," pp. 145–188 in J. Gray (Ed.), *The Symbolic Universe: Geometry and Physics, 1900–1915,* Oxford: Oxford University Press, 1999.

[Field 1980] Field, Hartry, *Science without Numbers: A Defense of Nominalism*, Princeton: Princeton University Press, 1980.

[Fine/Rosenberger 1997] Fine, B. & Rosenberger, G., *The Fundamental Theorem of Algebra*, New York: Springer-Verlag, 1977.

[Frege 1971] Frege, G., "On the Foundations of Geometry" (E.-H. Kluge, Trans.), pp. 22–37 in E.-H. Kluge (Ed.), *On the Foundations of Geometry and Formal Theories of Arithmetic,* New Haven: Yale University Press, 1971.

[Freudenthal 1962] Freudenthal, H., "The Main Trends in the Foundations of Geometry in the 19th Century," pp. 613–621 in E. Nagel, P. Suppes & A. Tarski (Eds.), *Logic, Methodology and Philosophy of Science: Proceedings of the 1960 International Congress,* Stanford: Stanford University Press, 1962.

[Gödel 1964a] Gödel, K., "Russell's Mathematical Logic," pp. 211–232 in P. Benacerraf & H. Putnam (Eds.), *Philosophy of Mathematics: Selected Readings,* Englewood Cliffs, NJ: Prentice-Hall, 1964.

[Gödel 1964b] Gödel, K., "What is Cantor's Continuum Problem?" pp. 258–273 in P. Benacerraf & H. Putnam (Eds.), *Philosophy of Mathematics: Selected Readings*, Englewood Cliffs, NJ: Prentice-Hall, 1964.

[Hardy 1941] Hardy, G. H., *A Mathematician's Apology*, Cambridge, England: Cambridge Univesity Press, 1941.

[Hardy 2002] Hardy, G. H., "Mathematical Proof," pp. 173–186 in D. Jacquette (Ed.), *Philosophy of Mathematics: an Anthology*, Oxford: Blackwell Publishers Inc., 2002.

[Hilbert 1971] Hilbert, D., *Foundations of Geometry* (L. Unger, Trans. Second English Translation ed.), La Salle, Illinois: Open Court, 1971.

[Hilbert 1980] Hilbert, D., Letter to Frege 29.12.1899 (H. Kaal, Trans.), pp. 38–41 in G. Gabriel, H. Hermes, F. Kambartel, C. Thiel & A. Veraart (Eds.), *Gottlob Frege: Philosophical and Mathematical Correspondence,* Chicago: The University of Chicago Press, 1980.

[Jaffe/Quinn 1993] Jaffe, A., & Quinn, F. (1993). "'Theoretical Mathematics': Toward a Cultural Synthesis of Mathematics and Theoretical Physics," *Bulletin of the American Mathematics Society* 29 (1993), pp. 1–13.

[Knopp 1945] Knopp, K., *Theory of Functions, Part I: Elements of the General Theory of Analytic Functions* (First American Edition ed.), New York: Dover Publications, Inc., 1945.

[Laubenbacher/Pengelley 1999] Laubenbacher, R., & Pengelley, D., *Mathematical Expeditions: Chronicles by the Explorers*, New York: Springer-Verlag, 1999.

[Mac Lane 1986] Mac Lane, S., *Mathematics: Form and Function*, New York: Springer-Verlag, 1986.

[Maddy 1990] Maddy, P, *Realism in Mathematics*, Oxford: Oxford University Press, 1990.

[Maki/Thompson 1973] Maki, D. P., & Thompson, M., *Mathematical Models and Applications*, Englewood Cliffs, New Jersey: Prentice-Hall, 1973.

[Mates 1972] Mates, B., *Elementary Logic* (2nd ed.), New York: Oxford University Press, 1972.

[Mostowski 1967] Mostowski, A. (1967), "Recent Results in Set Theory," pp. 82–96 in I. Lakatos (Ed.), *Problems in the Philosophy of Mathematics,* Amsterdam: North-Holland, 1967.

[Mueller 1981] Mueller, I., *Philosophy of Mathematics and Deductive Structure in Euclid's Elements*, Cambridge, Massachusetts: M. I. T. Press, 1981.

[Parsons 1995] Parsons, C, "Structuralism and the Concept of Set," pp. 14–92 in W. Sinnott-Armstrong, D. Raffman & N. Asher (Eds.), *Modality, Morality, and Belief: Essays in Honor of Ruth Barcan Marcus*, Cambridge: Cambridge University Press, 1995.

[Putnam 1967] Putnam, H., "The Thesis That Mathematics is Logic," pp. 273–303 in R. Schoenman (Ed.), *Bertrand Russell: Philosopher of the Century*, London: George Allen & Unwin Ltd, 1967.

[Resnik 1981] Resnik, M., "Mathematics as a Science of Patterns: Ontology and Reference," *Nous* 16 (1981), pp. 529–550.

[Resnik 1997] Resnik, M., *Mathematics as a Science of Patterns*, Oxford: Oxford University Press, 1997.

[Resnik 1998] Resnik, M., "Holistic Mathematics," pp. 227–246 in M. Schirn (Ed.), *Philosophy of Mathematics Today*, Oxford: Oxford University Press, 1998.

[Russell 1956] Russell, B., "The Philosophy of Logical Atomism," pp. 177–281 in R. C. Marsh (Ed.), *Logic and Knowledge*, London: George Allen & Unwin, 1956.

[Shapiro 1997] Shapiro, S., *Philosophy of Mathematics: Structure and Ontology*, Oxford: Oxford University Press, 1997.

[Singh 1997] Singh, S., *Fermat's Enigma*, New York: Walker and Company, 1997.

[Suppe 1974] Suppe, F., *The Structure of Scientific Theories*, Urbana: University of Illinois Press, 1974.

[Suppes 1967a] Suppes, P., "A Comparison of the Meaning and Uses of Models in Mathematics and the Empirical Sciences," *Synthese* 12 (1967), pp. 287–301.

[Suppes 1967b] Suppes, P., "What is a Scientific Theory?" pp. 55–67 in S. Morgenbesser (Ed.), *Philosophy of Science Today*, New York: Basic Books, 1967.

[Thurston 1994] Thurston, W., "On Proof and Progress in Mathematics," *Bulletin of the American Mathematical Society* 30 (1994), pp. 161–177.

8

Mathematical Objects

Stewart Shapiro
O'Donnell Professor of Philosophy
The Ohio State University

From the Editors

Stewart Shapiro, in this chapter, sets forth the questions that philosophers of mathematics have been trying to answer, dividing philosophers of mathematics along two axes: whether or not they believe mathematical objects exist objectively in some way (realism or nominalism of ontology), and whether or not they believe the theorems of mathematics are objectively true (realism or fictionalism in epistemology). He introduces the problems connected with each of these viewpoints, and describes how they developed. He then gives more details on several approaches that are receiving considerable attention currently, including neo-logicism (successors to Frege and Russell) and structuralism. Structuralism is immediately of interest because it appears that what we study in mathematics are structures—whether general structures such as topological spaces, or specific structures such as the real numbers. Generally, mathematicians are not very interested in what kind of thing a real number is (is it an object in some non-physical realm, a mark on a piece of paper, an idea in people's heads?), but in how it interacts with the rest of the real numbers. So in this sense, mathematicians study structures. Stewart Shapiro and some others (including Michael Resnik) have been trying to see whether that view of mathematics can resolve some of the philosophical problems that arise in a platonic approach to the philosophy of mathematics.

Stewart Shapiro is the O'Donnell Professor of Philosophy at The Ohio State University and a Professorial Fellow at the University of St. Andrews. His research interests include logic, philosophy of logic, epistemic logic, philosophy of language, and the philosophy of mathematics. Among his books and articles that readers of this volume are likely to be interested in are Thinking about mathematics: The philosophy of mathematics *(2000);* Philosophy of Mathematics: Structure and Ontology, *(1997); "Categories, structures, and the Frege-Hilbert controversy: the status of meta-metamathematics",* Philosophia Mathematica *(2005); "Simple truth, contradiction, and consistency", in* The law of non-contradiction, *edited by Graham Priest and J. C. Beall, (2004);*

"All sets great and small: and I do mean ALL", Philosophical Perspectives *(2003); "Space, Number, and Structure: A Tale of Two Debates,"* Philosophia Mathematica *(1996); "Reasoning, Logic and Computation,"* Philosophia Mathematica, *(1995); "Mathematics and the Philosophy of Mathematics,"* Philosophia Mathematica, *(1994); and "Mathematics and Reality,"* Philosophy of Science *(1983).*

1 Battle Lines

Examples of mathematical objects include natural numbers, real numbers, complex numbers, sets, geometric points, functions, topological spaces, groups, rings, and fields. These items are described by common nouns in ordinary mathematical discourse; some are referred to by proper names, such as '3' and 'π'. A number of philosophical questions come to mind almost immediately. Do mathematical objects exist? Do they exist independently of the language, mind, social contexts, or, to use a Wittgensteinian term, "form of life" of the mathematician? In short, do mathematical objects exist objectively? If mathematical objects exist, how do we know about them?

There is no agreed upon background framework for addressing questions like these. Early on in my education, I learned that in philosophy, an important part of each question—well over half of the battle—is to figure out what is being asked. This includes figuring out what the words in the question mean. What is it to *be* an object? What is it for an object, or for a type of object, to exist objectively? How do we know that objects exist? How do we know anything? Such questions can be asked for other sorts of objects, both mundane and exotic, such as rocks, baseballs, planets, people, nations, borders, galaxies, electrons, and quarks.

This, of course, is not the place to attack the general questions in a systematic manner, but it is clear that such questions must be addressed as we tackle the special cases of them that concern us here. As far as possible, we will keep focus on the specific questions concerning mathematics.

The view that mathematical objects—numbers, sets, points, etc.—exist objectively is sometimes called *realism in ontology.* Opponents of this view logically fall into two groups. *Nominalists* deny the existence of mathematical objects altogether. Members of the other group hold that mathematical objects exist, but not objectively. *Constructivists* hold that mathematical objects are constructions of a perhaps idealized mind. So mathematical objects are mind-dependent. If there were no minds, then there would be no mathematical objects. *Social constructivists* hold that mathematical objects are products of social mathematical activity. If there were no mathematicians, or no mathematical community, then there would be no mathematical objects (see Julian Cole's contribution to this volume).

Georg Kreisel is often credited with shifting attention from the nature of mathematical objects to the objectivity of mathematical discourse. Are the basic propositions of mathematics (nontrivially) true or false? Are they true or false independently of the judgments of mathematicians and their communities? Are there, or can there be, unknowable mathematical truths? These, too, are instances of more general questions. What is it for a given proposition to be true or false? What is it for a given area of discourse to be objective? Can there be *any* unknowable truths?

Realism in truth-value is the view that the basic propositions of mathematics—at least the bulk of the axioms and theorems of the major branches of mathematics—are objectively true. Here, again, the opposition falls into two camps. *Fictionalists* hold that mathematical assertions are, by and large, false. As indicated by the name, mathematical objects are likened to characters of fiction. Statements such as "$|3 + 2i|^2 = 13$" and "for each natural number n, there is a prime number $m > n$" have the same status as "Miss Marple's nosiness has brought several murderers to justice." In a sense, statements like these are true in the stories, but strictly and literally, they are false. Since Miss Marple does not exist, she is in no position to bring anyone (or at least any existing person) to (existing) justice. Other truth-value irrealists agree that basic mathematical propositions are true, but not objectively. In some way, the truth of basic mathematical propositions is due to the mind, language, or social life of the mathematician. A common theme of this second group of irrealists is that all mathematical truths are knowable.[1]

There is a natural alliance between realism in ontology and realism in truth-value. The latter holds that mathematical assertions are objectively true or false. To get to ontological realism from this, one just has to understand mathematical assertions literally, at face value. Consider the two examples above, "$|3 + 2i|^2 = 13$" and "for each natural number n, there is a prime number $m > n$." In the first, '3', '2', i', '13' are proper names, which at least purport to denote objects. The second proposition, taken literally, speaks of objects called "prime numbers," and it, together with a premise that there is at least one number and some other well-known facts about arithmetic, implies the existence of infinitely many prime numbers. According to truth-value realism, the statements are objectively true. If they are read literally, at face value, then it would seem that the objects referred to in these statements exist objectively.

There is a similar alliance between the respective irrealisms. A fictionalist about mathematical truth, for example, is likely to hold that mathematical objects do not exist. What is the point of saying that the mathematicians make fictional statements—assertions which are only "true in the story"—about objectively existing mathematical objects? Similarly, someone who holds that mathematical objects are the constructions of a mind is likely to also hold that mathematical propositions are likewise mind-dependent. How can we make objectively true statements about mind-dependent objects?

As natural as these alliances may be, they are not universal. Every combination of ontological realism/irrealism and truth-value realism/irrealism is articulated and defended by established and respected philosophers of mathematics. Kurt Gödel [1944], [1964], Michael Resnik [1997], Penelope Maddy [1990], and myself [1997] combine realism in ontology with realism in truth-value. Charles Chihara [1990] and Geoffrey Hellman [1989] develop programs to establish the combination of irrealism in ontology with realism in truth-value. The idea is to interpret mathematical statements in a non-face value manner. The result is that such statements enjoy objective truth-values even though (or even if) there are no distinctive mathematical objects. Hartry Field [1980], [1989] and Steven Yablo [2002], [2005] are fictionalists, adopting both forms of irrealism. Michael Dummett [1973], [1977] and the traditional intuitionists L. E. J. Brouwer (e.g., [1912], [1948]) and Arend Heyting (e.g., [1930], [1931], [1956]) hold that mathematical objects and mathematical truth are alike mind-dependent. Neil Tennant [1987], [1997] holds that mathematical objects exist objectively, but also that, in a sense, no truth—mathematical

---

[1] The word "anti-realism" is sometimes used for the view that it is impossible for there to be any unknowable truths. See, for example, Tennant [1997]. Here, I use "irrealism" for all views opposed to realism.

or otherwise—is completely independent of the human ability to determine its truth. In present terms, this is realism in ontology, irrealism in truth-value.

General questions and issues concerning objectivity, objecthood, and existence loom large in this taxonomy. It may be that the different authors do not accept the same conceptions of the basic metaphysical notions, and thus they may not disagree as much as it looks. Let us turn to the strengths, shortcomings, and issues of some of the major positions.

2 What Mathematical Objects are Like, or Would be Like if they Existed

It seems that mathematical objects, if such there be, do not have physical properties. No one wonders how much the number 2 weighs, or where it is located, or whether it can be moved. It is misguided to even wonder about such things. We might as well ask if the number 2 is brave, or funny. Of course, one can slaughter and eat two goats, and one can vaporize four ice cubes. So some physical relations invoke mathematical entities, or at least they are usually stated using mathematical language. But, it seems, one cannot destroy the number two itself and the number four itself. Realists in ontology typically also hold that mathematical objects, such as natural numbers, were not created. Mathematical objects thus have no beginning or end in time. They are eternal, or perhaps better, timeless. In like manner, mathematical objects do not have any causal relations with material reality, or anything else for that matter. Nothing that anyone or anything does can have any effect on the arithmetic properties of natural numbers, and numbers have no effects on anything else—or so it seems. Objects with properties like these are sometimes called *abstract*.

According to Plato, properties or Forms such as Beauty, Justice, and Goodness, are similarly eternal, unchanging, and acausal. For this reason, realism in ontology is sometimes called "Platonism."[2] Sometimes this is written "platonism," with a lowercase 'p', probably to mark some distance from the master. On views like these, mathematical objects are sometimes said, often pejoratively, to reside in "Plato's heaven." This, of course, is only a metaphor. According to the typical ontological realist, mathematical objects are not located anywhere.

3 A Dilemma

Much of the work in the philosophy of mathematics over the last 30 years can be traced to a dilemma posed by Paul Benacerraf [1973]. There are two desiderata for a philosophical account of mathematics. The first is that languages of mathematics and the languages of ordinary discourse and science be understood the same way. In other words, there should be a *uniform semantics* for the various languages. Ordinary discourse contains mathematical terms, and just about all of science does. Otherwise, mathematics would not be applicable to the material world, the world studied by science. If we had to understand mathematical language and non-mathematical language in different ways, then we would need an account of how the discourses interact, and a way to understand mixed statements, such as "one can slaughter and eat two goats" and "the force of gravity between two objects is proportional to the square of the distance between them."

---

[2] Plato's own views on mathematics are a matter of some scholarly dispute. For a brief overview, and some references to the extensive literature, see Shapiro [2000, Chapter 3].

Now consider the following pair of sentences:

There are filthy cities larger than Chicago.
There are prime numbers larger than 100.

These sentences have the same grammatical form. The first desideratum would suggest that they be understood in closely analogous ways. The first sentence has a proper name, "Chicago," a complex predicate "filthy city," and a term for the relation of largeness. The sentence is true if (and only if) there is a city denoted by "Chicago" and some cities are both filthy and larger than it. The second sentence has a name "100" that purports to refer to a natural number, a complex predicate symbol "prime number," and the relation symbol "larger." The sentence is true if (and only if) there is a number denoted by "100" and there are, in fact, some natural numbers that are both prime and larger than that number.

The first sentence is clearly true; there are such cities (although I will not name any, to avoid giving offence). The second sentence is an easy theorem of elementary arithmetic, and mathematicians assert it, without reservation. It is natural to hold that mathematicians know what they are talking about, that they mean what they say, and that they get things right, at least most of the time, especially when dealing with matters this elementary. So it would seem that the second sentence is true, like the first. Benacerraf's first desideratum thus suggests that mathematical objects exist. If we combine this with the above observations concerning the nature of mathematical objects, we are led to the thesis that mathematical objects are eternal, unchanging, and acausal. In other words, the first desideratum suggests realism in ontology. As above, this is allied with realism in truth-value.

A *second desideratum* for the philosophy of mathematics is a plausible (and uniform) *epistemology*. Mathematical knowledge should not be mysterious. It seems that we do know some mathematics. How? Benacerraf points out that the way we come to know about ordinary objects and the objects of science is by interacting with them. We see baseballs, trees, people, and the like. Of course, we do not literally see (or hear or touch) some of the theoretical posits of science, but we do see the effects of these objects, in Brownian motion and in cloud chambers, for example. In contrast, mathematical objects—if such there be—seem to have no effects on us, or on anything else. So even if mathematical objects do exist, we cannot interact with them, and thus know anything about them. We cannot even know that they exist.

This, then, is **Benacerraf's dilemma**. The *first desideratum* is to understand mathematical and non-mathematical discourse (as well as mixed discourse) the same way. This would have us take mathematical propositions at face value. If some such propositions are true, we are led to the existence of an acausal realm of mathematical objects. The second desideratum suggests that we cannot know anything about such objects, since we do not and cannot interact with them.

The second horn of the dilemma, as presented in [Benacerraf 1973], invokes a premise sometimes called the "causal theory of knowledge." The thesis is that there can be no knowledge about a type of object unless at least some knowers have at least some causal contact with at least some of the objects. This is impossible with abstract mathematical objects, as conceived above. Although causal theories have fallen into disrepute among epistemologists, the general theme of the Benacerraf dilemma continues to bother those working in the philosophy of mathematics, no matter what their views on mathematical objects and no matter what their views on knowledge. Hartry Field [1989, Essay 7] formulated a variation of the second horn of the dilemma that is substantially independent of the details of any particular theory of knowledge. According to realism in ontology, mathematical propositions are about a realm of objects that have no causal relations with the human mathematician (or anything else for that matter). Presumably, the realist also holds that at least the bulk of the beliefs of practicing mathematicians about these abstract objects are true: every natural number has a successor, the successor function is one-to-one, etc. So the ontological realist postulates a correlation between the beliefs of mathematicians and the facts about an eternal, acausal realm of abstract mathematical objects. The burden is to explain this correlation. It cannot be an accident that humans evolved to the point that they have correct beliefs about the mathematical realm—about Plato's heaven. How did we accomplish this feat? What *reason* do we have to think that we *did* accomplish this feat? If ontological realism is true, then what reason do we have for thinking that mathematicians are correct in their mathematical beliefs?

Benacerraf's dilemma thus leads to a pair of problems, one for the realist and one for the irrealist. Perhaps the most serious problem with ontological realism is to provide an account of mathematical knowledge, an account that makes it the case that mathematicians (and ordinary folk) have genuine knowledge about abstract objects. A serious problem for the various irrealisms is to give an account of how we understand mathematical language, and how mathematical knowledge figures in ordinary and scientific knowledge of the material world.

4 The Irrealist Horn

I turn to a sketch of some views that attack the first horn of Benacerraf's dilemma, and deny that mathematical objects enjoy an objective existence. As noted above, there are two versions of ontological irrealism. One has it that mathematical objects are mind-dependent, and the other holds that there are no mathematical objects at all.

The more radical version of ontological irrealism is *nominalism*, the view that mathematical objects do not exist—mind-independently or otherwise. As noted, one variety of this view is that mathematical objects are fictions. Accordingly, what passes for mathematical knowledge is just knowledge of the "stories" that mathematicians weave, and knowledge of the consequences of these stories. Presumably, that sort of knowledge is not problematical.

Other varieties of nominalism provide ways to interpret mathematical statements in a non-face value manner (e.g., [Chihara 1990], [Hellman 1989]). The basic axioms and theorems end up objectively true, but they do not imply the existence of distinctively mathematical objects. Typically, the statements of mathematics are about what is necessary and/or possible. Such statements are sometimes called "modal." According to views like this, mathematics is known the same ways that any modal propositions are known. That is, we know mathematics the same way that we know how anything is necessary or possible. It is not clear that such an epistemology for mathematics is any more tractable than one for the ontological realist (see [Shapiro 1997], Chapter 7).

In all these nominalistic accounts, the first desideratum of the Benacerraf dilemma is left unfulfilled: mathematical statements are not understood the same way as ordinary and scientific language is. The advocates of the various fictionalist and modal reconstructive programs take this burden seriously, and attempt to provide accounts of how mathematical propositions (whether fictional or modal) are useful, or indeed practically necessary, for understanding the physical world. A central focus of these projects is to show how to understand the statements of science, such as the laws of physics, that themselves invoke mathematics. Since the concern of this article is mathematical *objects*, I will go no further on the details of these accounts.[3]

We turn now to views that mathematical objects exist, but not objectively. The intuitionist Arend Heyting explicitly denied the mind-independence of mathematical objects:

> . . . we do not attribute an existence independent of our thought, i.e., a transcendental existence, to the integers or to any other mathematical objects . . . mathematical objects are by their very nature dependent on human thought. Their existence is guaranteed only insofar as they can be determined by thought. They have properties only insofar as these can be discerned in them by thought . . .   ([Heyting 1931], pp. 52–53])

> . . . [L. E. J.] Brouwer's program . . . consisted in the investigation of mental mathematical construction as such, . . . a mathematical theorem expresses . . . the success of a certain construction . . . In the study of mental mathematical constructions, "to exist" must be synonymous with "to be constructed" . . . . In fact, mathematics, from the intuitionist point of view, is a study of certain functions of the human mind.

> ([Heyting 1956], pp. 1, 8, 10)

Benacerraf argues that the philosophies of mathematics like this have a more tractable line on epistemology, on the second horn of the dilemma. If mathematical objects are mind-dependent, then mathematical knowledge is knowledge of our own minds. As mysterious as the mind is, it is not nearly as mysterious as Plato's heaven. Surely, we have some access to the mind, and to its productions. If nothing else, we have introspection.

If we take the statement of this philosophy literally, at face value, then it is not clear that it can make sense of the mathematics that we all know and love. It is an easy theorem of set theory that there are uncountably many real (and complex) numbers, and even more functions on real numbers. The universe of set theory itself is even more staggering. Surely, no one can claim that humans have constructed that many objects. When did we do so? The objects of advanced mathematics cannot be constructed one item at a time. There are too many of them.

Following Aristotle, the traditional intuitionist rejects the existence of an actual infinity. For example, there is no completed, existing set of natural numbers. Rather, we have a procedure for generating ever more natural numbers. The intuitionist understands the axiom that every natural number has a successor to mean that the procedure for constructing natural numbers does not run out. If given a natural number, one can construct its successor. But this is not to say that anyone ever could construct all of the natural numbers. As usual, one can think of an integer as a pair of natural numbers, a rational number as a pair of integers, and a real number as a Cauchy sequence of rational numbers (or a Dedekind cut). But here, again, the sequence (or cut) is not thought of as an actual infinity. It is not as if the members of a given Cauchy sequence exist all at once. Rather, we think of the sequence as generating its members, one at a time. The sequence is a potential infinity, in that it does not terminate.

Even the elementary parts of the arithmetic go beyond what humans can actually construct. We talk about gargantuan natural numbers when discussing the national debt, the distance between galaxies, and the numbers of atoms in a given substance. Avogadro's number is 6.02×10^{23}. No

---

[3] The interested reader is directed to the previously cited sources, and to Burgess and Rosen [1997] for a less than sympathetic account. See also Shapiro [2000, Chapter 9].

one has carried out the procedure for constructing natural numbers (one at a time) that far. Yet the intuitionist has no trouble countenancing the existence of numbers that large. It is a theorem of intuitionistic arithmetic that every number is either prime or composite. But clearly, there are numbers that are so large that no one can determine whether they are prime or composite before the sun goes cold, using the resources available in this limited universe.

The standard response here is to idealize. Even though Heyting speaks of mathematics as "a study of certain functions of the human mind," it is really an idealized mind that is being studied. And the study is hardly empirical—mathematics is not a branch of psychology. This might lessen the benefits conceded to the intuitionist on the second, epistemological, horn of the Benacerraf dilemma. We may have access to *our* finite and limited minds, via introspection, but do we have access to the postulated ideal mind?

The intuitionistic perspective has ramifications within mathematics itself. Brouwer showed that every function on (a closed and bounded set of) real numbers is (uniformly) continuous. Most intuitionists accept this. For example, they hold that there simply is no real valued function f such that $f(x) = 1$ if $x < 0$, and $f(x) = 2$ if $x \geq 0$. This is despite that fact that discontinuous functions are easily definable in classical mathematics—I just defined one—and that such functions have found application in science. In like manner, the intermediate value theorem cannot be proved in intuitionistic real analysis. An ongoing research program is to see how well science can get on with intuitionistic mathematics. A neutral observer might say that the jury is still out on that matter.

Brouwer, who is usually difficult to interpret, articulated a Kantian view that arithmetic concerns the ways that humans (or creatures with minds like humans) perceive the world in time. The natural numbers are the structure of our experience of events following each other, one at a time. Following Kant, he postulated that we have a faculty, called "intuition," which yields knowledge of the basic truths of arithmetic. This is the source of our knowledge of arithmetic. It is not clear whether such an epistemology is any more tractable than one for ontological realism.

What of the first horn of the Benacerraf dilemma, the desire that the languages of mathematics, ordinary discourse, and science be understood the same way? One option for the intuitionist would be to articulate a Kantian philosophy that ordinary objects and those of science are mind-dependent, at least to some extent. Then all discourses would receive a uniform treatment, and the first desideratum would be satisfied. Another option would be to develop contrasting and complementary accounts for the various discourses, including the mixed language of science, and show how the systems complement each other and function together.

Another variation of ontological irrealism is the view that mathematical objects are *social constructions*, along the lines of money, property, statutes, and borders. The idea, again, is that mathematical objects exist, but not independently of the mathematical and scientific community. A mathematical proposition is the same sort of statement as "the border between the USA and Canada is north of the equator" and "three strikes and you are out." Versions of this relatively fresh perspective are articulated in [Ernest 1998] and [Hersh 1997] (see Julian Cole's contribution to this volume).

5 The Realist Horn

We now take up attempts to grasp the first horn of the Benacerraf dilemma, and thus try to overcome the second, epistemological horn. These are usually tied to detailed, speculative accounts of the nature of mathematical objects.

5.1 An Intuitive Grasp

Plato held that the human mind has some sort of direct grasp of the world of eternal, unchanging Forms. Taking a cue from this, the ontological realist about mathematics might conjecture that there is a faculty for perceiving relations among mathematical objects, a faculty that gives us a glimpse into Plato's heaven, which Plato called the World of Being. The mathematician and logician Kurt Gödel is well-known for holding a view like this. In a much quoted philosophical essay, he spoke favorably of the philosopher who "considers mathematical objects to exist independently of our constructions and of our having an *intuition of them individually . . .*" ([Gödel 1964], p. 474, my emphasis). He elaborates:

> But, despite their remoteness from sense experience, we do have something like a perception also of the objects of set theory, as is seen from the fact that the axioms force themselves on us as being true. I don't see any reason why we should have less confidence in this kind of perception, i.e., in mathematical intuition, than in sense perception, which induces us to build up physical theories and to expect that future sense perceptions will agree with them . . .
>
> It should be noted that mathematical intuition need not be conceived of as a faculty giving an *immediate* knowledge of the objects concerned. Rather it seems that, as in the case of physical experience, we *form* our ideas also of those objects on the basis of something else which *is* immediately given. Only this something else here is *not*, or not primarily, the sensations.   ([Gödel 1964], pp. 483–484)

Gödel's use of the word "intuition" here is explicitly Kantian. Unlike the intuitionists, and Kant himself, however, Gödel was a realist in ontology, writing that the "given" underlying mathematics "may represent an aspect of objective reality, but, as opposed to the sensations, their presence in us may be due to another kind of relationship between ourselves and reality." That is, Gödel took mathematical intuition to be an intuition of an objectively existing realm.[4]

5.2 Holism

Gödel's views capture an experience described by many mathematicians. The view, however, is often derided by philosophers. It is difficult to accept the idea of a special intuitive grasp of a Platonic mathematical realm. Gödel's philosophy suggests that mathematical objects do have some sort of effect on us, since we intuit them, presumably as they are. Apparently, Gödel rejects the common thesis, noted above, that mathematical objects lack causal powers. In any case, the lack of a detailed account of mathematical intuition makes it tempting to both reject and overly simplify the view.

A popular view today, at least in North America, is *naturalism*, characterized by W. V. O. Quine [1981, p. 72] as "the abandonment of first philosophy" and "the recognition that it is within science itself . . . that reality is to be identified and described" (see also [Quine 1969]). The naturalistic philosopher sees the human knower as a thoroughly physical being within a physical universe. So any faculty that the philosopher invokes to explain knowledge must involve only natural processes amenable to ordinary scientific scrutiny: "The naturalistic philosopher begins

---

[4] James Robert Brown [1999] advocates a similar use of intuition on behalf of ontological realism.

his reasoning within the inherited world theory as a going concern;" and the "inherited world theory is primarily a scientific one, the current product of the scientific enterprise." It is hard to see what sort of empirical study could confirm the existence of a Kantian-cum-platonic intuition.

Naturalism exacerbates the epistemic problems with realism in ontology, the second horn of the Benacerraf dilemma. The challenge to the ontological realist is to show how a physical being in a physical universe can come to know about *abstracta* like mathematical objects, or the truths about such objects.

Besides being an unrelenting naturalist, Quine is an unrelenting empiricist, holding that all knowledge is ultimately based on sensory observation. He proposes a metaphor that our system of beliefs is a "seamless web." Each "node" (belief) has innumerable links to other nodes in the web, via logic and linguistic usage. Some nodes are fairly directly related to experience, so that they can be confirmed by direct observation. These lie at or near the "edges" of the web. To pursue the metaphor, sensory experience impinges on the web only at these "edges," through irritations on our nerve endings—observation. New observations bring about changes inside the web, via the innumerable links between the nodes, until some sort of equilibrium is achieved.

For Quine, "science is a tool . . . for predicting future experience in the light of past experience" ([Quine 1951], §6). Although the only evidence relevant to a theory is sensory experience,[5] Quine argues that experience does not bear on scientific statements considered one at a time. Our beliefs face the tribunal of experience only in groups. In light of recalcitrant experience, the scientist has many options on which of her beliefs to modify. In philosophy, the technical term for Quine's view is *holism*.

As noted above, several times, mathematics is a central part of the sciences. Thus, for Quine mathematics itself has a central place in the web of belief. He accepts mathematics as true for the same reason he accepts physics as true. Indeed, for Quine, mathematics has the same status as the more theoretical parts of science. It lies far from the "periphery" of the web, where observation has a more direct role. For Quine, the ultimate criterion for accepting anything—mathematics, physics, psychology, ordinary objects, myth—is that it play an essential role in the web of belief, in organizing and predicting experience. Physics, chemistry, and with those, mathematics, are entrenched in the web, and so we believe in the truth of the basic pronouncements of those fields. Mathematics, it seems, has a vast ontology, speaking of numbers, points, sets, and the like. So these objects exist.

One of the clearest articulations of the role of mathematics in Quinean holism is due to Hilary Putnam ([1971], Chapter 5). Define a "nominalist language" to be one that only refers to concrete, physical objects. Putnam invites us to "consider the best-known example of a physical law: Newton's law of gravitation,"[6] which

> . . . asserts that there is a force f_{ab} exerted by any body a on any other body b. The . . . magnitude F [of the force] is given by:
>
> $$F = gM_aM_b/d^2$$
>
> where g is a universal constant, M_a is the mass of a, M_b is the mass of b, and d is the distance which separates a and b.

---

[5] Quine allows other factors, like simplicity, to play a subsidiary role in developing theories.

[6] Putnam notes that it does not matter much that this particular law is not quite true. In any mature science, the laws of nature are thoroughly mathematical.

The point of the example is that Newton's law has a content which, although in one sense is perfectly clear (it says that gravitational "pull" is directly proportional to the masses and obeys an inverse-square law), quite transcends what can be expressed in nominalistic language. Even if the world were simpler than it is, so that gravitation were the only force, and Newton's law held exactly, still it would be impossible to "do" physics in nominalistic language. ([Putnam 1971], p. 37)

Since it seems that there is not much prospect of pursuing science without invoking real numbers, Putnam concludes that real numbers exist:[7]

If the numericalization of physical magnitudes is to make sense, we must accept such notions as function and real number; and these are just the notions the nominalist rejects. Yet if nothing really answers to them, then what at all does the law of gravitation assert? For that law makes no sense at all unless we can explain variables ranging over arbitrary distances (and also forces and masses, of course). ([Putnam 1971], p. 43)

Putnam ([1971], p. 57) sums things up:

. . . I have been developing an argument for realism along roughly the following lines: quantification over mathematical entities is indispensable for science . . . therefore we should accept such quantification; but this commits us to accepting the existence of the mathematical entities in question. This type of argument stems, of course, from Quine, who has for years stressed both the indispensability of quantification over mathematical entities and the intellectual dishonesty of denying the existence of what one daily presupposes.

The holistic perspective also attempts to provide a line on the epistemology of mathematics as well, to satisfy the second horn of the Benacerraf dilemma. According to Quine and Putnam, mathematics is known the same way as anything else in the web of belief is known, by playing a role in a highly successful way to organize and predict experience.

One would think, however, that articulations of the holistic picture should provide a careful analysis of the role of mathematics in science, rather than just noting the existence of this role. This analysis would shed light on the abstract, non-spatio-temporal, acausal nature of mathematical objects, and the relationships between such objects and ordinary and scientific material objects. How is it that talk of numbers and functions can shed light on rock formations, the stability of buildings, and the orbits of the planets? Such an analysis would go a long way toward defending the holistic picture.

Note also that Quine's view does not account for branches of mathematics, such as higher set theory, that have not found application in science. He takes a more hypothetical view toward unapplied branches of mathematics, claiming that he is not committed to the truth of these branches, or the existence of the indicated objects. Note also that mathematicians themselves do not follow the epistemology suggested by the Quinean picture. That is, mathematicians do not look for confirmation in science before believing (and publishing) their theorems. So Quine's picture does not account for mathematics as practiced.[7]

---

[7] See Colyvan [2001] and Resnik [2005] for a detailed articulation and defense of the holism and the indispensability argument. Penelope Maddy [1990] provides a defense of an ontological and truth-value realism that synthesizes aspects of Gödel's perspective and Quine's naturalism and empiricism.

6 A Matter of Meaning

I now present brief sketches of two specific accounts of the nature of mathematical objects, neo-logicism and structuralism. Each account cuts across the realist/irrealist divide of the past two sections, in the sense that each view has both realist and irrealist versions. The accounts are presented as a sample of the terrain, and are not even close to exhaustive of the rich and extensive literature in the philosophy of mathematics. The views were chosen, in part, because each has something specific to say about what it is to be a (mathematical) object.

One program in the philosophy of mathematics is to show that at least some basic principles of mathematics are what philosophers sometimes call "analytic." The idea is that anyone who properly understands the meanings of terms like "natural number," "successor function," "addition," and "multiplication," has all that she needs to see that the basic principles of arithmetic, such as the Peano postulates, are true.

The most prominent historical instance version of this program is *logicism*, the view that arithmetic truth, at least, is a species of logical truth. The most detailed developments are those of Gottlob Frege [1884], [1893] and Alfred North Whitehead and Bertrand Russell [1910]. Frege was a realist in ontology, in that he took the natural numbers to be objects, and to exist independent of the mathematician. Numbers are what may be called logical objects. Frege's program is thus an option to handle the second, epistemological horn of the Benacerraf dilemma. For Frege, arithmetical propositions are known the same way as any other logical truths (whatever way that may be).

Two concepts are *equinumerous* if they can be put in one-to-one correspondence. For example, the napkins on a table are equinumerous with the plates, provided that there is exactly one napkin corresponding to each plate. Frege showed how to define equinumerosity using the resources of what would later be called "second-order logic." Crucially, his definition does not explicitly invoke natural numbers. In attempting to define the natural numbers, and the general notion of *natural number*, Frege ([1884], §63) proposed the following principle, which has since become known as "Hume's principle":[8]

> For any concepts F, G, the number of F's is identical to the number of G's if and only if F is equinumerous with G.

For various reasons, Frege was not satisfied with Hume's principle as a characterization of *number*. He pointed out that the principle only determines the truth-values of identities of the form "the number of F's $=$ the number of G's," where 'F' and 'G' stand for concepts. Hume's principle does not determine the truth-value of sentences in the form "the number of F's $= t$," where t is an arbitrary name. In particular, Hume's principle does not determine whether the number 2 is identical with the empty set, or with Julius Caesar for that matter. The issue of "identifying" the individual natural numbers came to be known as the "Caesar problem."

In formulating this issue, and taking it seriously, Frege seems to be invoking a principle for numbers that applies to objects generally. Quine ([1981], p. 102) puts the criterion well: "We have an acceptable notion of class, or physical object, or attribute, or any other sort of object, only insofar as we have an acceptable principle of individuation for that sort of object. There is no

---

[8] This attribution is based on a passage from Hume that Frege quotes. Hume, however, made no substantial mathematical use of the principle. Since the name "Hume's principle" has stuck, we will go along with it here.

entity without identity." Hume's principle tells us when two terms for numbers, given a certain way, denote the same or different numbers. It does not tell us what it is for numbers in general to be identical or distinct from other sorts of objects, or even from numbers given in other ways.

To solve the Caesar problem, Frege provided an explicit definition of the natural numbers in terms of concepts and their extensions. The number three, for example, is the extension (or collection) of all concepts that hold of exactly three things. So the concept "child of Stewart Shapiro" is a member of the number three.

Frege's ontological realism was not compromised by this definition, since he took extensions to be objects. They are among the objects of logic. One of the principles governing extensions is Frege's Basic Law V:

> For any concepts F, G, the extension of F is identical to the extension of G if and only if every F is a G, and every G is an F.

This is an instance of a familiar style of definition by abstraction, used throughout mathematics. Basic Law V, however, is problematic, to say the least. Frege's logicist program came to a tragic end when his theory of extensions was shown to be inconsistent, via Russell's paradox: define x to be a "Russell-object" if there is a concept F such that x is the extension of F, and x is not an F (i.e., Fx is false). Let r be the extension of the concept of being a Russell-object. It follows, from Basic Law V, that r is a Russell-object if and only if r is not a Russell-object. This is (or leads to) a contradiction.

For his part, Russell ([1919], Chapter 2) was not deterred by this paradox. He held that Frege's account of the natural numbers is substantially correct:

> The question "What is number?" is one which has been often asked, but has only been correctly answered in our own time. The answer was given by Frege in 1884, in his *Grundlagen der Arithmetik*. Although this book is quite short, not difficult, and of the very highest importance, it attracted almost no attention, and the definition of number which it contains remained practically unknown until it was rediscovered by the present author [Russell] in 1901.

According to Russell, once Basic Law V is correctly understood, it is indeed a good definition of "extension" or "class." His diagnosis was that the derivation of the contradiction from Basic Law V invokes a fallacy. A definition of a mathematical entity is said to be *impredicative* if it refers to a collection that contains the defined entity. The usual definition of the "least upper bound" is impredicative since it refers to a set of upper bounds and characterizes a member of this set.

Russell ([1919], Chapter 17) argued that impredicative definitions are illegitimate, since they are circular:

> Whenever, by statements about "all" or "some" of the values that a variable can significantly take, we generate a new object, this new object must not be among the values which our previous variable could take, since, if it were, the totality of values over which the variable could range would be definable only in terms of itself, and we should be involved in a vicious circle. For example, if I say "Napoleon had all the qualities that make a great general," I must define "qualities" in such a way that it will not include what I am now saying, i.e., "having all the qualities that make a great general" must not be itself a quality in the sense supposed.

The "vicious circle principle" thus bans impredicative definitions and, in particular, the generation of Russell's paradox. Recall that we defined an object x to be a Russell-object just in case there is a concept F such that x is the extension of F and Fx is false. The definition of "Russell-object" thus refers to all concepts F, and "Russell-object" is just such a concept F. We then derived a contradiction from the assumption that the definition of "Russell-object" holds of its own extension. The ban on impredicative definitions precludes even making this assumption. Russell argues, from the vicious circle principle, that it "must under all circumstances be meaningless (not false) to suppose [that] a class [is] a member of itself or not a member of itself."

Russell proposed a *type theory*. Define an "individual" to be an object that is not a class. Individuals are of type 0, and classes of individuals are of type 1. Classes of classes of individuals are of type 2, and so on. So, for example, the people that make up a team are each individuals and so are type 0 objects. The team, regarded as a class of its players, is a type 1 object; and a league, regarded as a class of teams, is of type 2. A collection of leagues would be of type 3.

The move to classes allows a simplification of Frege's definitions of the natural numbers. For any class C, define the *number of C* to be the "class of all those classes (of the same type as C) that are" equinumerous with C (see [Russell 1919], Chapter 2). Let A be the class of my three children; so that A is of type 1. The number of A is the class of all three-membered type 1 classes. The number of my children is thus a type 2 class. Similarly, the number of a type 2 class is a type 3 class, etc. For Russell, a "number is anything which is the number of some class."

But what is mathematics *about*? What are numbers, function, etc., really? Do they exist objectively? Since Russell took the various sorts of numbers to be classes, the status of numbers turns on the status of classes. His mature writings deny the independent existence of classes: "the symbols for classes are mere conveniences, not representing objects called 'classes' . . . [C]lasses are in fact . . . logical fictions . . . [They] cannot be regarded as part of the ultimate furniture of the world" (Russell [1919, Chapter 18]). He called this view the "no class" theory. Talk of classes is only a "manner of speaking" about properties and relations of ordinary, non-mathematical objects (as well as properties of properties, etc.). Talk of classes and thus numbers is thus eliminable in principle. So at this period, Russell sharply departed from Frege's realism in ontology.

A variation on Frege's approach is pursued today, in the work of Crispin Wright, beginning with [Wright 1983], and Bob Hale [1987] (see [Hale/Wright 2001]) and other *neo-logicists*. The idea is to bypass the treatment of extensions and to work with Hume's principle, or something like it, directly. Hume's principle, recall, is:

> For any concepts F, G, the number of F's is identical to the number of G's if and only if F is equinumerous with G.

Frege's own technical development shows that the standard postulates for arithmetic, now known as the Peano-Dedekind axioms, can be derived from Hume's principle in a standard, higher-order logical system. This result has since been dubbed *Frege's theorem*. In Frege's foundational work on arithmetic, the only essential use of extensions, and Basic Law V, is in the derivation of Hume's principle. Everything else concerning natural numbers follows from that. Moreover, Hume's principle is consistent with second-order logic if second-order arithmetic is consistent (see [Boolos 1987] and [Hodes 1984]).

Wright and Hale hold that the right-hand side of Hume's principle gives the truth conditions for the left-hand side, but the left-hand side has the proper grammatical and logical form. In particular, locutions like "the number of F's" are terms that (purport to) denote objects. At least some instances of the right-hand side of Frege's principle are true on logical grounds alone. For example, it is a logical truth that the concept of "not identical to itself" is equinumerous with the concept "not identical to itself." Thus, from Hume's principle, we conclude that the number of non-self-identical things is identical to the number of non-self-identical things. Letting "0" denote the number of non-self-identical things, we conclude that $0 = 0$ and so zero exists: $\exists x(x = 0)$. Similarly, we can define the number one to be the number of the concept "identical to zero." This makes sense since it follows that exactly one thing is identical to zero. Hume's principle, and basic logic, implies that the number one exists: $\exists x(x = 1)$. The number two is the number of the concept "either identical to zero or identical to one." And it goes on from there: each natural number has a similar definition. It follows from Hume's principle that these natural numbers all exist, and are different from each other. In effect, it follows from Hume's principle that there are infinitely many natural numbers. If the neo-logicist holds that Hume's principle is objectively true, then he is a realist in ontology.[9]

Let us turn to epistemology, the second horn of the Benacerraf dilemma. If we know that Hume's principle is true, and if we know that the relevant principles of higher-order logic preserve truth, then we know the basic propositions of arithmetic. So, to provide an epistemology for basic arithmetic, the neo-logicist must show how Hume's principle is known. On most contemporary accounts, logical truths are known on the basis of their form alone, and, more importantly, logical truths alone have no substantial consequences concerning what exists. Accordingly, Hume's principle is not a truth of logic, since it implies the existence of the natural numbers. The neo-logicist claims that Hume's principle is analytic of the concept of natural number, or is an explanation of the concept of natural number.[10] Anyone who grasps the concept will accept, or ought to accept, or has the wherewithal to know, Hume's principle:

> Frege's theorem will ... ensure ... that the fundamental laws of arithmetic can be de-
> rived within a system of second-order logic augmented by a principle whose role is
> to *explain*, if not exactly to define, the general notion of identity of cardinal number,
> and that this explanation proceeds in terms of a notion which can be defined in terms
> of second-order logic. If such an explanatory principle ... can be regarded as *analytic*,
> then that should suffice ... to demonstrate the analyticity of arithmetic. Even if that term
> is found troubling, ... it will remain that Hume's principle—like any principle serving
> implicitly to define a certain concept—will be available without significant epistemolog-
> ical presupposition ... So one clear a priori route into a recognition of the truth of ... the
> fundamental laws of arithmetic ... will have been made out. And if in addition [Hume's
> principle] may be viewed as a *complete* explanation—as showing how the concept of
> cardinal number may be fully understood on a purely logical basis—then arithmetic will
> have been shown up by Hume's principle ... as transcending logic only to the extent
> that it makes use of a *logical* abstraction principle—one [that] deploys only logical

---

[9] The aforementioned Caesar problem remains an active and open research problem on the philosophical agenda of neo-logicism (see Hale and Wright [2001a]). Apparently, the neo-logicist is obligated to give an account of the truth conditions of any statement in the form " $n = t$," where n is a natural number and t is any object whatsoever.

[10] Neil Tennant [1997] provides a different neo-logicist program. He shows how to derive the basic propositions of arithmetic from certain principles, and then claims that we know that such principles are necessarily true.

notions. So, . . . there will be an a priori route from a mastery of second-order logic to a full understanding and grasp of the truth of the fundamental laws of arithmetic. Such an epistemological route . . . would be an outcome still worth describing as logicism . . .

([Wright 1997], pp. 210–211)

Recently, Hale and Wright [2000] have argued that Hume's principle is an implicit definition, and true by stipulation. Notice that Hume's principle has the same form as Basic Law V. The latter, of course, cannot be true by stipulation, since it is not consistent, and it is not analytic of, and does not explain, and is not an implicit definition of, anything. Clearly, one cannot stipulate anything one wants, and expect to provide a plausible epistemology of the consequences of these stipulations. A major item on the agenda of neo-logicism is to articulate plausible doctrines for when one can stipulate a principle, and thus provide an epistemological account of the ontological consequences of the principle.

The neo-logicist project, as developed thus far, only applies to basic arithmetic and the natural numbers. Another item on their agenda is to extend the treatment to cover other areas of mathematics, such as real analysis, functional analysis, geometry, and set theory. The program involves the search for principles rich enough to characterize more powerful mathematical theories, and yet be plausible to assert by stipulation, as an implicit definition.

7 Mathematics is the Science of Structure

Structuralism is a popular philosophy of mathematics that comes in both realist and irrealist versions. The main theme is that the subject matter of a branch of mathematics, such as arithmetic, is the pattern common to any system of objects that shares a given form. The *natural number structure*, for example, is the pattern common to any infinite system of objects that has a distinguished initial object and a one-to-one successor relation or operation that satisfies the induction principle. The natural number structure is exemplified by the Arabic numerals, the sequences of characters on a finite alphabet in lexical order, and an infinite sequence of distinct moments of time. From this perspective, a natural number, such as 4, is a place in the natural number structure, the fourth place (if the structure starts with one). The number 4 is an office, that can be occupied by any number of objects. In the system of Arabic numerals, the symbol '4' occupies the four-place. In like manner, real analysis is about the real number structure, set theory is about the set-theoretic-hierarchy structure, topology is about topological structures, etc. The slogan of structuralism is that mathematics is the science of structure.

Since the same structure can have many different instances, a structure is a "one-over-many", of sorts, much like a property or universal in traditional metaphysics. For example, the property "horse" is a single property that applies to the many horses in the universe, and "beauty" is a single property that applies to the many beautiful things. There are many competing philosophical accounts of universals that are readily adapted to structuralism. One view is that there is no more to the natural number structure, for example, than the systems of objects that exemplify this structure. Destroy the systems and the structure goes with them. From this perspective, either structures do not exist at all, or the existence of structures is tied to their "instances," the systems that exemplify them. This ontologically parsimonious view is a nominalism toward structures— sometimes called *eliminative structuralism* (see [Benacerraf 1965], [Parsons 1990]). It is a structuralism without structures.

The eliminative structuralist does not understand arithmetic statements at face value. Apparent names, such as numerals, are really variables. For example, "$2 + 3 = 5$" comes to something like "in any natural number system S, any object in the 2-place of S S-added to the object in the 3-place of S is the object in the 5-place of S." On the eliminative view, neither the structures, nor their places, exist as objects in their own right.

The physical universe does not seem to have enough objects to exemplify all of the structures of mathematics. So it would seem that the eliminative structuralist is committed to a Platonic universe after all, in order to give the propositions of mathematics substantial content. Another option is to think in terms of *possible* instantiations of the various systems. According to the *modal structuralist*, "$2 + 3 = 5$" comes to something like "necessarily, in any possible natural number system S, any object in the 2-place of S S-added to the object in the 3-place of S is the object in the 5-place of S" (see [Hellman 1989]). Since our topic here is mathematical objects, I won't go much further into these views.

The competing realist (or platonic) view is that mathematical structures exist independently of whether they have instances in the physical world, or any other world for that matter. In a metaphysical sense, the structure is prior to its instances. This view is sometimes called *ante rem* structuralism (see Shapiro [1983], [1997] and Resnik [1981], [1988], [1997]). On this view, the subject matter of mathematics is a realm of structures.

Benacerraf [1965] was a relatively early advocate of eliminative structuralism. He made much of the fact that the set-theoretic hierarchy contains many instances of the natural number structure. He concluded from this that numbers are not objects. The viability of this conclusion depends on the general philosophical issue concerning what it is to be an object. The ante rem structuralist argues that places within ante rem structures are themselves bona fide objects.

Consider a mundane pattern, such as an organizational chart for a certain corporation. The places in the pattern are the various positions in the organization: president, vice-president for human resources, chairman of the policy committee, etc. The chart specifies the relationships between the holders of the offices, indicating who reports to whom, and the like. For the structuralist, the natural number structure is understood in a similar manner. It consists of an infinite number of positions and the relations that the occupants of these positions bear to each other.

There are two ways to think about the places in a pattern, or structure. One can think of them as offices which can be held by different people or objects in different exemplifications of the pattern. Thus, we might say that Shirley Jones was the president in the previous administration, and she presently serves as vice-president for commerce. Similarly, one may say that the numeral 'IV' is in the four-place of the natural number structure under one exemplification, while the moment of time 12:00:04 EST on January 1, 2005, occupies the four place in a different exemplification of the same structure. In this sense, which we can call *places-are-offices*, numbers are more like properties than objects. They apply to, or hold of, different objects.

Sometimes, however, we talk about the places in a pattern or structure as objects in their own right. We may say that the president has the authority to veto the recommendation of the policy committee. In a sense, we are not talking about the person or people who occupy the various offices, but about the offices themselves. Call this the "places-are-objects" perspective. The ante rem structuralist understands the language of arithmetic to be about the places of the natural number structure, understood from this perspective. An apparent proper name, like "2," is a genuine proper name, denoting a place in the natural number structure. The referent of the numeral '2' is the indicated place in the natural number structure. So the ante rem structuralist takes mathematical statements at face value. A sum, like "$2 + 3 = 5$," is a statement concerning operations in this structure.

The ante rem structuralist easily accommodates Benacerraf's concerns about the many models of the natural number structure. The natural number structure, construed from the places-are-offices perspective, is exemplified by many systems of objects: the Roman numerals, a sequence of moments of time, etc. The view adds that there is a different perspective, one from which we can talk about the places of the structure as objects.

Ante rem structuralism also accounts for the fact that mathematical structures are exemplified by *other* mathematical objects. For example, the natural number structure is exemplified by the finite von Neumann ordinals, and by the following rational numbers 1, .5, .25, .125, . . . Indeed, the natural number structure is exemplified by various *systems of natural numbers*, such as the even numbers and the prime numbers. The ante rem structuralist explains this as follows. The natural numbers, as places in the natural number structure, and the rational numbers, as places in the rational number structure, exist as objects. Again, this is to adopt the places-are-objects perspective. Some of these objects can be organized into systems, and some of these systems exemplify various structures—including the natural number structure itself. So now we are taking the places-as-offices perspective. So, for example, in the system "even numbers," six occupies the three placé (since it is the third even number), and in the system of prime numbers five occupies the three place (since it is the third prime number). In the system of rationals described just above, .25 occupies the three-place.

So the ante rem structuralist satisfies the first desideratum of the Benacerraf dilemma. The languages of mathematics are understood and treated in the same way as the languages of ordinary and scientific discourse. The proper names of mathematics refer to places in structures, and the bound variables range over places in structures. Ante rem structuralists have developed several strategies for resolving the epistemic problems with mathematics, the second horn of the Benacerraf dilemma. The question becomes: how do we know about structures? The psychological mechanism of pattern recognition may be invoked for at least small, finite structures. By encountering instances of a given pattern, we obtain knowledge of the pattern itself ([Shapiro 1997], Chapter 4). Resnik [1997] proposes a strategy much like that of Quine. The success of patterns in organizing our experience and in the overall scientific enterprise—the web of belief—justifies us in postulating the instances of the patterns, or at least the places in patterns. More sophisticated structures are apprehended via implicit definition. One gives axioms that characterize the places and relationships of a given structure. If the axiomatization is coherent, then at least one structure satisfies it. We learn about these structure(s) by deducing the consequences of the axioms. The neo-logicist strategy of abstraction-principles can also be adapted to structuralist aims. The success of Hume's principle, for example, suggests that the natural number structure is coherent, and thus that the structure exists (again, see [Shapiro 1997], Chapter 4).

As noted, the above survey is not exhaustive of the views on the existence and nature of mathematical objects. The bulk of the space here was devoted to positions that endorse the existence of such objects, and this is perhaps not typical in the philosophy of mathematics. I do hope that the reader has been induced to delve further into the rich literature on this topic.

References and Further Reading

[Benacerraf 1965] Benacerraf, P., "What numbers could not be", *Philosophical Review 74* (1965), pp. 47–73; reprinted in [Benacerraf/Putnam 1983], pp. 272–294. Classic article, argues that numbers are not objects; early defense of eliminative structuralism.

[Benacerraf 1973] Benacerraf, P., "Mathematical truth", *Journal of Philosophy 70* (1973), pp. 661–679; reprinted in [Benacerraf/Putnam 1983], pp. 403–420. Sets a dilemma that dominated much of the debate since.

[Benacerraf and Putnam 1983] Benacerraf, P., and H. Putnam, *Philosophy of Mathematics*, second edition, Cambridge, Cambridge University Press, 1983. Comprehensive anthology of important works to date.

[Boolos 1987] Boolos, G., "The consistency of Frege's *Foundations of arithmetic*", pp. 3–20 in *On being and saying: Essays for Richard Cartwright*, edited by Judith Jarvis Thompson, Cambridge, Massachusetts, The MIT Press, 1987; reprinted in [Boolos 1998], pp. 183–201.

[Boolos 1998] ——, *Logic, logic, and logic*, Cambridge, Massachusetts, Harvard University Press, 1998. Collected works in logic, includes much in the philosophy of mathematics.

[Brouwer [1912] Brouwer, L. E. J., *Intuitionisme en Formalisme*, Gronigen, Noordhoof, 1912; translated as "Intuitionism and Formalism", pp. 77–89 in [Benacerraf/Putnam 1983].

[Brouwer 1948] ——, "Consciousness, philosophy and mathematics", pp. 90–96 in [Benacerraf/Putnam 1983].

[Brown 1999] Brown, James Robert, *Philosophy of mathematics: an introduction to the world of proofs and pictures*, New York, Routledge, 1999. Contemporary defense of a Platonist epistemology.

[Burgess/Rosen 1997] Burgess, J. and G. Rosen, *A subject with no object: Strategies for nominalistic interpretation of mathematics*, Oxford, Oxford University Press, 1997. Detailed account of nominalism, mostly negative.

[Chihara 1990] Chihara, C., *Constructibility and mathematical existence*, Oxford, Oxford University Press, 1990. Defense of nominalism.

[Chihara 2004] ——, *A structural account of mathematics*, Oxford, Oxford University Press, 2004. Sequel to Chihara [1990], replying to critics and supplying an account of the applicability of mathematics.

[Colyvan 2001] Colyvan, M., *The indispensability of mathematics*, Oxford, Oxford University Press, 2001. Extensive articulation and defense of the indispensability argument.

[Dummett 1973] Dummett, M., "The philosophical basis of intuitionistic logic", pp. 215–247 in [Dummett 1978]; reprinted in [Benacerraf/Putnam 1983], pp. 97–129. Classic defense of intuitionism.

[Dummett 1977] ——, *Elements of intuitionism*, Oxford, Oxford University Press, 1977.

[Dummett 1978] ——, *Truth and other enigmas*, Cambridge, Massachusetts, Harvard University Press. 1978.

[Ernest 1998] Ernest, Paul, *Social constructivism as a philosophy of mathematics*, New York, State University of New York Press, 1998. Articulation and defense of a social constructivist account of mathematics.

[Field 1980] Field, H., *Science without numbers*, Princeton, Princeton University Press, 1980. Detailed defense of nominalism; replies to indispensability considerations.

[Field 1989] ——, *Realism, mathematics and modality*, Oxford, Blackwell, 1989. Collected works, includes several follow-ups to [Field 1980].

[Frege 1884] Frege, G., *Die Grundlagen der Arithmetik*, Breslau, Koebner, 1884; *The foundations of arithmetic*, translated by J. Austin, second edition, New York, Harper, 1960.

[Frege 1893] ——, *Grundgesetze der Arithmetik 1*, Olms, Hildescheim, 1893.

[Gödel 1944] Gödel, K., "Russell's mathematical logic", pp. 447–469 in [Benacerraf/Putnam 1983]. Articulation and defense of realism.

[Gödel 1964] ——, "What is Cantor's continuum problem", pp. 470–485 in [Benacerraf/Putnam 1983]. Articulation and defense of realism.

[Hale 1987] Hale, Bob, *Abstract objects*, Oxford, Basil Blackwell, 1987. Key item in the neo-logicist program.

[Hale/Wright 2000] Hale, Bob, and Crispin Wright, "Implicit definition and the a priori", pp. 286–319 in *New essays on the a priori*, edited by Paul Boghossian and Christopher Peacocke, Oxford University Press, 2000; reprinted in [Hale/Wright 2001], pp. 117–152.

[Hale/Wright 2001] ——, *The reason's proper study*, Oxford, Oxford University Press, 2001. Collected papers on neo-logicism.

[Hale/Wright 2001a] ——, [2001a], "To bury Caesar . . . ", pp. 33–398 in [Hale/Wright 2001]. Treatment of the Caesar issue.

[Hellman 1989] Hellman, G., *Mathematics without numbers*, Oxford, Oxford University Press, 1989. Detailed articulation and defense of nominalism.

[Hersh 1997] Hersh, Reuben, *What is mathematics, really?*, Oxford, Oxford University Press, 1997. A defense of social constructivism.

[Heyting 1930] Heyting, A., "Die Formalen Regeln der Intuitionischen Logik", *Sitzungsberichte Preuss. Akad. Wiss. Phys. Math. Klasse* (1930), pp. 42–56.

[Heyting 1931] Heyting, A., "The intuitionistic foundations of mathematics", pp. 52–61 in [Benacerraf/Putnam 1983].

[Heyting 1956] ——, *Intuitionism, an introduction*, Amsterdam, North Holland.

[Hodes 1984] Hodes, H., "Logicism and the ontological commitments of arithmetic", *Journal of Philosophy 81* (1984), pp. 123–149.

[Maddy 1990] Maddy, P., *Realism in mathematics*, Oxford, Oxford University Press, 1990. Defense of realism.

[Parsons 1990] Parsons, C., "The structuralist view of mathematical objects", *Synthese 84* (1990), pp. 303–346.

[Putnam 1971] Putnam, H., *Philosophy of logic*, New York, Harper Torchbooks, 1971.

[Quine 1951] Quine, W. V. O., "Two dogmas of empiricism", *Philosophical Review 60* (1951), pp. 20–43.

[Quine 1969] ——, *Ontological relativity and other essays*, New York, Columbia University Press, 1969.

[Quine 1981] ——, *Theories and things*, Cambridge, Massachusetts, Harvard University Press, 1981.

[Resnik 1981] Resnik, M., "Mathematics as a science of patterns: Ontology and reference", *Nous 15* (1981), pp. 529–550.

[Resnik 1988] ——, "Mathematics from the structural point of view", *Revue Internationale de Philosophie 42* (1988), pp. 400–424.

[Resnik 1997] ——, *Mathematics as a science of patterns*, Oxford, Oxford University Press, 1997. Defense of a realist version of structuralism.

[Resnik 2005] ——, "Quine and the web of belief", pp. 412–436 in [Shapiro 2005]. Extensive articulation of the Quinean perspective, covering aspects of holism.

[Russell 1919] Russell, B., *Introduction to mathematical philosophy*, Dover edition, New York, Dover, 1993.

[Shapiro 1983] Shapiro, S., "Mathematics and reality", *Philosophy of Science 50* (1983), pp. 523–548.

[Shapiro 1997] ——, *Philosophy of mathematics: structure and ontology*, New York, Oxford University Press, 1997. Articulation and defense of structuralism.

[Shapiro 2000] ——, *Thinking about mathematics: the philosophy of mathematics*, Oxford, Oxford University Press, 2000. Textbook in the philosophy of mathematics; attempts broad coverage.

[Shapiro 2005] ——, *Oxford Handbook of philosophy of mathematics and logic*, New York, Oxford University Press, 2005. Contemporary essays on key issues in the philosophy of mathematics.

[Tennant 1987] Tennant, N., *Anti-realism and logic*, Oxford, Oxford University Press, 1987.

[Tennant 1997] ——, *The taming of the true*, Oxford, Oxford University Press, 1997. Articulation and defense of intuitionism.

[Whitehead/Russell 1910] Whitehead, A. N., and B. Russell, *Principia Mathematica 1*, Cambridge, Cambridge University Press, 1910.

[Wright 1983] Wright, C., *Frege's conception of numbers as objects*, Aberdeen, Aberdeen University Press, 1983. Classic launch of neo-logicism.

[Wright 1997] ——, "On the philosophical significance of Frege's theorem", pp. 201–244 in *Language, thought, and logic*, edited by Richard Heck, Jr., Oxford, Oxford University Press, 1997; reprinted in [Hale/Wright 2001], pp. 272–306.

[Wright 1998] ——, "On the harmless impredicativity of N$^=$ (Hume's principle)", pp. 339–368 in *The philosophy of mathematics today*, edited by Mathias Schirn, Oxford, Oxford University Press, 1998; reprinted in [Hale/Wright 2001], pp. 229–255.

[Wright 1999] ——, "Is Hume's Principle analytic", *Notre Dame Journal of Formal Logic 40* (1999), pp. 6–30; reprinted in [Hale/Wright 2001], pp. 307–334.

[Yablo 2001] Yablo, S., "Go figure: a path through fictionalism", *Midwest Studies in Philosophy 25* (2001), pp. 72–102.

[Yablo 2005] ——, "The myth of the seven", Chapter 3 in *Fictionalism in metaphysics*, edited by Mark Kalderon, Oxford, Oxford University Press, 2005.

9

Mathematical Platonism[1]

Mark Balaguer
Department of Philosophy
California State University, Los Angeles

From the Editors

"Platonism" as a philosophy of mathematics refers back to Plato's dialogues on the Forms, which have been represented as existing in some eternal, unchanging, non-physical realm. Platonism in mathematics locates mathematical objects there. Many mathematicians believe that platonism as a philosophy of mathematics has been discredited, in part due to the contradictions of naive set theory, in part because of the question of how we physical beings could contact such a realm. However, in fact platonism remains the default philosophy of mathematics among philosophers, one that few are willing to defend in the strong form attributed to Gödel, but which will not be replaced until a satisfactory alternative has been found.

This chapter summarizes over forty years of such discussion, between those defending some version of platonism (called platonists, or realists) and those opposing platonism, usually called nominalists. Because it is summarizing discussion that has developed in several hundred articles and dozens of books, this chapter is not one to read casually. However, this chapter is very clearly and comprehensively structured, so that those who take the effort it to read it will be rewarded with a thorough survey of the many different schools of thought that have developed among philosophers during this period. Thus this chapter provides an excellent introduction for anyone who would like to be able to start reading original work by philosophers of mathematics.

Mark Balaguer is a Professor of Philosophy at California State University, Los Angeles (www.calstatela.edu/faculty/mbalagu/). His interests are metaphysics, philosophy of mathematics, philosophy of language, and metaethics. He has published numerous articles in the philosophy of mathematics, including "A Theory of Mathematical Correctness and Mathematical Truth," Pacific Philosophical Quarterly *(2001), "A Fictionalist Account of the Indispensable Applications*

---

[1] Many of the ideas in this essay, and in some cases the wording of the ideas, come from my [1998a] and my [2004].

of Mathematics," Philosophical Studies (1996), "Towards a Nominalization of Quantum Mechanics," Mind (1996), "A Platonist Epistemology," Synthese (1995), and has an article soon to be published, "Realism and Anti-Realism in Mathematics," in a forthcoming book, Handbook of the Philosophy of Mathematics, *edited by Andrew Irvine. He has published one book,* Platonism and Anti-Platonism in Mathematics *(1998), which does in a much more extensive way what this chapter does: it explains the various platonist (realist) and anti-platonist (nominalist) approaches to the philosophy of mathematics, and attempts to show that there is only one plausible version of each, and that these two versions are effectively indistinguishable.*

1 Introduction

Philosophers of mathematics are interested in the question of what our mathematical sentences and theories are *about*. These sentences and theories *seem* to be making straightforward claims about certain objects. Consider, for instance, the sentence '3 is prime.' This sentence seems to be a simple subject-predicate sentence of the form 'The object *a* has the property *F*'—like, for instance, the sentence 'The moon is round.' This latter sentence seems to make a straightforward claim about the moon. Likewise, the sentence '3 is prime' seems to make a straightforward claim about the number 3. But this is where philosophers get puzzled. For it's not clear what the number 3 is supposed to *be*. What kind of thing is a *number*? Some philosophers (anti-realists, or nominalists) have responded here with disbelief: according to them, there are simply no such things as numbers. Others (the realists) think that there *are* such things as numbers (as well as other mathematical objects, such as sets). But among the realists, there are several different views of what *kind of thing* a number is. Some people have thought that numbers are mental objects (something like ideas in our heads). Others have claimed that numbers exist outside of our heads, as features of the physical world. The historically most popular view, however, called **platonism**, is the view that numbers are **abstract objects**. An abstract object is an object that exists outside of space and time. These objects (if there really are such things) are wholly non-physical, non-mental, and causally inert. In other words, they do not enter into causal relations with other objects. (The fact that abstract objects are non-causal in this way follows from the fact that they're nonspatiotemporal. Since they're not extended in space, and since they aren't made of physical *stuff*, they cannot enter into cause-and-effect relationships. They cannot cause other objects to *move* in the way that, say, a cue ball can.) So according to platonists, 3 is a real and objective thing that, like the moon, exists independently of us and our thinking (that is, it's not just an idea in our heads). But 3 is also *different* from the moon, according to platonists, because it's not a *physical* thing. That is, numbers exist (really and objectively and independently of us and our thoughts), but they do not exist in space and time.

   Given these remarks, it might seem that a "philosophy of mathematics" is essentially an **ontological** theory. (An ontological theory is a theory about what sorts of things really exist. Thus, for instance, the claim that there are unicorns is a false ontological theory, and the claim that there are tigers is a true ontological theory.) Now, there is often (though not always) an ontological *component* to a philosophy of mathematics; but if you want to understand what philosophers of mathematics are really up to, it is important to realize that the *first* thing they want to do (and sometimes the only thing) is to construct a ***semantic*** theory. A semantic theory is a presumably empirical theory about what certain expressions mean (or refer to) in ordinary discourse. So, for instance, the claim that the term 'George W. Bush' denotes (in English) the Empire State Building is a false semantic theory, and the claim that 'George W. Bush' denotes (in English) the forty-third president of the United States is a true semantic theory. A philosophy of mathematics involves a semantic theory because it tells us how to interpret the sentences of ordinary mathematical discourse. It tells us what sorts of objects, if any, terms like '3' are supposed to refer to. For instance, platonism tells us that numerals like '3' are supposed to refer to abstract objects. Another theory (psychologism) tells us that numerals are supposed to refer to mental objects.

Some philosophical views about mathematics also have ontological components. For instance, platonism tells us not just that numerals like '3' are supposed to refer to abstract objects, but that there really do exist such things. But other philosophical views don't have any ontological components. For instance, some views hold that numerals aren't supposed to refer to objects at all, and so, on views like this, ontological questions (about what kinds of objects really exist) are simply irrelevant to the philosophy of mathematics. And finally, there are other philosophical views that contain *entirely uncontroversial* ontological components. For instance, the psychologistic view that says that numerals like '3' refer to ideas in our heads does contain an ontological thesis (namely, that we really do have ideas of numbers in our heads). But this ontological thesis is not very controversial: of course we have such ideas. In contrast to this, however, any theory that could accurately be called a *philosophy of mathematics* is going to contain a (controversial) semantic component.

Given this, the relationship between a mathematician and a philosopher of mathematics is analogous to the relationship between a native speaker of French and a certain sort of *linguist*—in particular, a grammarian of French whose native tongue is English but who has learned a good deal of French in order to construct a grammar for that language. There is an obvious sense in which the native speaker of French knows her language better than the linguist does. But the linguist has been trained to construct syntactic theories, and most native speakers of French have not. Thus, while the linguist has to respect the linguistic intuitions of native speakers, he cannot very well ask them what the right theory is. Likewise, while it is obvious that mathematicians know mathematics (and the language of mathematics) better than philosophers do, most of them have not been trained to construct semantic theories in the way that philosophers have. So while philosophers of mathematics have to respect the intuitions of mathematicians, they cannot very well ask them what the right theory is.

In this essay, I will provide an overview of the most important views and arguments in the philosophy of mathematics, and at the same time, I will provide some (brief and incomplete) arguments for what I think is the *right* view. In section 2, I will provide a more precise statement of the platonistic view of mathematics, and then I will formulate the central argument in favor of this view. In the process of running through this argument and the various possible responses to it, I will also provide a description and critique of the various *alternatives* to platonism. (These two tasks naturally go together, because the main argument for platonism is centrally concerned with showing that none of the alternatives to platonism is plausible. And this should not be surprising. For while platonism seems to provide a very satisfying account of our mathematical theories, it can be pretty hard to swallow from an ontological point of view. After all, the platonist's ontological thesis—that there are wholly non-physical and non-mental objects that exist outside of space and time—seems rather weird and mysterious. One might simply find it hard to believe that there really exist such things as abstract objects. But the problem is that it's very hard to find a plausible alternative to platonism. And this is the basis of the central argument in favor of platonism. The idea is to try to show that none of the alternatives is capable of providing a satisfactory account of mathematical theory and mathematical practice.) After running through the central argument for platonism in section 2, I will move on in section 3 to a discussion of what is widely thought to be the best argument *against* platonism. My own view is that in the end, neither of these two arguments succeeds, and in discussing these arguments, I will explain why I think they fail. But since this essay is primarily a survey piece, my remarks will have to be rather sketchy and incomplete. (I will, however, refer the reader to other works that fill in the details.) Finally, in section 4, I will provide a few concluding remarks about what I think we ought to say about the question of whether platonism is true and, more generally, about the philosophy of mathematics.

2 The Fregean Argument for Mathematical Platonism

(and a Taxonomy of the Alternatives to Platonism)

2.1 The Argument

Mathematical platonism, formally defined, is the view that (a) there exist abstract mathematical objects—objects that are non-spatiotemporal and wholly non-physical and non-mental—and (b) our mathematical theories are true descriptions of such objects. This view has been endorsed by a number of different people, including Plato, Frege, Gödel, and in some of his writings, Quine.[2]

The central argument for platonism is due mainly to Frege ([1884] and [1893–1903]), although I will present it somewhat differently than he did. The argument can be put like this:

(1) Our mathematical theories are extremely useful in empirical science—indeed, they seem to be indispensable to our empirical theories—and the only way to account for this is to admit that our mathematical theories are true. Therefore,

(2) The sentences of our mathematical theories—sentences like '3 is prime'—are true. Moreover, it seems that

(3) Sentences like '3 is prime' should be read at face value. (Philosophers would put this by saying that the logical form of '3 is prime' is '*a* is *F*,' where '*a*' is a constant and '*F*' is a predicate. Thus, the claim here is that '3 is prime' has the same logical form as, e.g., 'Mars is red.' Both sentences just make straightforward claims about the nature of certain objects. The one makes a claim about the nature of Mars, and the other makes a claim about the nature of the number 3.) But

(4) If we allow that sentences like '3 is prime' are true, and if moreover we allow that they should be read at face value, then we are committed to believing in the existence of the objects that they're about. For instance, if we read '3 is prime' as making a straightforward claim about the nature of the number 3, and if we allow that this sentence is literally true, then we are committed to believing in the existence of the number 3. But

---

[2] See, e.g., Plato's *Meno* and *Phaedo*; [Frege 1893-1903]; [Gödel 1964]; and [Quine 1948], [Quine 1951].

(5) If there are such things as mathematical objects (i.e., things that our mathematical theories are about), then they are abstract objects. For instance, if there is such a thing as the number 3, then it is an abstract object, not a physical or mental object. Therefore,

(6) There are such things as abstract mathematical objects, and our mathematical theories provide true descriptions of these things. In other words, mathematical platonism is true.

Arguments similar to this are sometimes called the ***Quine-Putnam indispensability argument***.[3] However, just about all of this argument, including the part in (1) about the applicability of mathematics, has roots in the work of Frege. In any event, when I talk in this essay about the "Quine-Putnam argument," I will have in mind only the subargument contained in (1)–(2).

There are a number of ways that anti-platonists can respond to the above argument, and different kinds of anti-platonists will respond in different ways. The most important divide in the anti-platonist camp is between the realists on the one hand and the anti-realists, or the nominalists, on the other. Nominalists reject the existence of mathematical objects like numbers and sets (or as philosophers would say, nominalists deny that mathematics has an *ontology*). So they deny that our mathematical theories provide true descriptions of some part of the world. Realistic anti-platonists, on the other hand, maintain that our mathematical theories *do* provide true descriptions of objects that exist in the world, but they deny that these objects are abstract. Thus, in connection with the above argument, realistic anti-platonists reject premise (5), and nominalists reject either (4), (3), or (2), depending on the kind of nominalism they endorse. In particular, neo-Meinongians reject (4); paraphrase nominalists reject (3); and fictionalists reject (2), as well as the argument for (2) contained in (1). Thus, the argument in (1)–(6) is set up so that as I run through the different responses to the argument, from the rejection of (5) back to the rejection of (2), we will get something of a taxonomy of the various kinds of anti-platonism. In what follows, I will run through these anti-platonist responses to the Fregean argument, indicating what (if anything) I think is wrong with the various views and responses. In particular, I will discuss realistic versions of anti-platonism in section 2.2 and nominalistic versions of anti-platonism in section 2.3.

2.2 Realistic Anti-Platonism

Realistic anti-platonism is the view that (a) our mathematical theories provide true descriptions of objects that exist in the world, but (b) these objects are not abstract. Now, if mathematical objects are not abstract—if, that is, they are concrete—then it seems that they must be either mental objects of some kind or (non-mental) physical objects of some kind. (One might think that they could also be social objects, or perhaps social constructions. It seems, though, that social objects would ultimately have to reduce to either mental objects or abstract objects.[4]) Thus, there

---

[3] See, e.g., [Quine 1948], [Quine 1951], [Putnam 1971], and [Colyvan 2001].

[4] This does not mean that we should think of social construction views of mathematics as either psychologistic or platonistic. It's easy to imagine versions of this view which are such that (a) the social objects in question could only be abstract objects, but (b) the view clearly rejects the existence of abstract objects, and so (c) in the end, the view is best thought of as a sort of anti-realism, perhaps a kind of fictionalism. I think this is probably the best way to understand Hersh's [1997] view, although I doubt that he would want to put it this way.

are two kinds of realistic anti-platonism: physicalism and psychologism. I will now discuss these in turn.

2.2.1 Physicalism

Advocates of physicalism maintain that our mathematical theories are about (non-mental) things that exist in the physical world. Thus, they agree with platonists that our mathematical theories provide true descriptions of things that exist independently of us and our thinking, but they reject the platonist idea that mathematical objects are abstract. There are a few different ways that one might try to develop a physicalist view of mathematics. The most famous strategy here is due to John Stuart Mill ([1843], book II, chapters 5 and 6). On his view, mathematics is about ordinary physical objects and it is just a very general empirical science. For instance, Mill takes arithmetical sentences like '2 + 3 = 5' to make very general claims about piles of objects. Thus, on this view, '2 + 3 = 5' does not tell us something about specific entities (numbers). Rather, it tells us that whenever we push a pile of two objects together with a pile of three objects, we get a pile of five objects—or something along these lines.

One problem with this view is that in order to account for contemporary mathematics in this general way, a contemporary Millian would have to take set theory to be about physical piles as well. This, however, is untenable. One argument here is that sets could not be piles of physical stuff, because corresponding to every physical pile—or, indeed, every individual physical object—there are infinitely many sets. Corresponding to a ball, for instance, is the set containing the ball, the set containing its molecules, the set containing that set, and so on. Clearly, these sets are not purely physical objects, because (a) they are all distinct from one another, and (b) they all share the same physical base (i.e., they're all made of the same matter and have the same spatiotemporal location). Thus, there must be something non-physical about these sets, over and above the physical base that they all share. So sets cannot be purely physical objects.

A second problem with physicalism is that it seems to imply that mathematics is an empirical science, contingent on physical facts and susceptible to empirical falsification. This seems to fly in the face of the facts about actual mathematical methodology. Of course, there's a *sense* in which mathematical thinking is sometimes "empirical": mathematicians often proceed by thinking of examples and counterexamples. But Millian physicalism implies that *all* mathematical claims—e.g., '2 + 3 = 5'—could in principle be falsified by discoveries about *physical* objects. This seems not just implausible (because most mathematical assertions cannot be empirically falsified by discoveries about the nature of the physical world), but also out of step with mathematical practice. It is just not plausible to interpret ordinary utterances of sentences like '2 + 3 = 5' as being about the physical world in a way that makes them contingent upon physical facts.

A third problem is that physicalism seems incapable of accounting for the truth of mathematical sentences that require the existence of infinitely many objects. Consider, e.g., the sentence, 'There are infinitely many transfinite cardinals.' It is hard to believe that this is a true claim about purely physical objects. (For a more thorough argument against the Millian view of mathematics, see my [1998a], chapter 5, section 5.)

Philip Kitcher [1984] has developed a view that might *seem* like a contemporary version of Millian physicalism. On Kitcher's view, our mathematical theories should be interpreted as (or paraphrased into) sentences about the activities of an ideal agent (a creature who pushes blocks around, adding them to piles of blocks, taking them away from piles, and so on).[5] But according to Kitcher, there aren't really any such things as ideal agents, and so, on his view, our mathematical theories are vacuous—that is, in the end, they are not about anything. So Kitcher isn't really a physicalist at all; he is, rather, an anti-realist. In particular, he is what I will call a paraphrase nominalist. Thus, Kitcher's view falls prey to the argument I give below (section 2.3.2) against paraphrase nominalism.[6]

Another way to develop physicalism would be to claim that mathematical objects are properties of some kind and to adopt a physicalistic view of properties. For example, one might take natural numbers to be properties of piles of physical objects. (This sort of view has been defended by Armstrong [1978].) There are numerous problems with views like this. To name just one, they seem to encounter problems with branches of mathematics that deal with things not found in the physical world. For instance, talk of certain large transfinite cardinals is not plausibly interpreted as being about properties that are found in the physical world. One might try to interpret such talk as being about convoluted properties that *are* found in the physical world. But any such interpretation would involve a significant departure from actual mathematical practice, and it would be susceptible to the argument given below (section 2.3.2) against paraphrase nominalism.

2.2.2 Psychologism

This is the view that there do exist mathematical objects like numbers and sets but that they do not exist independently of us. Instead, they are mental objects; in particular, the claim is usually that they are something like ideas in our heads. Thus, for instance, on this view, '3 is prime' is about a certain mental object, namely, the idea of 3. Today, most philosophers of mathematics think that psychologism is completely untenable, but the view was popular in the late nineteenth and early twentieth centuries (see, e.g., the early Husserl [1891]).

(It is often thought that intuitionism is a form of psychologism, but this is a mistake. What's true is that many intuitionists—most notably, Brouwer ([1912] and [1948]), and Heyting [1956]—have also endorsed psychologism. But intuitionism is perfectly consistent with platonism and nominalism, and psychologism is consistent with a rejection of intuitionism. See my [forthcoming] for more on the difference between psychologism and intuitionism.)

We can obtain a better understanding of psychologism, and of why philosophers find the view implausible, by distinguishing two different theses that are inherent in this view. As we saw above (section 1), we can think of a philosophy of mathematics as a *semantic* theory (an empirical theory about what certain words mean) and (usually) an *ontological* theory (a theory about what sorts of things really exist). In particular, psychologism is the conjunction of a certain ontological thesis

---

[5] The reason we can't take these sentences to be about actual agents is that most of the operations in question have never been performed. For instance, it seems likely that no one has ever pushed a pile of 17,312 blocks together with a pile of 8,643,912 blocks.

[6] One might argue that Kitcher's view is actually worse off than other paraphrase nominalist views, because it fails to deliver what is ordinarily taken to be the benefit of paraphrasing, namely, salvaging the truth of mathematics while claiming that our mathematical theories aren't about anything. For while some mathematical claims come out vacuously true on Kitcher's view, some of them come out untrue. Thus, his view is actually an odd sort of cross between paraphrase nominalism and fictionalism.

(namely, that there do exist such things as number-ideas, and set-ideas, and so on, in human heads) and a certain semantic thesis (namely, that sentences like '3 is prime' and '8,937,461.1174 > 67,491.398,' in ordinary mathematical discourse, are best interpreted as being *about* these ideas in our heads). Thus, taking the two claims together, psychologism is the claim that mathematical sentences like the above are *true* claims about *actual objects* that *literally exist* in human heads. Now, psychologism's ontological thesis—that we have ideas of mathematical objects in our heads—is obviously true: we do have ideas of numbers and so on. So *whatever* philosophy of mathematics you endorse, you should endorse the psychologist's ontological thesis. Thus, it is the *semantic* thesis of psychologism that really sets that view apart from other views. And it is this semantic thesis that other philosophers (i.e., platonists and physicalists and anti-realists) deny. That is, they deny that terms like '67,491.398' should be interpreted as denoting ideas in our heads.

There are a number of well-known arguments against the psychologist's semantic thesis. Many of these were articulated by Frege ([1884, introduction and section 27], [1893–1903, introduction], [1894], and [1919]). I will present three of these arguments here (although I should note that while the arguments I provide are all Fregean in spirit, my formulations are a bit different from Frege's). The first argument is that psychologism makes mathematical truths contingent upon psychological truths. So, for instance, if we all died, '4 is greater than 2' would suddenly become untrue. But this seems wrong: it seems that mathematics is true independently of us. Whether 4 is greater than 2 has nothing at all to do with how many humans are alive. Second, psychologism seems incompatible with the fact that there are infinitely many numbers, since clearly, there are not infinitely many number-ideas in human brains. (This is not to say that humans can't *conceive* of an infinite set. The point is, rather, that it's not the case that there *is* an infinite set of actual objects (i.e., distinct number-ideas) *actually residing* in human heads. Standard arithmetical theories like Peano Arithmetic (PA) imply that there is such a thing as the number 1, and there is such a thing as the number 2, and 2 is not identical to 1, and so on. Thus, psychologism together with PA implies that there actually exist infinitely many distinct number-ideas in human heads. Thus, either psychologism or PA is false. Third, psychologism suggests that the proper methodology for mathematics is that of empirical psychology. That is, if psychologism were true, the proper way to discover whether, say, there is a prime number between 10,000,000 and 10,000,020, would be to do an empirical study of humans and ascertain whether there is, in fact, an *idea* of such a number in one of our heads. But, of course, this is not the proper methodology for mathematics; the proper methodology involves mathematical proof, not empirical psychology. As Frege says [Frege 1884, section 27], "Weird and wonderful . . . are the results of taking seriously the suggestion that number is an idea."

Given the implausibility of the semantic thesis inherent in psychologism, one might ask whether there's any way to endorse a psychologistic view of mathematics without endorsing the strong semantic thesis that leads to such obvious problems. The answer is no. One might *try* to do this by taking psychologism's ontological thesis (i.e., the thesis that humans do have ideas of numbers and sets and so on) and strengthening it in an effort to generate a thesis that isn't so trivial and that would separate psychologism from platonism and anti-realism. But I think it can be argued that this is hopeless. To appreciate this point, consider the following thesis:

> (**COR**) Mathematical correctness is ultimately settled by the ideas that we have in our heads. More specifically, a mathematical sentence is correct just in case it is "built into," or follows from, the notions, conceptions, intuitions, and so on that we have in

connection with the given branch of mathematics (or if we have no substantive pre-theoretic conception of the objects being studied, then correctness is determined by the axioms that we happen to be working with). Thus, for instance, an arithmetical sentence is correct if and only if it is built into our full conception of the natural numbers (FCNN), where FCNN is just the sum total of all of our "natural-number thoughts" and everything that follows from these thoughts. And a set-theoretic sentence is correct if and only if it is built into our full conception of the universe of sets, or our notion of set. And so on.[7]

This might seem, at first, like a pretty psychologistic thesis. But I have argued elsewhere [Balaguer 2001] that platonists and anti-realists can—indeed *should*—endorse (COR). I will say a few words about this below: we'll see in section 3.3.4 that the best versions of platonism imply (COR), and we'll see in section 2.3.3 that the best versions of anti-realism imply (COR). So (COR) is not a particularly psychologistic thesis. Likewise, it is not a particularly "social constructivist" thesis. It is perfectly consistent with platonism and anti-realism to claim that social and psychological facts are crucially important to the determination of mathematical truth. So, again, what separates psychologism from other more plausible views (e.g., platonism and anti-realism) is not anything like (COR); what sets psychologism apart is the *semantic* thesis that our mathematical sentences are about ideas in our heads. But as we've seen, there are strong arguments against this thesis. And this is why philosophers of mathematics roundly reject psychologism.

2.3 Anti-Realistic Anti-Platonism (aka Nominalism)

If arguments like the ones discussed above are correct, then sentences like '3 is prime' are *not* about physical or mental objects. So it would seem that premise (5) in the Fregean argument for platonism is true. We turn now to the nominalistic responses to the Fregean argument. For platonists, this is the hard part. There is a good deal of agreement among philosophers of mathematics that psychologism and physicalism are untenable. Hence, most philosophers of mathematics are either platonists or nominalists. But there is very little agreement as to whether platonism or nominalism is correct.

Mathematical nominalism is the view that there are no such things as mathematical objects like numbers and sets. So on this view, our mathematical theories do not provide true descriptions of some part of the world. Thus, while nominalists would admit that there are such things as piles of three stones, and thoughts of the number 3 in people's heads, they would deny that the '3' of ordinary mathematical discourse can plausibly be interpreted as denoting any of these things. Nominalists think that when we get the *correct* interpretation of the term '3,' as it's used in ordinary mathematical discourse, it turns out that there is simply no such thing as the number 3. There are at least three different major subcategories within the nominalist camp. The first, which I will call *neo-Meinongianism*, involves a rejection of premise (4) in the Fregean argument. The second, which I will call *paraphrase nominalism*, involves a rejection of premise (3). And the third, *fictionalism*, involves a rejection of premise (2) and hence also a rejection of the Quine-Putnam argument for (2), i.e., the argument contained in (1). I will now discuss these views in turn.

---

[7] I assume here, and in what follows, that our various full conceptions are consistent. Anyone who wanted to endorse a view like (COR) would obviously have to tell a different story in connection with cases involving inconsistent full conceptions. I won't go into this here, but see my [2001] for more on this.

2.3.1 Neo-Meinongianism

Let traditional Meinongianism be the view that

(a) every *singular term*—e.g., 'Clinton,' '3,' and 'Sherlock Holmes'—denotes an object that has some sort of being (that *subsists*, or that *is*, in some sense); but

(b) only some of these objects have full-blown existence; and

(c) mathematical sentences like '3 is prime' express truths about objects that don't exist (but that still have some sort of being).

This view has been almost universally rejected. The standard argument against it (see, e.g., [Quine 1948], p. 3) is that it does not provide a view that is clearly distinct from platonism and merely creates the illusion of a different view by altering the meaning of the term 'exist.' On the standard meaning of 'exist,' any object that has any being at all exists. So according to standard usage, traditional Meinongianism implies that numbers exist. But this view clearly doesn't take such things to exist in spacetime. Therefore, traditional Meinongianism implies that numbers are abstract objects—which, of course, is just what platonism says.

There is a second version of Meinongianism, however, that is not refuted by the above argument. This view, which we can call *neo-Meinongianism*, holds that

(a) '3 is prime' should be given a face-value interpretation according to which it makes a straightforward claim about the nature of the number 3 (in particular, it says that this object is prime); and

(b) there is no such thing as the number 3 (i.e., it has no sort of being whatsoever); and yet

(c) '3 is prime' is literally true.

Thus, the idea here is that we can make true claims about things that don't exist at all. We can say (truly) of the number 3 that it is prime, even though there is no such thing as 3. (It is standardly thought that Meinong [1904] held the traditional Meinongian view described above, but one might argue that he actually endorsed what I am calling neo-Meinongianism. In any event, neo-Meinongianism has been endorsed by Richard Routley [1980]—or Richard Sylvan, as he later became known—and by Priest [2003], Azzouni ([1994], [2004]), and Bueno [2005].

One problem with neo-Meinongianism is that just as traditional Meinongians seem to alter the meaning of 'exist,' so neo-Meinongians seem to alter the meaning of 'true.' It seems that, on the standard meaning of 'true,' if there is no such thing as the object a, then sentences of the form 'a is F' cannot be *literally* true. In other words, if you believe that 'a is F' is literally true, then you also have to believe in the existence of the object a. Given this, it seems that neo-Meinongianism is untenable. If we want to maintain that mathematical sentences like '3 is prime' are literally true (and that they are of the form 'a is F'), then we have to admit that there are mathematical objects like the number 3. (See also David Lewis [1990] for an argument against Routley's view.)

Another view that might be mentioned here—a view that's related to neo-Meinongianism but also importantly different—is *conventionalism* (see, e.g., [Ayer 1946, chapter 4], [Hempel 1945], [Carnap 1934], and [Carnap 1956]). This view holds that sentences like '3 is prime' and 'There are infinitely many prime numbers' are *analytic*. That is, they are true in virtue of meaning, or linguistic conventions—along the lines of, say, 'All bachelors are unmarried' or 'All warlocks are warlocks.' But again, if sentences like '3 is prime' imply that mathematical objects like 3 exist, then it's hard to see how these sentences could be true by convention, or true in virtue of meaning. It's hard to see how the *existence* of infinitely many numbers could follow from our accepting a set of linguistic conventions—unless we're talking about some sort of psychologism, a view that we've already dispensed with.

2.3.2 Paraphrase Nominalism

Paraphrase nominalists reject premise (3) of the Fregean argument. That is, they claim that we should *not* read sentences like '3 is prime' at face value, i.e., as being of the form '*a* is *F*.' Instead, they claim, sentences like this should be read as having different logical forms. In particular, paraphrase nominalists think that these sentences have logical forms that do not imply the existence of any objects. Thus, to back this claim up, paraphrase nominalists have to specify what they think sentences like '3 is prime' *are* saying. Or in other words, they have to give *paraphrases* of these sentences that reveal their real logical forms. There are several different strategies in the literature for paraphrasing mathematics. One view here, known as *if-thenism*, holds that '3 is prime' can be paraphrased by 'If there were numbers, then 3 would be prime.'[8] For an early view of this general kind, see the early Hilbert ([1899] and his letters to Frege in [Frege 1980]); for later versions, see [Putnam 1967a], [Putnam 1967b] and [Hellman 1989]. For other paraphrase views, see, e.g., [Curry 1951] and [Chihara 1990].

The main problem with paraphrase nominalist views is that they're committed to implausible empirical hypotheses about the intentions of mathematicians and ordinary folk. For instance, if-thenism is committed to the thesis that when mathematicians and ordinary folk utter sentences like '3 is prime,' what they really mean to say is that if there were numbers, then 3 would be prime. But there is no evidence for the thesis that this is what is meant, and what's more, it seems obviously false. The same point can be made in connection with all of the other versions of paraphrase nominalism. In short, it just seems wrongheaded to claim that when mathematicians and ordinary folk utter sentences like '3 is prime,' they are speaking non-literally and really *mean* to be saying something other than what they *seem* to be saying, e.g., that 3 is prime. Now, one might question whether paraphrase nominalists are really committed to the empirical hypothesis that their nominalist paraphrases capture what mathematicians really mean to be saying when they utter sentences like '3 is prime.' But it's easy to see that they *are* so committed. For if they admit that ordinary mathematical utterances should be interpreted at face-value, then their view will collapse into fictionalism (see 2.3.3). They will be committed to saying that when we interpret our mathematical theories as saying what they actually mean in the mouths of mathematicians, they imply the existence of abstract objects. But since paraphrase nominalists deny that there are any such things as abstract objects, they will have to say that our mathematical theories, interpreted literally, are not true. And this is just what fictionalism says.

---

[8] One might point out here that '3 is prime' can be paraphrased by the sentence 'Aside from 3 and 1, there does not exist a pair of natural numbers, n and m, such that $n \times m = 3$.' But this sort of mathematical paraphrase is not helpful to nominalists, because many such paraphrases will still imply the existence of mathematical objects. For instance, if we switch the example to '4 is composite,' then the relevant mathematical paraphrase would be something like this: 'In addition to 4 and 1, there exists a pair of natural numbers, n and m, such that $n \times m = 4$.' But this sentence obviously implies the existence of mathematical objects. So this kind of paraphrase won't work. What nominalists need is a general method of paraphrasing that always delivers sentences that don't imply the existence of any mathematical objects.

2.3.3 *Mathematical Fictionalism*

Mathematical fictionalism (or simply fictionalism, as I'll call it) is the view that

(a) our mathematical sentences and theories should be interpreted at face value; (e.g., '3 is prime' should be interpreted as making—or purporting to make—a straightforward claim about the number 3); but

(b) there are no such things as abstract objects such as the number 3; and so

(c) mathematical sentences like '3 is prime' are not true.

Or, equivalently, we can say that fictionalists endorse the *semantic* theory of platonists but reject the *ontological* theory of platonists. Thus, according to fictionalists, our mathematical theories are not literally true for the same reason that, say, *Alice in Wonderland* is not literally true. Just as there are no such things as talking rabbits and hookah-smoking caterpillars and so on, so too there are no such things as numbers and sets and so on. Fictionalists agree with platonists that premises (3)–(5) in the Fregean argument are correct. They admit that if our mathematical theories are true, then there are abstract objects and platonism is correct. But fictionalists reject (2), and they reject the Quine-Putnam argument for (2)—i.e., the argument contained in (1).[9] Fictionalism was first introduced by Hartry Field ([1980], [1989]) (see below). He saw the view as being wedded to the thesis that empirical science can be nominalized. That is, on Field's view, scientific theories can be restated so that they don't contain any reference to, or quantification over, mathematical objects. In my [1996a] and [1998a], I defend a version of fictionalism that is divorced from the nominalization program, and similar versions of fictionalism have been endorsed by Rosen [2001] and Yablo [2002].

The most important objection to fictionalism is the Quine-Putnam indispensability objection. The argument here can be put like this:

Our mathematical theories are extremely useful in empirical science; indeed, they seem to be indispensable to our empirical theories. Therefore, assuming that we want to claim that our empirical theories provide accurate pictures of the world, it seems that we also have to maintain that our mathematical theories provide accurate descriptions of the world. Therefore, it seems that fictionalism is false.

Before saying how fictionalists can respond to this argument, it should be noted that fictionalism is not the only philosophy of mathematics that encounters a problem here. Every philosophy of mathematics has to account for the fact that mathematics is applicable (and perhaps indispensable) to empirical science. And for most of the standard views in the literature, there is some initial reason to think that they might not be able to provide the required explanation. The only exception here is physicalism (a view that, as we've seen, is untenable for other reasons). Indeed, I think it can be argued that the problem of applications is essentially equivalent for all non-physicalistic

---

[9] In addition to fictionalism, there is a second (much more radical) way to claim that our mathematical theories are not true and, hence, that (2) is false. One could endorse a *non-cognitivist* view of mathematics, claiming that sentences like '3 is prime' don't really say anything at all and, hence, aren't the sorts of things that have truth values. One such view is *game formalism*, which holds that mathematics is a game of symbol manipulation. According to this view, '3 is prime' is one of the "legal results" of the game of arithmetic. This view was defended by Heine and Thomae and attacked vigorously by Frege (see Frege [1893–1903], sections 88–131). One might also interpret Wittgenstein's [1956] philosophy of mathematics as non-cognitivist, although this is controversial.

views of mathematics. And I also think it can be argued that either all of these views can solve the problem or else none of them can. In what follows, I will discuss the question of how fictionalists can respond to this worry; but the view of applications and indispensability that I favor can be conjoined with other views as well—most notably, with platonism and other (non-fictionalistic) versions of nominalism.

Fictionalists have developed two different responses to the Quine-Putnam argument. The first was developed by Field [1980]. He argues that

(a) mathematics is in fact not indispensable to empirical science; and

(b) the fact that it is applicable to empirical science in a dispensable way can be explained without abandoning fictionalism.

Claim (b) is fairly plausible and has not been subjected to much criticism,[10] but claim (a) is highly controversial. In order to establish thesis (a), we would have to argue that all of our empirical theories can be *nominalized*, i.e., reformulated in a way that avoids reference to, and existential quantification over, abstract objects. Field [1980] tried to motivate this by carrying out the nominalization for one empirical theory, namely, Newtonian Gravitation Theory. And in my [1996b] and [1998a], I show how to extend Field's strategy to quantum mechanics. However, philosophers have raised several objections to Field's nominalization program—see, e.g., [Malament 1982], [Resnik 1985], and [Chihara 1990, chapter 8, section 5]. The consensus opinion seems to be that this program cannot be made to work, although this is far from established.

The second fictionalist response to the Quine-Putnam argument is

(a) to grant for the sake of argument that mathematics is hopelessly and inextricably woven into some of our empirical theories; and

(b) to simply *account* for these indispensable applications from a fictionalist point of view.

I developed this strategy in my [1996a], [1998b], and [1998a, chapter 7]; the idea has also been pursued by Rosen [2001] and Yablo [2002]. The central ideas are as follows. Because abstract objects are entirely non-causal, and because our empirical theories don't assign any causal role to abstract objects, it follows that the truth of empirical science depends upon two sets of facts that are entirely independent of one another. That is, the two sets of facts hold or don't hold independently of one another. One of these sets of facts is purely platonistic and mathematical, and the other is purely physical (or more precisely, purely nominalistic). Consider, for instance, the sentence

(A) The physical system S is forty degrees Celsius.

This sentence says that the physical system S stands in the Celsius relation to the number 40. But, trivially, it does not assign any causal role to the number 40. It is not saying that the number 40 is *responsible* in some way for the fact that S has the temperature it has. Rather, what's going on here is that we are using the numeral '40' to help us say what we want to say about S. In essence, what we're doing is using '40' as a *name* of a certain temperature state. (It is *convenient* to use numerals here, instead of ordinary names like 'Ralph' and 'Jane,' because the real numbers are structured in the same way that the possible temperature states are structured.) Thus, given

---

[10] See, however, [Shapiro 1983a] for one objection; and for a response, see [Field 1989, essay 4].

this, it follows that if (A) is true, it is true in virtue of facts about S and 40 that are entirely independent of one another. And the same point seems to hold for all of empirical science. Since no abstract objects are causally relevant to the physical world, and since empirical science never says that they are, it follows that if empirical science is true, then its truth depends upon two entirely independent sets of facts: a set of purely physical or nominalistic facts and a set of purely platonistic facts.

Now, since these two sets of facts hold or don't hold independently of one another, it could be that (a) there does exist a set of purely physical facts of the sort required for the truth of empirical science, but (b) there *doesn't* exist a set of purely platonistic facts of the sort required for the truth of empirical science (because there are no such things as abstract objects). Therefore, mathematical fictionalism is perfectly consistent with the claim that empirical science paints an essentially accurate picture of the physical world. In other words, fictionalists can endorse what I call **nominalistic scientific realism**. This is just the view that (a) and (b) above are true. In other words, it's the view that the physical world holds up *its end* of the "empirical-science bargain." (This view is different from standard scientific realism, because it doesn't imply that our empirical theories are *strictly* true. Nonetheless, this view is still a realist view, for according to this view, the physical world is *essentially just the way empirical science makes it out to be.* After all, this view says that there does exist a set of purely physical facts of the sort needed for the truth of empirical science.) Therefore, fictionalism is consistent with whatever role mathematics plays in empirical science, indispensable or not. For even if mathematics can't be eliminated from our empirical theories, and even if there are no such things as mathematical objects (and hence our empirical theories aren't literally true), the picture that empirical science paints of the physical world could still be essentially accurate.

Now, one might wonder what mathematics is doing in empirical science, if it doesn't need to be true in order for empirical science to be essentially accurate. The answer, I think, is that mathematics appears in empirical science as a *descriptive aid*. That is, it provides us with an easy way of saying what we want to say about the physical world.[11] For a more complete formulation of this second fictionalist response to the Quine-Putnam argument, see my [1996a], [1998b], and [1998a, chapter 7].

Even if fictionalists can successfully respond to the Quine-Putnam argument in this way, they are not out of the woods. For while the Quine-Putnam argument is the most important worry about fictionalism, there are other objections that one might raise against the view. I do not have the space to discuss all of these objections here, or to respond in full to the ones I do discuss, but I would like to say a few words about some of the more obvious of these objections.

Objection 1: *Fictionalism seems incapable of accounting for the objectivity of mathematics. In particular, it seems inconsistent with the fact that there is an important difference between sentences like '3 is prime' on the one hand and sentences like '3 is composite' on the other. It seems that the difference here is that '3 is prime' is true whereas '3 is composite' is false. But*

---

[11] One might wonder how it could be that mathematics is indispensable to an empirical theory T if the mathematics in T functions merely as a descriptive aid in that theory. The answer is that it may be impossible to formulate a theory that doesn't use any mathematics and yet still counts as a "version of T." There might be theories that don't use any mathematics and that are *empirically equivalent* to T in the sense that they imply the same predictions about the physical world. But they might be so unlike T, in "look and feel," that it would be implausible to treat them as "alternate versions of T."

fictionalists can't say this, because they think that both of these sentences are untrue. So what can *fictionalists say about this?*

Response: As Field [1980] pointed out when he first introduced the view, fictionalists can say that the difference between '3 is prime' and '3 is composite' is analogous to the difference between 'Oliver Twist lived in London' and 'Oliver Twist lived in sin.' In other words, the difference is that '3 is prime' is part of a certain well-known mathematical story, whereas '3 is composite' is not. We might express this idea by saying that while neither '3 is prime' nor '3 is composite' is literally true, there is another truth predicate (or pseudo-truth predicate, as the case may be)— viz., 'is true in the story of mathematics'—that applies to '3 is prime' but not to '3 is composite.' This seems to be the view that Field endorses, but there is more that needs to be said on the topic. According to fictionalism, there are alternative mathematical "stories" consisting of sentences that are not part of standard mathematics. Thus, the real difference between '3 is prime' and '3 is composite' is that the former is part of *our* story of mathematics, whereas the latter is not. (Of course, there is no *consistent* mathematical story that contains the sentence '3 is composite' and in which that sentence means what it does in English; but that's not relevant here.)

Objection 2: *OK, this will enable fictionalists to account for the difference between '3 is prime' and '3 is composite,' but there is more than this to the objectivity of mathematics. For instance, it could turn out that mathematicians are going to* discover *an objectively correct answer to the question of whether the continuum hypothesis (CH) is true or false; but it's not clear how fictionalists could account for this. Given that both CH and its negation (~CH) are consistent with the standard Zermelo-Fraenkel axiomatization of set theory (ZF), how could fictionalists claim that one of these sentences is true in the story of mathematics whereas the other is not?*

Response: In order to respond to this objection, fictionalists need to get more precise about what determines whether a sentence is true in the story of mathematics, or part of our story of mathematics. According to the version of fictionalism I favor, our story of mathematics goes beyond the axioms systems that we currently accept. It covers what I call the *full conceptions* that we have of the objects, or purported objects, in the various branches of mathematics. That is, it covers the sum total of the intentions that we, as a community, have regarding those objects. So the story of arithmetic includes everything that follows from our full conception of the natural numbers. And the story of set theory includes everything that follows from our full conception of the universe of sets, or our notion of set. And if we have no substantive pretheoretic conception of the objects being studied, then the given "full conception" is exhausted by the axiom system in question.

Given this, fictionalists can account for how mathematicians could *discover* an objectively correct answer to the CH question. Suppose, for instance, that some mathematician thought of some new axiom candidate A such that (i) all mathematicians agreed that A was intuitively obvious, and (ii) ZF + A implied CH. Then mathematicians would claim that we had discovered that CH was correct (and that it had been correct all along, that we hadn't just made this up). Fictionalists of the kind I have in mind can account for this. They can claim that the fact that A was intuitively obvious to all mathematicians shows that it was inherent in, or followed from, our notion of set, and hence that CH followed from our notion of set (even before we discovered A). So fictionalists can claim in this case that CH was part of the story of mathematics all along, even though we hadn't noticed this.

Now, this is not to say that fictionalists are committed to saying that there *is* an objectively correct answer to the CH question. In fact, they're not—fictionalism of this kind is consistent with the claim that it might be that there is no objectively correct answer to the CH question. For it *may* be that neither CH nor ∼CH follows from our notion of set. In this case, according to my version of fictionalism, neither CH nor ∼CH would be true in the story of mathematics. And so there would be no objectively correct answer to the CH question. Fictionalism is thus consistent with whatever mathematicians end up saying about CH. And this, I think, is a very attractive feature of fictionalism. For the question of what we ought to say about CH is a *mathematical* question. We don't want our philosophy of mathematics dictating what mathematicians ought to say about this.[12]

(By the way, the above considerations suggest that fictionalists should endorse (COR) (see section 2.2.2). From this it follows that fictionalism has more in common with "social constructivist" views than one might have thought. Below, we will see that the best versions of platonism also imply (COR), and so they too have something in common with social constructivist views. Indeed, depending on what is meant by "social constructivism," one might even conclude that, surprisingly, the best versions of platonism and fictionalism *are* social constructivist views. But if "social constructivism" is taken to involve a non-standard view of the *meanings* of mathematical sentences, then these views are not social constructivist views. For according to both platonism and fictionalism, our mathematical sentences and theories should be read at face value, as being about abstract mathematical objects (or as fictionalists would put it, as *purporting* to be about such objects).)

I have just scratched the surface of the topic of fictionalism and objectivity. There is much more to say about this. I cannot say any more about it here, but I have discussed it at length elsewhere, in my [2001] and my [1998a].

Objection 3: *Fictionalism is just wildly implausible on its face. Mathematics and fiction are radically different enterprises; there are numerous obvious disanalogies between the two.*

Response: Fictionalists can simply grant that there are deep and important disanalogies between mathematics and fiction, because they aren't committed to there being any deep similarities between the two enterprises. Mathematical fictionalism is a view about mathematics only. It doesn't say anything at all about fictional discourse, and so it is not committed to there being any deep similarities between mathematics and fiction. Likewise, fictionalists aren't committed to there being any deep similarities between mathematics and metaphor, or between mathematics and anything else. (For this reason, the name 'fictionalism' might be a bit misleading; a less misleading name might be 'lack-of-reference-ism,' or 'not-true-ism.')

Let's summarize what we've found so far. There are two realistic alternatives to platonism, namely, physicalism and psychologism. And there are three main anti-realist alternatives to platonism, namely, neo-Meinongianism, paraphrase nominalism, and fictionalism. I have argued (albeit rather briefly) that, aside from fictionalism, none of these views is tenable. Thus, returning to the Fregean argument for platonism with which we began, it seems that premises (3)–(5) are correct. This means that if (2) is true then (6) is also true. That is, it means that if our mathematical

---

[12] The remarks in the text suggest that fictionalists should say that a mathematical sentence is "correct" iff it follows from the full conception that we have of the objects, or purported objects, in the given branch of mathematics. But this isn't quite right, for if the given full conception is *inconsistent*, fictionalists need to say something else. See my [2001, note 5] for more on this.

theories are literally true then platonism is true. But my own view is that we don't have any good reason to believe that our mathematical theories are literally true; in particular, I don't think the Quine-Putnam argument contained in (1) gives us a good reason to believe (2). As we will presently see, however, this is *not* to say that I think we have good reason to endorse fictionalism.

3 The Epistemological Argument Against Platonism

There are a number of arguments against platonism in the literature, but one of these arguments stands out as the strongest, namely, the epistemological argument. This argument goes all the way back to Plato, but it has received renewed interest since 1973, when Paul Benacerraf presented a version of the argument. The argument can be put in the following way (see my [1998a]):

(E1) Human beings exist entirely within spacetime.

(E2) If there exist any abstract mathematical objects, then they exist outside of spacetime. Therefore, it seems very plausible that

(E3) If there exist any abstract mathematical objects, then human beings could not attain knowledge of them. Therefore,

(E4) If mathematical platonism is correct, then human beings could not attain mathematical knowledge.

(E5) Human beings have mathematical knowledge. Therefore,

(E6) Mathematical platonism is not correct.

The argument for (E3) is everything here. If it can be established, then so can (E6), because (E3) trivially implies (E4), (E5) is beyond doubt, and (E4) and (E5) trivially imply (E6). Now, (E1) and (E2) do not strictly imply (E3), and so there is room for platonists to maneuver here. And as we'll see, this is precisely how most platonists have responded. However, it is important to notice that (E1) and (E2) *seem* to provide strong motivation for (E3). They seem to imply that mathematical objects (if there are such things) are totally inaccessible to us, i.e., that information cannot pass from mathematical objects to human beings. But given this, it's hard to see how human beings could acquire knowledge of mathematical objects. Thus, we should think of this argument not as *refuting* platonism but as issuing a challenge to platonists to explain how human beings could acquire knowledge of abstract mathematical objects.

There are three strategies that platonists can use in trying to respond to this argument, and I will now discuss these in turn.

3.1 Rejecting the View that the Human Mind is Purely Physical

First, platonists can try to argue that (E1) is false and that the human mind is capable of somehow forging contact with abstract objects and thereby acquiring information about them. This strategy has been pursued by Plato (see *The Meno* and *The Phaedo*) and Gödel [1964]. Plato's idea is that our immaterial souls acquired knowledge of abstract objects before we were born and that mathematical learning is really just a process of coming to remember what we knew before we were born. On Gödel's version of the view, we acquire knowledge of abstract objects in much the same way that we acquire knowledge of concrete physical objects. Just as we acquire information about physical objects via the faculty of sense perception, so we acquire information about abstract objects by means of a faculty of *mathematical intuition*. Now, other philosophers have endorsed the idea that we possess a faculty of mathematical intuition. But Gödel's version of this view—and he seems to be alone in this—seems to involve the idea that the mind is non-physical in some sense and that we are capable of forging contact with, and acquiring information from, abstract objects.[13] This view has been almost universally rejected. One problem is that denying (E1) doesn't seem to help. The idea of an immaterial mind receiving information from an abstract object seems just as mysterious and confused as the idea of a physical brain receiving information from an abstract object.

3.2 Rejecting the Thesis that Abstract Mathematical Objects Exist Outside of Spacetime

The second strategy that one might pursue in responding to the epistemological argument is to argue that (E2) is false and that human beings can acquire information about mathematical objects via ordinary perceptual means. The early Maddy [1990] pursued this idea in connection with set theory, claiming that sets of physical objects can be taken to exist in spacetime and, hence, that we can perceive them. For instance, on Maddy's view, if there are two books on a table, then the set containing these books exists on the table, in the same place that the books exist, and we can see the set and acquire information about it in this way. Now, according to the definitions I've been using here, views like Maddy's—i.e., views that reject (E2)—are not versions of platonism at all, because they do not take mathematical objects to be non-spatiotemporal. Nonetheless, there is some rationale for thinking of Maddy's view as a sort of non-traditional platonism. First of all, Maddy's view implies that there is an infinity of sets associated with every ordinary physical object, all sharing the same spatiotemporal location and the same physical matter. (For example, corresponding to a book, there is the set containing the book, the set containing that set, the pair containing those two sets, and so on and so forth.) But since these sets all share the same *physical base* (i.e., the same location and matter), and since they are all distinct objects, Maddy has to allow that they differ from one another in some non-physical way. Hence, on Maddy's view, there must be something about these sets that is non-physical, or abstract, in some sense of these terms. Moreover, if Maddy *didn't* take this line, her view would be untenable, because it would collapse into a version of physicalism along the lines of Mill's view, which we've already rejected. In any event, regardless of whether Maddy's view counts as version of "platonism," it is an available response to the above epistemological argument.

Maddy's early view has been subjected to much criticism, including arguments from the later Maddy [1997]. Others to attack the view include Lavine [1992], Dieterle and Shapiro [1993], myself ([1994], [1998a]), Milne [1994], Riskin [1994], and Carson [1996]. One strategy here is to argue as follows:

(a) there is more to an early-Maddian set than the aggregate of physical stuff with which it shares its location (in particular, as we've seen, there is something abstract about the set, over and above the physical aggregate); but

---

[13] This interpretation of Gödel is a bit controversial. Evidence for it comes not just from his [1964], but also from his [1951]. See my [1998a, section 4.2] for a discussion.

(b) human beings don't receive any sensory data about any such sets that go beyond the data that they receive about physical aggregates; therefore,

(c) there is still an unexplained epistemic gap between the information we receive in sense perception and the relevant facts about sets.

In fact, one might push this line of thought a bit further and argue as follows:

(a) traditional platonists can grant that humans receive sensory information about physical aggregates; and

(b) traditional platonists can also claim that humans can use this information in coming to knowledge of sets; therefore,

(c) Maddian platonists are no better off here, epistemologically speaking, than traditional platonists.

For a fuller version of this argument, see my [1998a].

3.3 Explaining How We Could Have Knowledge of Abstract Mathematical Objects Without Any Contact With Such Objects

The third and final strategy that platonists can pursue is to accept (E1) and (E2) and explain why (E3) is nonetheless false. This strategy is different from the first two in that it doesn't involve the postulation of an information-transferring contact between human beings and abstract objects. The idea here is to grant that human beings do not have any such contact with abstract objects and to explain how they can nonetheless acquire knowledge of such objects. This has been the most popular strategy among contemporary platonists. Its advocates include Quine [1951, section 6], Steiner [1975, chapter 4], Parsons ([1980], [1994]), Katz ([1981], [1998]), Resnik ([1982], [1997]), Wright [1983], Lewis [1986, section 2.4], Hale [1987], Shapiro ([1989], [1997]), myself ([1995], [1998a]), and Linsky and Zalta [1995]. There are several different versions of this view. We will look very briefly at the most prominent of them.

3.3.1 Justification via Empirical Confirmation?

One version of the third strategy, implicit in the writings of Quine [1951, section 6] and developed by Steiner [1975, chapter four, especially section IV] and Resnik [1997, chapter 7], is to argue as follows:

(a) our mathematical theories are embedded in our empirical theories; and

(b) these empirical theories (including their mathematical parts) have been confirmed by empirical evidence; moreover,

(c) when an empirical theory is confirmed by empirical evidence, the *entire* theory is confirmed; therefore,

(d) even though we don't have any contact with mathematical objects, we have empirical evidence for believing that our mathematical theories are true and hence for believing that there do exist abstract mathematical objects.

Given what I said above about how mathematical fictionalists can account for the role that mathematics plays in empirical science, this view seems implausible. Since abstract objects do not enter into any causal relations with anything in the physical world, it follows that we humans would receive the same perceptual information—i.e., we would have the same set of empirical data—whether there were any such things as mathematical objects or not. Thus, empirical data can provide reason for believing only that there exist purely physical facts of the sort needed for the truth of empirical science. Empirical data do *not* provide any good reasons for believing that our empirical theories (including their implications about the existence of mathematical objects) are literally true.

A second problem with the Quine-Steiner-Resnik view is that it leaves unexplained the fact that mathematicians acquire knowledge of their theories before these theories are applied in empirical science. (For a more complete argument against the Quinean view, see my [1998a].)

3.3.2 Justification via Necessity?

A second version of the third strategy, developed by Katz ([1981], [1998]) and Lewis [1986, section 2.4], is to argue that we can know that our mathematical theories are true, without any sort of information-transferring contact with mathematical objects, because these theories are *necessarily* true. It may be that in order to know that fire engines are red, we need some sort of information-transferring contact with fire engines. But according to the Katz-Lewis view, we don't need any such contact with the number 3 in order to know that it's prime, because it couldn't have been composite, i.e., because the sentence '3 is prime' is *necessarily* true. For criticisms of this view, see Field [1989, pp. 233–38] and my [1998a, chapter 2, section 6.4]. One problem here is that there doesn't seem to be any epistemologically relevant sense in which our mathematical theories are necessarily true. Since sentences like '3 is prime' and 'There is a null set' assert that certain objects exist, they don't seem to be logically or conceptually necessary. (One might try to argue that they're *metaphysically* necessary, but there are serious problems with this suggestion. See my [1998a, chapter 2] for more on this.)

3.3.3 Structuralism

A third version of the third strategy has been developed by Resnik [1997] and Shapiro [1997]. Both of these philosophers endorse (platonistic) *structuralism*, a view that holds that our mathematical theories provide true descriptions of mathematical structures, which, according to this view, are abstract. Moreover, Resnik and Shapiro both claim that human beings can acquire knowledge of mathematical structures (without coming into any sort of information-transferring contact with such things) by simply constructing mathematical axiom systems. For these axiom systems provide *implicit definitions* of structures. There are a few different problems with this view. I discuss these problems in my [1998a], but I will just mention one of them here: Resnik-Shapiro structuralism doesn't explain how human beings could know *which* of the various axiom systems that we might formulate actually pick out real structures that exist in the mathematical realm.

3.3.4 Full-Blooded Platonism

A fourth and final version of the third strategy, developed in my own writings (see my [1992], [1995], and [1998a], and see Linsky and Zalta [1995] for a related view), is based upon the adoption of a particular version of platonism that can be called *plenitudinous platonism*, or as I call it, *full-blooded platonism* (FBP). FBP can be intuitively but sloppily expressed with the slogan, 'All possible mathematical objects exist.' More precisely, the view is that all the mathematical objects that possibly *could* exist actually *do* exist. I argue that if platonists endorse

FBP, then they can explain how human beings could acquire knowledge of abstract mathematical objects without the aid of any sort of information-transferring contact with such objects. If FBP is true, then all consistent purely mathematical theories accurately describe some collection of abstract mathematical objects. Thus, to attain knowledge of abstract mathematical objects, *all we need to do* is acquire knowledge that some purely mathematical theory is *consistent*. (It doesn't matter how we come up with the theory; some creative mathematician might simply "dream it up.") But knowledge of the consistency of a theory doesn't require any sort of contact with, or access to, the objects that the theory is about. Thus, the epistemological problem has been solved. We can acquire knowledge of abstract mathematical objects without the aid of any sort of information-transferring contact with such objects.

There are a number of objections that one might raise against FBP and the above FBP-based epistemology. Here, for instance, are six different objections that one might raise:

1. Your view seems to assume that humans are capable of *thinking about* abstract objects, or *referring to* them, or *formulating theories about* them. But it's not clear how humans could do these things.

2. The above sketch of your epistemology seems to assume that it will be easy for FBP-ists to account for how human beings could (without the aid of any contact with mathematical objects) acquire knowledge that certain mathematical theories are *consistent*. But it's not clear how FBP-ists could do this.

3. You may be right that if FBP is true, then all consistent purely mathematical theories truly describe *some* collection of mathematical objects, or some part of the mathematical realm. But *which* part? How do we know that it will be true of the part of the mathematical realm that its authors intended to characterize? Indeed, it seems mistaken to think that such theories will characterize *unique* parts of the mathematical realm at all. (For instance, if FBP is true, then there are infinitely many ω–sequences in the mathematical realm. Can FBP-ists maintain that some unique one of these sequences is *the* sequence of natural numbers?)

4. All your theory can explain is how humans could know that *if* FBP is true, then our mathematical theories truly describe parts of the mathematical realm. It doesn't explain how humans could acquire genuine knowledge of the mathematical realm, because it doesn't explain how humans could know that FBP is true.

5. How can FBP-ists account for the applications of mathematics to empirical science? FBP implies that our mathematical theories are about objects that are causally isolated from the physical world. So why do our physical theories make use of these mathematical theories?

6. FBP seems to be inconsistent with the *objectivity* of mathematics. It seems to imply that, for example, the continuum hypothesis (CH) has no determinate truth value because CH and $\sim$CH both accurately describe parts of the mathematical realm. Indeed, one might argue that because of this, FBP leads to the contradictory result that CH and $\sim$CH are both true.

I respond to all of these objections, as well as a few others, in my [1998a] and my [2001]. Moreover, in my [forthcoming], I respond to some objections that have been put forward recently by other philosophers, most notably Colyvan and Zalta [1999] and Restall [2003]. I do not have the space to address these objections here, but it is worth noting, in connection with objection 6, that the FBP-ist account of mathematical objectivity is (surprisingly) virtually identical to the fictionalist account of objectivity. According to FBP, whether CH (or any other mathematical sentence) is *true* (not just true in some model, or some part of the mathematical realm) depends on whether it's true in the *intended model* (or more precisely, in *all* intended models). But given this, it can be argued that FBP implies that a mathematical sentence is true if and only if it follows from *our intentions*, or from the full conception that *we have* of the objects in the given branch of mathematics. (And it should be noted that if we have no substantive pretheoretic conception of the objects being studied, then the given "full conception" is exhausted by the axiom system in question.) What this means is that FBP-ists are going to endorse the thesis that I called (COR) in section 2.2.2. They will not take (COR) to provide a *definition* of mathematical correctness. Like all platonists, they think that, by definition, correctness has to do with accurately describing the intended objects. But on the FBP-ist view, it *turns out* that (COR) is true.

So according to FBP, mathematical truth is ultimately determined by what follows from our "full conceptions." For instance, if CH follows from our full conception of the universe of sets, or our notion of set, then it will be true in all intended parts of the mathematical realm, and hence it will be *true*. Similarly, if ~CH follows from this full conception, then it will be true and, hence, CH will be false. And if neither CH nor ~CH follows from our full conception of the universe of sets, then they will both be true in some intended hierarchies and false in others. In this case, there will be no fact of the matter as to whether either of them is true or false. But this is essentially equivalent to what fictionalists say about how mathematical correctness—or truth in the story of mathematics—is ultimately determined. (For a full discussion of this issue, see my [2001].)

There are many other appealing features of FBP, aside from the fact that it enables platonists to solve the epistemological problem with their view. For instance, as I have argued elsewhere ([Balaguer 1998a], [Balaguer 2001]), it is only by adopting FBP that platonists can provide a plausible account of how our mathematical intuitions could be accurate indicators of mathematical truth.

In sum, then, the epistemological argument against platonism is not entirely successful. It succeeds in refuting all traditional versions of platonism, but it does not refute FBP.

4 Concluding Remarks

So we seem to be left with just one version of platonism (namely, FBP) and one version of anti-platonism (namely, fictionalism). Now, I have argued elsewhere ([Balaguer 1998a], [Balaguer 2001]) that FBP and fictionalism (interestingly and surprisingly) agree on almost everything about the interpretation of mathematical practice. I just said a few words about this in connection with the question of what ultimately determines mathematical truth. But it turns out that FBP and fictionalism agree on much more than this. It can be argued that FBP-ists and fictionalists should say essentially the same things in response to virtually *all* questions about mathematical practice. The reason for this is two-pronged. First, FBP-ists think that mathematical objects are causally inert, so that the existence or nonexistence of mathematical objects is irrelevant to the practice of mathematics. And second, FBP-ists think that every consistent purely mathematical theory accurately describes some collection of mathematical objects, so that, like fictionalists, they are committed to the thesis that from a purely *ontological* point of view, all consistent purely mathematical theories are equally good. Because of these two points, FBP-ists end up agreeing with fictionalists on almost all important questions about mathematical practice. They agree on questions about how and why mathematics is applicable to empirical science, what mathematical knowledge ultimately consists in, the semantics of mathematical discourse, the roles of creation and discovery in mathematics, and many other things. In short, FBP and fictionalism offer almost identical views of mathematics. The only questions about which they disagree are the question of whether there actually exist any abstract mathematical objects and (as a result) the question of whether our mathematical theories are literally true. That is, they disagree on the question of whether the mathematical statements that we all agree are *good* (or *correct*, or *acceptable*) are distinguished by being *literally true* or *true in the story of mathematics*.

Now, in my [1998a], I argue that while FBP and fictionalism can be defended against all of the standard objections to those views in the literature, there are no good positive arguments for either view. That is, there are no good arguments for the claim that there are abstract objects and hence that our mathematical theories are literally true, and there are no good arguments for the claim that there are no abstract objects and hence that our mathematical theories are strictly speaking untrue. Indeed, I argue that for a variety of reasons (most notably because we cannot obtain any information about whether there are any mathematical objects), we could *never* have any good reason to endorse FBP over fictionalism or vice versa. And finally, I argue that there is actually no fact of the matter about whether FBP or fictionalism is true, because there is no fact of the matter about whether there are any such things as abstract objects.

Bibliography

[Armstrong 1978] Armstrong, D.M., 1978, *A Theory of Universals*, Cambridge: Cambridge University Press.

[Ayer 1946] Ayer, A.J., *Language, Truth and Logic*, 2nd ed., New York: Dover Publications, 1946. (First published 1936.)

[Azzouni 1994] Azzouni, J., *Metaphysical Myths, Mathematical Practice*, Cambridge: Cambridge University Press, 1994.

[Azzouni 2004] ———, *Deflating Existential Consequence: A Case for Nominalism*, Oxford: Oxford University Press 2004.

[Balaguer, M., 1992] Balaguer, M., "Knowledge of Mathematical Objects," Ph.D. dissertation, CUNY Graduate Center, New York, 1992.

[Balaguer 1994] ———, "Against (Maddian) Naturalized Platonism," *Philosophia Mathematica* 2 (1994), pp. 97–108.

[Balaguer 1995] ———, "A Platonist Epistemology," *Synthese* 103 (1995), pp. 303–25.

[Balaguer 1996a] ———, "A Fictionalist Account of the Indispensable Applications of Mathematics," *Philosophical Studies* 83 (1996), pp. 291–314.

[Balaguer 1996b] ———, "Towards a Nominalization of Quantum Mechanics," *Mind* 105 (1996), pp. 209–26.

[Balaguer 1998a] ———, *Platonism and Anti-Platonism in Mathematics*, Oxford: Oxford University Press, 1998.

[Balaguer 1998b] ———, "Attitudes Without Propositions," *Philosophy and Phenomenological Research* 58 (1998), pp. 805–26.

[Balaguer 2001] ———, "A Theory of Mathematical Correctness and Mathematical Truth," *Pacific Philosophical Quarterly* 82 (2001), pp. 87–114.

[Balaguer 2004] ———, "Platonism in Metaphysics," *Stanford Encyclopedia of Philosophy*, plato.stanford.edu/entries/platonism/ (2004).

[Balaguer Forthcoming] ——, "Realism and Anti-Realism in Mathematics," forthcoming in the *Philosophy of Mathematics* volume of *The Handbook of the Philosophy of Science*, Amsterdam: North-Holland.

[Benacerraf 1973] Benacerraf, P., "Mathematical Truth," *Journal of Philosophy* 70 (1973), pp. 661–79.

[Benacerraf/Putnam 1983] Benacerraf, P. and Putnam, H. (eds.), *Philosophy of Mathematics*, Cambridge: Cambridge University Press (1983).

[Brouwer 1912] Brouwer, L.E.J., "Intuitionism and Formalism," reprinted in [Benacerraf/Putnam 1983], pp. 77–89.

[Brouwer 1948] ——, "Consciousness, Philosophy, and Mathematics," reprinted in [Benacerraf/Putnam 1983], pp. 90–96.

[Bueno 2005] Bueno, O., "Dirac and the Dispensability of Mathematics," *Studies in History and Philosophy of Modern Physics* 36 (2005), pp. 465–490.

[Burgess/Rosen 1997] Burgess, J. and Rosen, G., *A Subject With No Object*, Oxford: Oxford University Press, 1997.

[Carnap 1934] Carnap, R., *Logische Syntax der Sprache*, 1934. Translated by A. Smeaton as *The Logical Syntax of Language*, New York: Harcourt Brace, 1937.

[Carnap 1956] ——, "Empiricism, Semantics, and Ontology," reprinted in [Benacerraf/Putnam 1983], pp. 241–257.

[Carson 1996] Carson, E., "On Realism in Set Theory," *Philosophia Mathematica* 4 (1996), pp. 3–17.

[Chihara 1990] Chihara, C., *Constructibility and Mathematical Existence*, Oxford: Oxford University Press, 1990.

[Colyvan 2001] Colyvan, M., *The Indispensability of Mathematics*, New York: Oxford University Press, 2001.

[Colyvan/Zalta 1999] Colyvan, M. and Zalta, E., "Mathematics: Truth and Fiction?," *Philosophia Mathematica* 7 (1999), pp. 336–49.

Curry, H.B., 1951, Outlines of a Formalist Philosophy of Mathematics, Amsterdam: North-Holland.

[Dieterle/Shapiro 1993] Dieterle, J. and Shapiro, S., Review of Maddy, *Realism in Mathematics*, *Philosophy of Science* 60 (1993), pp. 659–660.

[Field 1980] Field, H., *Science Without Numbers*, Princeton, NJ: Princeton University Press, 1980.

[Field 1989] ——, *Realism, Mathematics, and Modality*, New York: Basil Blackwell, 1989.

[Field 1991] ——, "Metalogic and Modality," *Philosophical Studies* 62 (1991), pp. 1–22.

[Frege 1884] Frege, G., *Der Grundlagen die Arithmetik* (1884). Translated by J.L. Austin as *The Foundations of Arithmetic*, Oxford: Basil Blackwell, 1953.

[Frege 1893–1903] ——, *Grundgesetze der Arithmetik* (1893-1903). Translated (in part) by M. Furth as *The Basic Laws of Arithmetic*, Berkeley, CA: University of California Press, 1964.

[Frege 1894] ——, Review of E. Husserl's *Philosophie der Arithmetik*, *Zeitschrift für Philosophie und phil. Kritik*, 103 (1894), pp. 313–332.

[Frege 1919] ——, "The Thought: A Logical Inquiry," reprinted in *Essays on Frege*, E.D. Klemke (ed.), Urbana, IL: University of Illinois Press, 1968, pp. 507–35.

[Frege 1980] ——, *Philosophical and Mathematical Correspondence*, Chicago: University of Chicago Press 1980.

[Gödel 1951], Gödel, K., "Some Basic Theorems on the Foundations of Mathematics and Their Implications," 1951, pp. 304–23 in Gödel's *Collected Works, Volume III*, S. Peterman, J.W. Dawson, Jr., W. Goldfarb, C. Parsons, and R.N. Solovay (eds.), Oxford: Oxford University Press, 1995.

[Gödel 1964] Gödel, K., "What is Cantor's Continuum Problem?," 1964, reprinted in [Benacerraf/Putnam 1983], pp. 470–85.

[Hale 1987] Hale, R., *Abstract Objects*, Oxford: Basil Blackwell, 1987.

[Hellman 1989] Hellman, G., *Mathematics Without Numbers*, Oxford: Clarendon Press, 1989.

[Hempel 1945] Hempel, C., "On the Nature of Mathematical Truth," 1945, reprinted in [Benacerraf/Putnam 1983], pp. 377–393.

[Hersh 1997] Hersh, R., *What is Mathematics, Really?*, New York: Oxford University Press, 1997.

[Heyting 1956] Heyting, A., *Intuitionism*, Amsterdam: North-Holland, 1956.

[Hilbert 1899] Hilbert, D., *Grundlagen der Geometrie,* 1899. Translated by E. Townsend as Foundations of Geometry, La Salle, IL: Open Court, 1959.

[Husserl 1891] Husserl, E., *Philosophie der Arithmetik*, Leipzig: C.E.M. Pfeffer 1891.

[Katz 1981] Katz, J., *Language and Other Abstract Objects*, Totowa, NJ: Rowman and Littlefield 1981.

[Katz 1998] ——, *Realistic Rationalism*, Cambridge, MA: MIT Press 1998.

[Kitcher 1984] Kitcher, P., *The Nature of Mathematical Knowledge*, Oxford: Oxford University Press, 1984.

[Kreisel 1967] Kreisel, G., "Informal Rigor and Completeness Proofs," in I. Lakatos (ed.), *Problems in the Philosophy of Mathematics*, Amsterdam: North-Holland 1967.

[Lavine 1992] Lavine, S., Review of Maddy, *Realism in Mathematics*, *Journal of Philosophy*, 89 (1992), pp. 321–26.

[Lewis 1986] Lewis, D., *On the Plurality of Worlds*, Oxford: Basil Blackwell, 1986.

[Lewis 1990] ——, "Noneism or Allism?" *Mind* 99 (1990), pp. 23–31.

[Linsky/Zalta 1995] Linsky, B. and Zalta, E., "Naturalized Platonism vs. Platonized Naturalism," *Journal of Philosophy* 92 (1995), pp. 525–55.

[Maddy 1990] Maddy, P., *Realism in Mathematics*, Oxford: Oxford University Press, 1990.

[Maddy 1997] ——, *Naturalism in Mathematics*, Oxford: Oxford University Press, 1997.

[Malament 1982] Malament, D., Review of Field, *Science Without Numbers*, *Journal of Philosophy*, 79 (1982), pp. 523–34.

[Meinong 1904] Meinong, A., "Über Gegenstandstheorie," in *Untersuchungen zur Gegenstandstheorie und Psychologie*, A. Meinong (ed.), Leipzig: Barth, 1904.

[Mill 1843] Mill, J.S., *A System of Logic*, London: Longmans, Green, and Company, 1843.

[Milne 1994] Milne, P., "The Physicalization of Mathematics," *British Journal for the Philosophy of Science* 45 (1994), pp. 305–340.

[Parsons 1980] Parsons, C., "Mathematical Intuition," *Proceedings of the Aristotelian Society* 80 (1980), pp. 145–168.

[Parsons 1994] ——, "Intuition and Number," pp. 141–57 in *Mathematics and Mind*, A. George (ed.), Oxford: Oxford University Press, 1994.

[Priest 2003] Priest, G., "Meinongianism and the Philosophy of Mathematics," *Philosophia Mathematica* 11 (2003), pp. 3–15.

[Putnam 1967a] Putnam, H., "Mathematics Without Foundations," 1967, reprinted in [Benacerraf/Putnam 1983], pp. 295–311.

[Putnam 1967b] ——, "The Thesis that Mathematics is Logic," in *Bertrand Russell, Philosopher of the Century*, R. Schoenman (ed.), London: Allen and Unwin, 1967.

[Putnam 1971] Putnam, H., *Philosophy of Logic*, New York: Harper and Row, 1971.

[Quine 1948] Quine, W.V.O., "On What There Is," 1948, reprinted in [Quine 1961], pp. 1–19.

[Quine 1951] ——, "Two Dogmas of Empiricism," 1951, reprinted in [Quine 1961], pp. 20–46.

[Quine 1961] ——, *From a Logical Point of View*, 2nd ed., New York: Harper and Row, 1961.

[Resnik 1982] Resnik, M., "Mathematics as a Science of Patterns: Epistemology," *Nous* 16 (1982), pp. 95–105.

[Resnik 1985] ——, "How Nominalist is Hartry Field's Nominalism?" *Philosophical Studies*, 47 (1985), pp. 163–181.

[Resnik 1997] ——, *Mathematics as a Science of Patterns*, Oxford: Oxford University Press, 1997.

[Restall 2003] Restall, G., "Just What *Is* Full-Blooded Platonism?" *Philosophia Mathematica*, 11 (2003), pp. 82–91.

[Riskin 1994] Riskin, A., "On the Most Open Question in the History of Mathematics: A Discussion of Maddy," *Philosophia Mathematica* 2 (1994), pp. 109–121.

[Rosen 2001] Rosen, G., 2001, "Nominalism, Naturalism, Epistemic Relativism," in *Philosophical Topics* XV (Metaphysics) (2001), J. Tomberlin (ed.), pp. 60–91.

[Routley 1980] Routley, R., *Exploring Meinong's Jungle and Beyond*, Canberra: RSSS, Australian National University, 1980.

[Shapiro 1983] Shapiro, S., "Conservativeness and Incompleteness," *Journal of Philosophy* 80 (1983), pp. 521–31.

[Shapiro 1989] ——, "Structure and Ontology," *Philosophical Topics* 17 (1989), pp. 145–71.

[Shapiro 1997] ——, *Philosophy of Mathematics: Structure and Ontology*, New York: Oxford University Press, 1997.

[Steiner 1975] Steiner, M., *Mathematical Knowledge*, Ithaca, NY: Cornell University Press, 1975.

[Wittgenstein 1956] Wittgenstein, L., *Remarks on the Foundations of Mathematics*, Oxford: Basil Blackwell, 1956.

[Wright 1983] Wright, C., *Frege's Conception of Numbers as Objects*, Aberdeen, Scotland: Aberdeen University Press, 1983.

[Yablo 2002] Yablo, S., "Abstract Objects: A Case Study," *Nous* 36, supplementary volume 1 (2002), pp. 220–240.

10

The Nature of Mathematical Objects

Øystein Linnebo
Lecturer in Philosophy
University of Bristol, England

From the Editors

As we work on a mathematical problem, the mathematical objects we are working with seem very real and concrete. For example, when they do not behave as we had expected, they certainly seem quite objective and separate from our internal thoughts. As a result, many mathematicians tend toward a platonist/realist view of mathematical objects, at least until the problems with platonism are voiced. In this chapter, Øystein Linnebo attempts to resolve some of those problems by looking carefully at the language we use when referring to mathematical objects.

In ordinary speech, when we say "that's merely a semantic distinction," we mean that there is no real difference—they are just two different ways of saying the same thing. But for philosophers, questions of semantics—what words mean and how they are used—are essential to unraveling many apparent disagreements or contradictions. Mathematicians, for example, generally are indifferent to the distinction between numbers and numerals (or tend to resolve them, as Errett Bishop did, by saying "I identify a number with its numeral"). But paradoxes appear when one is not careful about the distinction between an object and its name. In this chapter, Linnebo uses this distinction in his attempt to resolve some of the problems plaguing those who would like to view mathematical objects as objective.

Øystein Linnebo is a Lecturer in Philosophy at the University of Bristol, England, (seis.bris.ac.uk/~plxol/). He received his Ph.D. in 2002 from Harvard under the direction of Charles Parsons. His main research interests are in the philosophies of logic and mathematics, metaphysics and the philosophy of language. He is particularly interested in questions concerning ontology, individuation, essence, reference (especially to abstract objects), necessity and knowledge of necessary truths. Among his articles that are likely to be of interest to readers of this volume are "Frege's Context Principle and Reference to the Natural Numbers," forthcoming in S. Lindström (ed.), Logicism, Formalism, Intuitionism—What Has Become of Them?, *"Against*

Limitation of Size," in Paradox: Logical, Cognitive and Communicative Aspects *(2006), and "Epistemological Challenges to Mathematical Platonism," Philosophical Studies (2006).*

1 *Frege's Argument for Mathematical Platonism*

Philosophers classify objects as either concrete or abstract. Roughly speaking, an object is *concrete* if it exists in space-time and is involved in causation. Otherwise the object is *abstract*.[1] Someone who believes that there exist abstract objects is said to be a *platonist*, and someone who denies this is called a *nominalist*.

On the face of it, platonism seems very far removed from the scientific world view that dominates our age. Nevertheless many philosophers and mathematicians believe that modern mathematics requires some form of platonism. The defense of mathematical platonism that is both most direct and has been most influential in the analytic tradition in philosophy derives from the German logician-philosopher Gottlob Frege (1848–1925).[2] I will therefore refer to it as *Frege's argument*. This argument is part of the background of any contemporary discussion of mathematical platonism.[3]

Frege's argument begins with the observation that the language of mathematics contains expressions that are supposed to refer to abstract mathematical objects such as numbers, functions, spaces, and geometrical figures. We see this already from a casual inspection of the language of mathematics: this language has its own stock of proper names that are supposed to denote mathematical objects (for instance '5' and 'π'), and it contains quantifier phrases that are supposed to to range over mathematical objects ('for any natural number *n*' and 'there is a real number *x*'). Frege's argument continues by claiming that a lot of mathematical statements are true. Evidence for this claim is that lots of such statements are asserted in complete earnest by everyone from lay people to expert mathematicians, and that such statements are employed in everything from quotidian reasoning to advanced science. Combining this second premise about truth with the first premise about the semantic purpose of various mathematical expressions, Frege's argument concludes that there exist mathematical objects. For a sentence containing expressions that are supposed to refer to a certain kind of object cannot be true unless there really exist objects of the kind in question.

The two premises of Frege's argument take at face value certain apparent features of mathematical language and mathematical practice. But as we all know, appearances can sometimes deceive. Both premises have therefore been challenged.

Consider the first premise that the language of mathematics is supposed to refer to mathematical objects. Here one may challenge the classification of certain mathematical expressions as

---

[1] This distinction between abstract and concrete is different from the one used in mathematics, where it means something like the distinction between general and particular.

[2] See in particular [Frege 1884].

[3] What I call "Frege's argument" abstracts from certain aspects of Frege's own defense of mathematical platonism, most importantly, from his view that arithmetic is reducible to logic. These aspects are in my opinion best seen as providing further support for and explanation of the premises of what I call "Frege's argument." By thus detaching Frege's argument from optional add-ons, we get an argument with very broad appeal. For instance, the so-called "indispensability argument" for mathematical platonism, deriving from W.V. Quine and Hilary Putnam, can be seen as just another way of supporting and explaining the premises of Frege's argument.

singular terms and quantifiers. For instance, one may argue that the adjectival use of the numerals (as in "there are five apples") is more fundamental than the substantival use (as in "the number of apples is five"). Indeed, it is a commonplace that the surface structure of natural language can deceive. Consider for instance a sentence like 'Tom did it for John's sake'. Although at a superficial level the expression 'John's sake' appears to be analogous to the expression 'John's car', a more careful logico-linguistic analysis reveals that these two expressions function differently. Frege himself was acutely aware of the danger of being misled by the surface structure of natural language. For this reason, he gave a sustained defense of his claim that certain mathematical expressions function logically as singular terms and quantifiers. He did this by developing the logical analysis of the language of mathematics which is now standard in philosophy and in mathematical logic. On this analysis, the numerals function logically as singular terms, and what look like quantifiers do indeed function as such. If this logical analysis is acceptable, then so will be the first premise of Frege's argument.

The second premise of Frege's argument states that the theorems of mathematics for the most part are true. Why, one may wonder, should we accept this claim? Can't we just regard these theorems as part of a useful game or a convenient fiction, and in this way avoid assigning any truth-values to them? On this sort of view, mathematical theorems would play a role that is strictly internal to a game or a fiction, and we wouldn't in full earnest have to accept them as true. Again, Frege has a response.[4] Unless we accept the theorems of mathematics as true, Frege says, we won't be able to account for their applicability. On the formalist and fictionalist interpretations of mathematics it remains a mystery how mathematics can be applied. For if mathematics isn't true, why should an empirical statement deduced from a true empirical theory along with a body of mathematics itself be true? The conclusion of a valid argument is guaranteed to be true only if all the premises are true, or so one would think. So Frege's response has great force.[5]

Although this defense of the premises of Frege's argument is less than conclusive, it at least shows that they enjoy great plausibility. No wonder, then, that Frege's argument has so profoundly changed the nature of the debate about mathematical platonism. The argument identifies two premises that are eminently natural and plausible, and it shows that mathematical platonism follows from these two premises. Frege's argument therefore succeeds in shifting the burden of proof onto the nominalist. Since our starting point is to believe in these two premises, we now need a reason *not* to believe in mathematical platonism.

2 Two Challenges to Mathematical Platonism

To examine such reasons, I will now discuss two challenges to mathematical platonism.

The first challenge is that mathematical platonism appears to make mathematical knowledge impossible.[6] How can the human mind reach out to the platonist's universe of abstract mathematical objects? Any causal relation is obviously out of the question, given that abstract objects

---

[4] See in particular [Frege 1903], Section 91 and the surrounding discussion.

[5] But his response isn't conclusive as it stands. For it is not ruled out that mathematical theorems have some property *other than truth* which guarantees the truth of all empirical statements that are deduced from a true empirical theory along with these mathematical theorems. For instance, Hartry Field has argued that the *semantic conservativeness* of pure mathematics is such a property. See [Field 1980] and [Field 1989].

[6] The classical source is [Benacerraf 1973], but see also [Field 1989], chapters 1 and 7.

aren't involved in causation. How then can our mathematical beliefs be sensitive to truths about this universe of abstract objects? In fact, it seems that our mathematical beliefs are *completely insensitive* to such truths. For people's beliefs are determined by facts about their brains and their physical environments. So the causal processes that take place in the physical world would have produced in us precisely the same mathematical beliefs regardless of the universe of abstract objects! This abstract universe appears to contribute nothing to the fact that we believe as we do. Even if this mathematical universe had not existed at all, our mathematical beliefs would have been precisely the same. The contrast with knowledge of the physical world is stark. My belief that there is a computer in front of me is caused by there actually being a computer in front of me. Had there not been a computer in front of me, I would not have believed that there is one. But mathematical platonists place the subject matter of mathematics outside of space-time and deny that it is involved in causation, and in so doing they foreclose the possibility of any causal explanation of mathematical knowledge. It therefore looks like the platonist's conception of mathematics makes mathematical knowledge impossible.

It may be responded that it is illegitimate to require, as this challenge does, that mathematical beliefs depend on or be sensitive to mathematical truths. For to say that X depends on or is sensitive to Y is, at the very least, to say that X co-varies with Y. But this claim makes no sense when Y consists of *necessary* truths. For a necessary truth could not have been any other way. Since mathematical truths are traditionally taken to be necessary, it therefore makes no sense to ask how other truths depend on them or are sensitive to them.[7]

This response is correct as far as it goes. But our first challenge to mathematical platonism can be stated so as to avoid this problem. We arrive at our mathematical beliefs by undergoing certain processes and by following certain methods. We would therefore like an account of how the processes and methods by which we arrive at our mathematical beliefs are *relevant* to what these beliefs are about. These processes and methods must somehow be appropriate for finding out about this subject matter. For surely it isn't just an *accident* that beliefs arrived at in these ways tend to be true.[8]

Can this challenge be met? The challenge confronts mathematical platonists with the following exercise of "solving for the unknown." Hold fixed our assumption that there is such a thing as mathematical knowledge and that this knowledge has an explanation. Then mathematical objects must be such that the methods by which we arrive at our mathematical beliefs are conducive to finding out about such objects. It is far from obvious that this equation has a solution. Since we cannot causally interact with abstract mathematical objects, the model that is appropriate to empirical knowledge doesn't apply to pure mathematics. And it remains a wide open question whether an alternative model exists which *is* appropriate to pure mathematics.

The second challenge to mathematical platonism is concerned with "Occam's razor," which instructs us not to postulate entities beyond necessity. Other things being equal, we should prefer lean scientific theories to ones with excess fat. By and large, this seems to be good scientific methodology. But not in mathematics! For mathematical objects are cheap.[9] In general,

---

[7] A response of this sort is developed in [Lewis 1986], pp. 111–12.

[8] For more on the ideas of this paragraph, see my [2006].

[9] At least according to the norms that are standard in contemporary mathematics. But even on more restrictive views on mathematical method, such as constructivism, there will be a contrast between pure mathematics and empirical science of the sort I am calling attention to.

if mathematical objects answering to some natural description *can* exist, then they *do*. Given that imaginary numbers can exist, we assume (following Cardano and others) that they do. Given that "ideal numbers" can exist, we assume (following Dedekind and others) that they do. Given that various large cardinal numbers can exist, we assume (following past and present set-theorists) that they do. This view of correct mathematical methodology is nicely expressed in the following passage from the inventor of modern infinitary set theory, Georg Cantor.

> Mathematics is in its development entirely free and is only bound in the self-evident respect that its concepts must both be consistent with each other and also stand in exact relationships, ordered by definitions, to those concepts which have previously been introduced and are already at hand and established.
>
> In particular, in the introduction of new numbers it is only obligated to give definitions of them which will bestow such a determinacy and, in certain circumstances, such a relationship to the older numbers that they can in any given instance be precisely distinguished. As soon as a number satisfies all these conditions it can and must be regarded in mathematics as existent and real. [. . .] for the *essence* of *mathematics* lies precisely in its *freedom*.   ([Cantor 1883], p. 896)

Most extravagant of all branches of mathematics is set theory. For arguably, the guiding norm in set theory is to maximize, to postulate as many sets as possible, stopping just short of inconsistency.[10] But this norm is diametrically opposite to Occam's razor!

This challenge confronts mathematical platonists with another exercise of "solving for the unknown." Hold fixed that the extravagant methods of modern mathematics are suited to finding out about mathematical objects. What is it about mathematical objects that makes these methods appropriate? Again it is unclear whether the platonists' equation has a solution. For it seems to follow from our common sense idea of an object that objects should not be postulated lightly.

3 From Objects to Semantic Values

We have seen that different considerations pull in different directions. On the one hand, Frege's argument for mathematical platonism has great plausibility. On the other hand, mathematical platonism faces two very serious challenges. To make progress, let's consider a radical proposal. Perhaps the two challenges appear insuperable only because we are operating with a wrong model of what "mathematical objects" are. Our model of objecthood has been physical bodies. According to this model, objects are like sticks and stones, apples and oranges. They are nuggets of stuff, lumps of reality. Being abstract rather than concrete, mathematical objects obviously cannot be *entirely* like sticks and stones. But they are supposed to be pretty much like sticks and stones except for being "outside of space-time" and causally isolated from us. Now, if *this* is what mathematical objects are, how on earth can our mathematical methods be conducive to our finding out about them? And how can such utterly exotic things be postulated so lightly?

Can we find a better model of what it is to be an object? I believe progress can be made by returning to Frege's argument. The argument's first premise, we recall, says that the language of mathematics contains expressions that contribute to sentences in which they occur in a way

---

[10] For documentation, see [Maddy 1997], esp. p. 131.

that is similar to how more familiar proper names contribute to sentences in which *they* occur. The similarity is that both kinds of expressions refer to objects. But in our present context, this characterization of the similarity isn't very helpful, given that our goal is precisely to explicate what it is to be an object. More helpful is the notion of a *semantic value*, which plays a fundamental role in contemporary semantics and philosophy of language. Very briefly, this notion can be explained as follows. Each component of a sentence makes some definite contribution to the truth or falsity of the sentence. This contribution is its *semantic value*. Consider a simple sentence such as 'George W. Bush is president.' The proper name 'George W. Bush' makes a definite contribution to the truth of this sentence, namely its referent, George W. Bush. Likewise, the predicate 'is president' makes a definite contribution, namely a specification of what is required of an object for the predicate to be true of it; in this case, that the object be president (as opposed to, say, being king). Moreover, the truth or falsity of a sentence is determined as a *function* of the semantic values of its constituents. This is known as *the principle of compositionality*. For instance, in our toy sentence, it doesn't matter how the semantic value George W. Bush is picked out. If the proper name 'Dubya' has the same semantic value as 'George W. Bush', then 'Dubya is president' must have the same semantic value (in this case, truth-value) as our toy sentence.

The second premise of Frege's argument is that many mathematical sentences are true. For this to be the case, all expressions involved in these sentences must succeed in making their appropriate semantic contributions. Using our new terminology, this means that all of these expressions must have semantic values, and that these semantic values must combine so as to make the relevant sentences true. The conclusion of Frege's argument can now be re-stated as the claim that mathematical singular terms have semantic values. As we have seen, these semantic values (unlike those of proper names such as 'George W. Bush') cannot be identified with any concrete objects. Perhaps we can shed light on mathematical objects by explaining how mathematical singular terms manage to have semantic values and what the nature of these semantic values is.

To develop this idea, we need a better understanding of what is involved in our reference to various sorts of objects. Since human beings are very complex organisms who stand in very complex social relations, we will have to simplify. What we want is a good model, not necessarily a description that is accurate in every detail. I therefore propose that we develop a model of our reference to various sorts of objects based on robots (or computers embedded in, and interacting with, a physical environment). I will thus investigate under what conditions it makes sense to ascribe to robots a semantics involving reference to different sorts of objects. I will focus on two fundamental cases: reference to physical bodies and reference to natural numbers.

4 Reference to Physical Bodies

What is required of a robot for it to refer to physical bodies, such as sticks and stones, in its environment? I believe we get a better understanding of the problem by focusing on the senses of sight and touch, and on some very fundamental thought processes. Other senses, such as smell, taste, and hearing, appear to play a less fundamental role in our reference to physical bodies. Consciousness too (in the sense of awareness of what it is like to have various sorts of experiences) will be put to one side, as it appears inessential to our core notion of reference.[11]

———

[11] This section draws on my [2005], especially Section 4.

So consider a robot equipped with senses of sight and touch similar to our natural human senses. Such a robot must interact with its environment by detecting light reflected by surrounding surfaces and by having a capacity for touching and grasping things in its vicinity. What is required for such a robot to make reference to a physical body in its environment? Obviously the robot must "perceive" the body in the sense that it must receive light from some part of its surface or touch some part of it. The robot will thus receive information from some spatiotemporal part of the body. These parts need not have natural boundaries in either space or time; they are simply the sum-totals of the particle-instants with which the robot causally interacts in this perception-like way.

But it is not sufficient, in order for a robot to refer to a body, that it should receive such information from some part of the body. The robot also needs a mechanism for determining when two such pieces of information belong to *the same* body. This task is far from trivial. For we are surrounded by bodies that are partially hidden, that are occluded by other bodies, and that move in and out of view. For instance, a stick can be partially buried, and a stone can be partially covered by other stones piled up around it. So there will always be different ways of "getting at" one and the same physical body, both from different spatial points of view and at different moments of time. It is therefore essential that the robot have some mechanism for grouping together pieces of perceptual information that belong to one and the same body.

I claim that what matters for this task is that the chunks of stuff from which the robot receives perceptual information be *spatiotemporally connected* (in some suitable way to be spelled out shortly). Assume for instance that the robot establishes visual contact with part of a stick that emerges from the ground and that one of its "arms" is simultaneously probing into the ground nearby and encountering something hard. What should we "teach" the robot about the conditions under which the two chunks of stuff it interacts with belong to the same body? Roughly, the kind of connectedness that matters has to do with solidity and motion: the two chunks must be related through a continuous stretch of solid[12] stuff, all of which belongs to the same unit of independent motion (roughly in the sense that, if you wiggle one chunk, the other chunk follows along).

To produce a more precise answer, think of this as an exercise in robotics. I submit that the following fundamental principles are part of an analysis of the concept of a physical body, and will therefore have to be implemented in the robot.[13]

(B1) Bodies are three-dimensional, solid objects.

(This principle holds because any two parts of a three-dimensional solid are naturally connected in space.) Thus, a cloud of gas doesn't qualify as a body in the present sense. This means that not all spatiotemporal objects are *bodies*.

(B2) Bodies have natural and relatively well distinguished spatial boundaries.

---

[12] I here mean 'solid' in the ordinary sense in which a stick or a stone is said to be solid. Of course, physics tells us that even sticks and stones aren't solid in the stricter sense of filling up all space at an atomic level.

[13] These principles are also constitutive of the concept of what psychologists sometimes call "Spelke-objects." This concept corresponds closely to my concept of a physical body. See e.g. [Spelke 1993] and [Xu 1997].

For instance, an undetached half-rock fails to be a body because it lacks sufficiently natural boundaries, and a mountain fails because its boundaries are insufficiently well distinguished.[14]

(B3) Bodies are units of independent motion.

Thus, although a book is a body, a pile of papers is not.

(B4) Bodies move along continuous paths.

Consider the object that came into being with the birth of Bill Clinton, coincided with Clinton until the end of his presidency, and thenceforth coincides with George W. Bush. By (B4) this object cannot be a body.

(B5) Bodies have natural and relatively well distinguished temporal boundaries.

So arbitrary temporal parts of bodies are not themselves bodies.

I believe this relatively simple model captures the core of the phenomenon of reference to physical bodies. What matters is that our agents (whether human or robot) receive sensory information from parts of bodies and that they have a capacity for grouping together such pieces of information just in case these pieces derive from parts that are spatiotemporally connected in the way spelled out above. Let $\sim$ be this relation of spatiotemporal connectedness. This is an equivalence relation on parts of bodies. This equivalence relation determines a (partial) function B that maps a part u to the physical body, if any, that u picks out. That is, $B(u)$ is the body that u is part of. Bodies are then subject to the following criterion of identity:

(Id-B) $\forall u \forall v (B(u) = B(v) \leftrightarrow u \sim v)$

5 Reference to Natural Numbers

Might it be the case that the structure involved in reference to physical bodies is just an instance of a more general phenomenon? Perhaps reference *always* consists in some relation to parts or aspects of objects, accompanied by some mechanism for determining when two such parts or aspects pick out the same object.[15]

Let's attempt to apply this idea to the natural numbers. Instead of information causally linked to some part of a body, a natural number is presented by means of a numeral. In fact, the most immediate ways in which a natural number is presented to people from contemporary Western culture is by means of an ordinary decimal numeral. So assume that our robots too operate with this system of numerals. However, no system of numerals can be *identified with* the natural numbers. For it is part of ordinary arithmetical competence that the natural numbers are

---

[14] Precisely how well distinguished must the boundaries of a body be? Presumably, a shedding cat is still a physical body despite all the hairs that are in the process of falling off. Although I doubt that our question allows of any precise answer, I am hopeful that an approximate answer can be given by empirical investigation of ordinary people's concept of a body.

[15] This section and the next draw on my [forthcoming], especially Sections 3 and 4. The most influential attempt to account for reference to natural numbers by means of equivalence relations on other entities is due to Wright ([1983]); see also [Hale/Wright 2001], especially the Introduction. Their attempt differs from mine in two main respects. Firstly, they take natural numbers to be presented by means of *concepts*, and they take the equivalence relation on such presentations to be that of one concept's being in one-to-one correspondence with another. Secondly, they deny that their account brings with it any sort of reductionism, whereas I argue in the next section that mine does.

"notation independent," in the sense that they can be denoted by different systems of numerals. Even people with a very rudimentary knowledge of arithmetic know that the natural numbers can be denoted not only by ordinary decimal numerals but also by their counterparts in written and spoken English (and in other natural languages) and by sequences of strokes (perhaps grouped in fives). Many people also know alternative systems of numerals such as the Roman numerals and the numerals of position systems with bases other than ten, such as binary and hexadecimal numerals.

I will here take a numeral to be any object that occupies a position in a well-ordering. In fact, since it is convenient to make the well-ordering explicit, I will take a numeral to be an ordered pair $\langle u, R \rangle$, where u is the numeral proper and R the well-ordering in which u occupies a position. On this very liberal view of the matter, the numeral proper u need not be a syntactical object, at least not in any traditional sense. (For instance, if a pre-historic shepherd counts his sheep by matching them with cuts in a stick, then these cuts count as numerals.) Moreover, since R can be any well-ordering, these numerals refer to ordinal numbers but not necessarily to finite ones.

Next, we need to equip our robot with a general condition for when two numerals determine the same number. A moment's reflection shows that two numerals $\langle u, R \rangle$ and $\langle u', R' \rangle$ determine the same number just in case u and u' occupy analogous positions in their respective orderings; for instance, that both occupy the 17th position. This can easily be given a precise mathematical definition and (at least in principle) implemented in our robot. Let $\approx$ be the resulting equivalence relation on numerals.[16] This equivalence relation determines a function N that maps a numeral to the number that it determines:

(Id-N) $N\langle u, R \rangle = N\langle u', R' \rangle \leftrightarrow \langle u, R \rangle \approx \langle u', R' \rangle$

The numbers to which the numerals are mapped are not equivalence classes of numerals but form their own category of objects. (Compare the physical bodies to which the function B from the previous section maps parts of physical stuff. These physical bodies are not equivalence classes of their parts but form a distinct category of objects.) And the criterion of identity (Id-N) tells us how these objects are identified, just like the criterion of identity (Id-B) tells us how physical bodies are identified. Let O (for "ordinal") be a predicate that holds of all and only objects in the range of N.

Next we want to define a relation $P^{\#}$ that holds between two numerals $\langle u, R \rangle$ and $\langle u', R' \rangle$ just in case the number determined by the former immediately precedes the number determined by the latter. One easily sees that the right definition is that u' has an R'-predecessor v such that $\langle u, R \rangle \approx \langle v, R' \rangle$. It is then easy to verify that $P^{\#}$ doesn't distinguish between numerals that are equivalent under $\approx$. This means that the relation $P^{\#}$ on numerals induces a predecessor relation P on the ordinals themselves, defined by

(Def-P) $P(N\langle u, R \rangle, N\langle u', R' \rangle) \leftrightarrow P^{\#}(\langle u, R \rangle, \langle u', R' \rangle)$.

Finally, following the ordinary practice of counting, we let 1 be the first number. We may for instance define 1 as $N\langle \text{'1'}, D \rangle$, where D is the familiar well-ordering of base 10 numerals for positive integers.

---

[16] The relation $\approx$ may be taken to be a linguistic object, not a properly mathematical one, by identifying it with a formula with two free variables in an interpreted language.

With these definitions, it is easy to establish some of the basic axioms for ordinal numbers, for instance:

(O1) $O(1)$
(O2) $\neg \exists x\, P(x, 1)$
(O3) $P(x, y) \wedge P(x', y) \rightarrow x = x'$
(O4) $P(x, y) \wedge P(x, y') \rightarrow y = y'$

For instance, (O3) follows from the fact that any two numerals that precede a third are equivalent under $\approx$, which means that they determine the same ordinal.

However, I have not yet said anything very substantial about *how many* ordinals there are. For the purpose of describing the natural numbers (which I identify with the finite ordinals), the only such principle we need is an axiom that ensures the existence of successors:

(O5) $\forall x(O(x) \rightarrow \exists y\, P(x, y))$

This axiom doesn't follow from what has been said so far about the numerals. But the axiom can be motivated as follows. Begin with the extremely plausible principle that for any numeral, there *could be* (roughly, that it is consistent that there is) another numeral directly succeeding it. By (Def-P), this means that for any ordinal, there *could be* another ordinal directly succeeding it. From this we get (O5) by invoking the principle that any ordinal that could exist, does exist.

Finally, we need to specify some condition of finitude with which to restrict the ordinals such that we get all and only the natural numbers but no infinite ordinals. I claim that this condition is simply that mathematical induction should be valid of the natural numbers. That is, an ordinal n is a *natural number* just in case the following schema holds for any predicate ϕ:

(MI) $\phi(1) \wedge \forall x \forall y[\phi(x) \wedge P(x, y) \rightarrow \phi(y)] \rightarrow \phi(n)$.

Our characterization of the natural numbers has thus allowed us to derive all the familiar axioms of Peano Arithmetic.

6 The "Thinness" of the Natural Numbers

I will now explain a fundamental difference between physical bodies and natural numbers that has to do with the ways in which these objects possess properties. Consider the question whether a physical body x has some property, say being round. To answer this question, it isn't sufficient to consider any proper part of x. Whether a body is round isn't determined by any of its proper parts but information is needed about the entire body. And there is nothing unusual about this case. It is in general true that, in order to determine whether a body x has some property G (such as a particular shape or mass), one needs information about many or all parts of x. The question whether a body has some property G cannot in general be reduced to a question about any *one* of its proper parts. This means that a body can have properties in an irreducible way, that is, in a way that isn't reflected in any property of any *one* of its proper parts.[17]

---

[17] The converse is true as well (although less important to the present discussion): a proper part can have properties that aren't reflected in any properties of the body of which it forms a part.

The situation is very different with the natural numbers. Consider the question whether a natural number n has some arithmetical property G, say the property of being even. In this case, unlike the situation of roundness of physical bodies, a standard presentation of n by some numeral (say a standard decimal numeral) suffices to answer the question. There is no need to examine other presentations of the number n or the number itself. In fact, the question whether the natural number n possesses an arithmetical property G can always be reduced to a question about the numeral by which the number n is presented. For all the usual arithmetical properties are definable (in second-order logic) from the predecessor relation P. And as (Def-P) shows, the question of whether P holds between two natural numbers is itself reducible to the question whether the relation $P^{\#}$ holds between certain numerals.

This means that on the view I have defended, the natural numbers are "impoverished" compared to numerals. For whenever a natural number n possesses some arithmetical property, its doing so is inherited from the fact that the numerals that present n possess some related property. Natural numbers are therefore "thinner" than the numerals that present them. In fact, since questions about natural numbers can be reduced to questions about the numerals that present them, this opens the possibility of a form of reductionism about natural numbers.

Given the possibility of this reductionism, does it still make sense to say that numerals *refer to natural numbers*? In light of Section 3, this question is best understood as asking whether it still makes sense to ascribe *semantic values* to numerals. I will now argue that this *does* still make sense. One observation that supports this claim is the following. The default assumption is that expressions that belong to the same syntactic category—in this case, the class of singular terms—should belong to the same semantic category as well. And indeed, when we analyze English and the language of arithmetic, singular terms such as '5' and '1001' seem to function just like singular terms such as 'Alice' and 'Bob.' But it is uncontroversial that singular terms such as 'Alice' and 'Bob' have semantic values, namely the physical bodies that they refer to. This provides at least some reason to think that arithmetical singular terms such as '5' and '1001' have semantic values as well.

It may be objected that this default assumption is overridden by our discovery that questions about natural numbers can be reduced to questions about the associated numerals. Since this reduction shows that it suffices to talk about the numerals, the objection continues, there is no need to ascribe any sort of semantic values to numerals. However, this objection assumes *that the structure responsible for the reduction that we have discovered is also the kind of structure that matters for semantic analysis*. I argue in other work that this condition is not met and that the objection therefore fails.[18] Taking a broader look at the issue, this shouldn't be very surprising. For not every kind of structure that is involved in the phenomenon of reference is semantic structure. For instance, reference is often based on perception, and perception is undoubtedly a complicated process that involves all kinds of structuring of sensory information. But this structure will typically not be semantic structure. Although perception is often *presupposed by* the relation of reference and thus also by semantics, perception and its structure aren't thereby *included in* semantics.

If I am right that the objection fails, then it still makes sense to ascribe semantic values to numerals. And since these semantic values are nothing other than the natural numbers, this means that the numerals do after all refer to natural numbers.

---

[18] See [Linnebo forthcoming], Section 4.

7 Back to the Two Challenges

I have argued that natural numbers are fundamentally different from physical bodies by being so "thin." Can this be used to answer the two challenges discussed in Section 2? I proceed in reverse order.

The second challenge was to explain why it is reasonable to operate with such ontologically "extravagant" methods as those of modern mathematics rather than the more "parsimonious" methods found in the empirical sciences. What is it about mathematical objects that makes it appropriate to postulate such objects so much more lightly than we postulate physical objects? When we translate talk about objects into talk about semantic values, the question becomes why so much less is required for a mathematical singular term to have a semantic value than for a physical singular term to have one. This is a question that we are now well equipped to answer. If pure mathematics is at all like arithmetic, then very little is required for a mathematical singular term to have a semantic value. All that is required is that the term be associated with some (possibly syntactic) object that serves as a presentation of some semantic value, and that we have some principled and well-founded way of telling when two such presentations determine the same semantic value. This supports a view of mathematics like the one expressed in the passage from Cantor quoted in Section 2.

The first challenge was to explain how knowledge of mathematical objects is possible without any causal interaction with such objects. How can the methods by which we arrive at our mathematical beliefs be sensitive to the facts that are involved in making these beliefs true? Let's begin by considering a simple belief about the physical world such as 'This body is round' (pointing at a near-perfect globe). The truth of this belief[19] depends on two things: first, that the belief has some particular proposition as its semantic content, and second, that the world is such as to make this proposition true. I will now describe these two kinds of dependence and explain how both contribute to the formation of true beliefs.

Let's begin with the first kind of dependence. Why does the belief have this particular proposition as its content? On the account I have been developing, this question can be reduced to the question why the various *simple constituents* of the belief have the semantic values that they happen to have. This is a question about which I have had quite a lot to say. I have proposed a model of how the expression 'this body' comes to refer to a particular body, in this case the globe. This involves facts about the causal transmission of information from the globe to our sense organs and about this information's being put together in a way that is sensitive to the natural spatiotemporal connectedness of the chunks of physical stuff from which it derives.[20] Since these facts make it the case that the belief has some particular proposition as its content, they contribute *semantically* to the truth of the belief in question.

The second kind of dependence requires less comment. Since the content of the belief in question is the proposition that the relevant globe is round, this globe's actually being round obviously contributes to making the belief true. Now, the fact that the globe is round isn't among

---

[19] Here and in what follows I use the word 'belief' to mean a particular sort of internal psychological state, considered in abstraction from any propositional content that this state may have. This is thus a syntactic, rather than semantic, notion of belief.

[20] A story can also be told about how 'is round' comes to have *its* semantic value. This story will crucially involve the fact that our subject takes this predicate to apply to all and only round things.

the facts that contribute semantically to the truth of the belief in question. I will therefore say that it contributes *non-semantically*. Note that we are not distinguishing between two kinds of facts—the semantic and the non-semantic—but rather between two kinds of *contributions* that a unique realm of facts can make to the truth of a belief.

Facts that contribute to making a belief true in either of these ways typically also contribute to an agent's formation of this belief. Let's begin with the facts that contribute non-semantically—in our example, the globe's actually being round. This fact obviously contributes to the agent's formation of the belief in question. Had this globe been seriously dented, say, the agent would have noticed and therefore not formed the belief in question. What about the facts that contribute semantically? Recall that these are facts about the agent's being in perceptual contact with the globe and about the resulting perceptual information's being put together in accordance with the principle (Id-B) (see the end of section 4). These facts too contribute to the formation of the belief. Had the agent been in perceptual contact with another body, or had he put together pieces of perceptual information in accordance with some principle other than (Id-B), he would most likely not have formed the belief in question.[21] So in this example, a complete account of why the agent formed the belief in question must appeal both to the facts that contribute semantically to the truth of the belief and to the facts that contribute non-semantically.

I turn now to a very simple mathematical example, involving the mathematical belief that 2 directly precedes 3. (Once this example is in place, more complex examples can be given by exploiting the fact that other arithmetical relations are definable from the predecessor relation.) In this example too there are facts that ensure that the constituents of the belief have the semantic values that they happen to have. First there is the fact that the numerals '2' and '3' occupy the second and third positions of the standard sequence of decimal numerals. Then there is the fact that the agent takes two numerals to determine the same number just in case they stand in the equivalence relation $\approx$. Finally there is the fact that the agent takes the predecessor relation P's holding of two natural numbers to be a matter of the associated numerals' standing in the relation $P^{\#}$, as described in (Def-P). Unlike the previous example, however, there are no facts whose contribution to the truth of the belief that 2 directly precedes 3 is completely non-semantic. This kind of contribution to truth has vanished entirely.[22]

Fortunately, the facts that contribute semantically to the truth of the belief that 2 directly precedes 3 also suffice to explain why an agent formed this belief. Because the agent treats the predicate 'directly precedes' in accordance with (Def-P), he regards the belief that 2 directly precedes 3 as true just in case the associated numerals stand in the relation $P^{\#}$. And because he regards the associated numerals '2' and '3' as ordinary decimal numerals, he deems that they indeed stand in the relation $P^{\#}$. Consequently he regards the belief as true. Had the agent not treated the predicate 'directly precedes' in accordance with the definition (Def-P), or had he not taken the numerals '2' and '3' to be ordered as decimal numerals, he would not have formed the belief.

---

[21] Much the same goes for the facts involved in giving the predicate 'is round' its semantic value.

[22] Whether or not this makes the belief in question analytic will depend on how one understands the notion of analyticity. I would deny that the belief is analytic in the traditional sense that anyone who grasps the belief can see it to be true by conceptual analysis alone. For the semantic facts that I have been talking about need not be consciously accessible even to people with a perfect grasp of the belief. (Similarly, the semantic facts involved in reference to physical bodies need not be consciously accessible even to people who are fully competent with such reference.)

Summing up, it turns out that the very same facts that make our sample mathematical belief true are also responsible for making the agent form the belief. The agent's belief is therefore appropriately sensitive to the truth of the belief, which answers the first challenge.

8 Conclusion

I began by outlining two conflicting views on mathematics: first Frege's argument that there exist abstract mathematical objects, and then two serious challenges to the idea that there could exist such objects. To make progress, I suggested that we must reject the standard conception of objecthood (which is modeled on the notion of a physical body) and instead use the technical notion of a *semantic value* to explicate the notion of objecthood. I then gave an account of reference to physical bodies, based on the idea that we perceptually interact with parts of such bodies and that we operate with an equivalence relation which tells us when two such parts belong to the same body. I next suggested that a similar account is possible of reference to natural numbers: natural numbers are presented to us by means of numerals, and we operate with an equivalence relation that tells us when two such numerals determine the same number. Natural numbers are on this view much "thinner" than physical bodies, in the sense that all properties of a natural number can be reduced to properties of the corresponding numeral, whereas not all properties of a physical body can be reduced to properties of its individual parts. I finally observed that on this conception of numbers (and of mathematical objects more generally) as "thin," we are able to *both* agree with Frege's argument *and* answer the two challenges to which this argument gives rise.

This means that we may continue to use platonistic language when thinking, talking and teaching about mathematics. For there is a perfectly legitimate sense in which we succeed in referring to mathematical objects. And there is nothing scientifically suspect about this form of platonism—at least not when the mathematical objects are understood as "thin."

Acknowledgements Thanks to Matti Eklund, Frode Kjosavik, and the editors of this volume for very valuable comments on an earlier version.

References

[Benacerraf 1973] Benacerraf, Paul, "Mathematical Truth," repr. in P. Benacerraf and H. Putnam (eds.), *Philosophy of Mathematics: Selected Readings* , Cambridge: Cambridge University Press, 1983.

[Cantor 1883] Cantor, Georg, *Foundations of a General Theory of Manifolds*. Transl. and repr. in William Ewald (ed.), *From Kant to Hilbert*, two volumes, Oxford: Clarendon, 1996.

[Field 1980] Field, Hartry, *Science without Numbers: A Defense of Nominalism*, Princeton, NJ: Princeton University Press, 1980.

[Field 1989] ——, *Realism, Mathematics, and Modality*, Oxford: Blackwell, 1989.

[Frege 1884] Frege, Gottlob, *Grundlagen der Arithmetik*. Transl. by J.L. Austin as *Foundations of Arithmetic*, Oxford: Blackwell, 1953.

[Frege 1903] ——, *Grundgesetze der Arithmetik*, Vol. 2. Excerpts transl., pp. 139–224 in P. Geach and M. Black (eds.), *Translations from the Philosophical Writings of Gottlob Frege*, Oxford: Blackwell, 1952.

[Hale/Wright 2001] Hale, Bob and Crispin Wright, *The Reason's Proper Study*, Oxford: Clarendon, 2001.

[Lewis 1986] Lewis, David K., *On the Plurality of Worlds*, Oxford: Blackwell, 1986.

[Linnebo 2005] Linnebo, Øystein, "To Be Is to Be an *F*," *Dialectica* 59(2) (2005), pp. 201–222.

[Linnebo 2006] ——, "Epistemological Challenges to Mathematical Platonism," *Philosophical Studies* 129(3) (2006), pp. 545–574

[Linnebo Forthcoming] ——, "Frege's Context Principle and Reference to Natural Numbers," to appear in Sten Lindström (ed.), *Logicism, Formalism, Intuitionism: What Has Become of Them?* New York: Springer.

[Maddy 1997] Maddy, Penelope, *Naturalism in Mathematics*, Oxford: Clarendon, 1997.

[Spelke 1993] Spelke, Elizabeth, "Object Perception," in A. Goldman (ed.), *Readings in Philosophy and Cognitive Science*, Cambridge, MA: MIT Press, 1993.

[Wright 1983] Wright, Crispin, *Frege's Conception of Numbers as Objects*, Aberdeen: Aberdeen University Press, 1983.

[Xu 1997] Xu, Fei, "From Lot's Wife to a Pillar of Salt: Evidence that *Physical Object* is a Sortal Concept," *Mind and Language* 12 (1997), pp. 365–92.

11

When is One Thing Equal to Some Other Thing?

Barry Mazur
Gerhard Gade University Professor
Harvard University

From the Editors

You may wonder why this chapter by Barry Mazur is in this section on the nature of mathematical objects. We have put it here because, in order to have an object, you need to be able to distinguish it from other objects that it is not identical to—as Cantor said, when defining transfinite numbers, objects must be "definite, well-distinguished objects of our perception or of our thought." We must be able to tell when a mathematical object given one way is the same as that given another way: "the integer that is the successor of 1" and "the only even prime," for example. The (general) solution to this problem is less obvious than it might initially appear, and this chapter is certainly the most mathematically sophisticated contribution to the question that we have seen.

In a way it is a continuation of the discussions on structuralism by the philosophers in this section, but now via category theory. It also leads naturally into the final section of this book, on the nature of mathematics.

Barry Mazur is the Gerhard Gade University Professor in the Department of Mathematics at Harvard University (www.math.harvard.edu/~mazur/) and a member of the National Academy of Sciences. His early work was in geometric topology, but he soon started working in algebraic geometry and number theory. In addition to his many mathematical papers, he has written several articles about the philosophy of mathematics, including "On the Absence of Time in Mathematics," For the Learning of Mathematics *(2004) and "Conjecture,"* Synthese *(1997). His mathematical expository work has won him prizes, including the Chauvenet prize in 1994 for "Number Theory as Gadfly" (*American Mathematical Monthly, *1991).*

Many of his expository articles include substantial philosophical content, particularly "How did Theaetetus prove his Theorem?" to be published in a Festschrift in honor of Eva Brann, "Perturbations, deformations, and variations (and 'near-misses') in geometry, physics, and number theory" in the issue of the Bulletin of the A.M.S. *in honor of René Thom (2004), an article with*

Federica la Nave, "Reading Bombelli," The Mathematical Intelligencer *(2002), and his book* Imagining Numbers (particularly the square root of minus fifteen) *(2003).*

In memory of Saunders Mac Lane

1 The Awkwardness of Equality

One can't do mathematics for more than ten minutes without grappling, in some way or other, with the slippery notion of *equality*. Slippery, because the way in which objects are presented to us hardly ever, perhaps never, immediately tells us—without further commentary—when two of them are to be considered equal. We even see this, for example, if we try to define real numbers as decimals, and then have to mention aliases like $20 = 19.999\ldots$, a fact not unknown to the merchants who price their items $19.99.

The heart and soul of much mathematics consists of the fact that the "same" object can be presented to us in different ways. Even if we are faced with the simple-seeming task of "giving" a large number, there is no way of doing this without also, at the same time, giving a hefty amount of extra structure that comes as a result of the way we pin down—or the way we present—our large number. If we write our number as 1729 we are, sotto voce, offering a preferred way of "computing it" (add one thousand to seven hundreds to two tens to nine). If we present it as $1 + 12^3$ we are recommending another mode of computation, and if we pin it down—as Ramanujan did— as the first number expressible as a sum of two cubes in two different ways, we are being less specific about how to compute our number, but have underscored a characterizing property of it within a subtle diophantine arena.

The issue of "presentation" sometimes comes up as a small pedagogical hurdle—no more than a pebble in the road, perhaps, but it is there—when one teaches young people the idea of *congruence mod N*. How should we think of 1, 2, 3, ... mod 691? Are these ciphers just members of a *new* number system that happens to have similar notation as some of our integers? Are we to think of them as equivalence classes of integers, where the equivalence relation is congruence mod 691? Or are we happy to deal with them as the good old integers, but subjected to that equivalence relation? The eventual answer, of course, is: all three ways—having the flexibility to adjust our viewpoint to the needs of the moment is the key. But that may be too stiff a dose of flexibility to impose on our students all at once.

To define the mathematical objects we intend to study, we often—perhaps always—first make it understood, more often implicitly than explicitly, how we intend these objects to be presented to us, thereby delineating a kind of *super-object;* that is, a species of mathematical objects garnished with a repertoire of *modes of presentation*. Only once this is done do we try to erase the scaffolding of the presentation, to say when two of these super-objects—possibly presented to us in wildly different ways—are to be considered *equal*. In this oblique way, the objects that we truly want enter the scene only defined as *equivalence classes of explicitly presented objects*. That is, as specifically presented objects with the specific presentation ignored, in the spirit of "ham and eggs, but hold the ham."

This issue has been with us, of course, forever: the general question of *abstraction*, as separating what we want from what we are presented with. It is neatly packaged in the Greek verb *aphairein,* as interpreted by Aristotle[1] in the later books of the *Metaphysics* to mean simply *separation*: if it is *whiteness* we want to think about, we must somehow separate it from *white horse, white house, white hose,* and all the other white things that it invariably *must* come along with, in order for us to experience it at all.

The little trireme of possibilities we encounter in teaching congruence mod 691 (i.e., is 5 mod 691 to be thought of as a symbol, or a stand-in for any number that has remainder 5 when divided by 691, or should we take the tack that it (i.e., "5 mod 691") *is* the (equivalence) class of all integers that are congruent to 5 mod 691?) has its analogue elsewhere—perhaps everywhere—in mathematics. Familiarity with a concept will allow us to finesse, or ignore, this, as when we are happy to deal with a fraction a/b ambiguously as an equivalence class of pairs of integers (a, b) with $b \neq 0$, where the equivalence relationship is given by the rule $(a, b) \sim (a', b')$ if and only if $ab' = a'b$, or as a particular member of this class. Few mathematical concepts enter our repertoire in a manner other than ambiguously a *single object* and at the same time an *equivalence class of objects*. This is especially true for the concept of *natural number*, as we shall see in the next section where we examine the three possible ways we have of coming to terms with the number 5.

One of the templates of modern mathematics, category theory, offers its own formulation of *equivalence* as opposed to *equality*; the spirit of category theory allows us to be content to determine a mathematical object, as one says in the language of that theory, *up to canonical isomorphism*. The categorical viewpoint is, however, more than merely "content" with the in-evitability that any particular mathematical object tends to come to us along with the contingent scaffolding of the specific way in which it is presented to us, but has this inevitability built in to its very vocabulary, and in an elegant way, makes profound use of this. It will allow itself the further flexibility of viewing any mathematical object "as" a *representation* of the theory in which the object is contained to the proto-theory of modern mathematics, namely, to *set theory*.

My aim in this article is to address a few points about *mathematical objects* and *equality of mathematical objects* following the line of thought of the preceding paragraph. I see these "points" borne out by the doings of working mathematicians as they go about their daily business thinking about, developing, and communicating mathematics, but I haven't found them specifi-cally formulated anywhere.[2] I don't even see how questions about these issues can even be raised within the framework of vocabulary that people employ when they talk about *the foundations of mathematics* for so much of the literature on philosophy of mathematics continues to keep to certain staples: formal systems, consistency, undecidability, provability and unprovability, and rigor in its various manifestations.

To be sure, people have exciting things to talk about, when it comes to the list of topics I have just given. These issues have been the focus of dramatic encounters—famous "conversations," let us call them—that represent turning points in our understanding of what the very mission of mathematics should be. The ancient literature—notably, Plato's comment about how the

---

[1] Aristotle first uses this term in Book XI Chap 3 1061a line 29 of the *Metaphysics*; his discussion in Book XIII, Chap 2 begins to confront some of the puzzles the term poses.

[2] The faintest resonance, though, might be seen in the discussion in Books 13 and 14 of Aristotle's *Metaphysics* which hits at the perplexity of whether the so-called *mathematicals* (that ostensibly play their role in the platonic theory of forms) occur *uniquely* for each mathematical concept, or *multiply*.

mathematicians bring their analyses back to the *hypotheses* that they frame, *but no further*—already delineates this mission[3]. The early modern literature—epitomized by the riveting use that Kant made of his starkly phrased question "how is pure mathematics possible?"—offers a grounding for it. In the century just past, we have seen much drama regarding the grounds for mathematics: the Frege-Russell correspondence, the critique that L.E.J. Brouwer made of the modern dealings with infinity and with Cantor's set theory, Hilbert's response to this critique, leading to his magnificent invention of formal systems, and the work of Gödel, itself an extraordinary comment on the relationship between the mission of mathematics and the manner in which it formulates its deductions.

Formal systems remain our lingua franca. The general expectation is that any particular work we happen to do in mathematics should be, or at least should be capable of being, packaged within some formal system or other. When we want to legitimize our modes of operation regarding anything, from *real numbers* to *set theory* to *cohomology*—we are in the habit of invoking *axiomatic systems* as standard-bearer. But when it comes to a crisis of rigorous argument, the open secret is that, for the most part, mathematicians who are not focussed on the architecture of formal systems per se, mathematicians who are *consumers* rather than *providers*, somehow achieve a sense of utterly firm conviction in their mathematical doings, without actually going through the exercise of translating their particular argumentation into a brand-name formal system.

If we are shaky in our convictions as to the rigor of an argument, an excursion into formal systems is rarely the thing that will shore up faith in the argument. To be sure, it is often very helpful for us to write down our demonstrations very completely using pencil and paper or our all-efficient computers. In any event, no matter how wonderful and clarifying and comforting it may be for mathematician X to know that all of his or her proofs have, so far, found their expression within the framework of Zermelo-Frankel set theory, the chances are that mathematician X, if quizzed on what—exactly—those axioms are, might be at a loss to answer.

Of course, it is vitally important to understand, as fully as we can, what tools we need to assemble in order to justify our arguments. But to appreciate, and discuss, a grand view of the nature of mathematical objects that has taken root in mathematical culture during the past half-century, we must also become conversant with a language that has a thrust somewhat different from the standard fare of *foundations*. This newer vocabulary has phrases like *canonical isomorphism, unique up to unique isomorphism, functor, equivalence of category* and has something to say about every part of mathematics, including the definition of the natural numbers.

2 Defining Natural Numbers

Consider natural numbers; for instance, the number 5. Here are three approaches to the task of defining the number 5.

- We could, in our effort to define the number 5, deposit five gold bars in, say, Gauss's observatory in Göttingen, and if ever anyone wants to determine whether or not their set

---

[3] "I think you know that students of geometry, calculation, and the like hypothesize the odd and the even, the various figures, the three kinds of angles, and other things akin to these in each of their investigations, as if they know them. They make these their hypotheses and don't think it necessary to give any account of them, either to themselves or others, as if they were clear to everyone."   —[Plato 1997] *Republic* Book VI 510c.

has cardinality five, they would make a quick trip to Göttingen and try to put the elements of their set in one-one correspondence with the bullion deposited there. Of course, there are many drawbacks to this approach to defining the number "five," the least of which is that it has the smell of contingency. Let us call this kind of approach the *bureau of standards attitude towards definition:* one chooses a representative exemplar of the mathematical object one wishes to define, and then gives a criterion for any other mathematical object to be viewed as *equal* to the exemplar. There is, after all, something nice and crisp about having a single concrete exemplar for a mathematical concept.

- The extreme opposite approach to this is Frege's: define a **cardinality** (for example, five) as an equivalence class in the set of all sets, the equivalence relation $A \sim B$ being the existence of a one-one correspondence between A and B. The advantage, here, is that it is a criterion utterly devoid of subjectivity: no set is preferred and chosen to govern as benchmark for any other set; no choice (in the realm of sets) is made at all. The disadvantage, after Russell, is well known: the type of universal quantification required in Frege's definition, at least when the equivalence classes involved are considered to be sets, leads to paradox. The Frege-Russell correspondence makes it clear that one cannot, or at least one should not, be too greedy regarding unconditional quantification. To keep clear of immediate paradox, we introduce the word *class* into our discussion, amend the phrase *set of all sets* to *class of all sets*, and hope for the best.

- A fine compromise between the above two extremes is to do what we all, in fact, do: a strategy that captures the best features of both of the above approaches. What I mean here, by the way, is to indicate what we *do*, rather than what we *say we do* when quizzed about our foundations. I allow my notation $1, 2, 3, 4, 5, 6, \ldots$ to play the role of my personal bureau-of-standards within which I happily make my calculations. I think of the set $\{1, 2, 3, 4, 5\}$, for example, as a perfectly workable exemplar for quintuples. Meanwhile you use your notation $1', 2', 3', 4', 5', 6', \ldots$ (or whatever it is) to play a similar role with respect to your work and thoughts, the basic issue being whether there is a faithful translation of structure from the way in which you view natural numbers to the way I do.

Equivalence (of structure) in the above "compromise" is the primary issue, rather than *equality* of mathematical objects. Furthermore, it is the *structure* intrinsic to the whole gamut of natural numbers that plays a crucial role there. For only in terms of this structure (packaged, perhaps, as a version of Peano's axioms) do we have a criterion to determine when your understanding of "natural numbers," and mine, admit "faithful translations" one to another. A consequence of such an approach—which is the standard modus operandi of mathematics ever since Hilbert—is that any single mathematical object, say the number 5, is understood primarily in terms of the structural relationship it bears to the other natural numbers. Mathematical objects are determined by–and understood by—the network of relationships they enjoy with all the other objects of their species.

3 Objects versus Structure

Mathematics thrives on going to extremes whenever it can. Since the "compromise" we sketched above has "mathematical objects determined by the network of relationships they enjoy with all the other objects of their species," perhaps we can go to extremes within this compromise, by taking the following further step. Subjugate the role of the *mathematical object* to the role of its network of relationships—or, a further extreme—simply *replace* the mathematical object by this network.

This may seem like an impossible balancing act. But one of the elegant–and surprising—accomplishments of category theory is that it performs this act, and does it with ease.

4 Category Theory as Balancing Act Rather Than "Foundations"

There are two great modern formulas—as I'll call them—for packaging entire mathematical theories. There is the concept of **formal system**, following David Hilbert, as discussed above. There is also the concept of **category**, the great innovation of Samuel Eilenberg and Saunders MacLane. Now, these two formulas have vastly different intents.

A **formal system** representing a mathematical theory has, within it, all of the mechanics and vocabulary necessary to discuss proofs, and the generation of proofs, in the mathematical theory; indeed, that is mainly what a formal system is all about.

In contrast, a **category** is quite sparse in its vocabulary: it can say nothing whatsoever about proofs; a category is a mathematical entity that, in the most succinct of languages, captures the essence of what a mathematical theory consists: *objects of the theory*, allowable *transformations* between these objects, and a *composition law* telling us how to compose two transformations when the range of the first transformation is the domain of the second.

It stands to reason, then, that the concept of category cannot provide us with anything that goes under the heading of "foundations." Nevertheless, in its effect on our view of mathematical objects it plays a fine balancing role: it extracts—as I hope you will see—the best elements from both a Fregean and a bureau-of-standards attitude towards the formulation of mathematical concepts.

5 Example: The Category of Sets

Even before I describe *category* more formally, it pays to examine *the category of sets* as an example. The category of sets, though, is not just "an" example, it is the proto-type example; it is as much *an* example of a category as Odette is *un* amour de Swann.

The enormous complexity to set theory is one of the great facts of life of mathematics. I suppose most people before Cantor, if they ever had a flicker of a thought that *sets* could occur at all as mathematical objects, would have expected that a rather straightforward theoretical account of the notion would encompass everything that there was to say about those objects. As we all know, nothing of the sort has transpired.

The famous attitude of St. Augustine towards the notion of "time," (i.e., "What then is time? If no one asks me, I know what it is. If I wish to explain it to him who asks, I do not know.") mirrors my attitude—and I would suppose, most people's attitude—toward sets. If I retain my naive outlook on *sets,* all is, or at least seems to be, well; but once I embark on formulating the notion rigorously and specifically, I am either entangled, or else I am forced to make very contingent choices.

Keeping to the bare bones, a *set theory* will consist of

- the repertoire of **elements** of the theory, and if I wish to refer to one of them I will use a lower case symbol, e.g., $a, b, \ldots$

- the repertoire of **sets** of the theory, and for these I will use upper case symbols, e.g., $X, Y, \ldots$
- the relation of **containment** telling us when an element x is contained in a set X ($x \in X$); each set X is extensionally distinguished by the elements, $x \in X$, that are in it.
- the **mappings** $f : X \to Y$ between sets of the theory, each mapping f uniquely characterized by stipulating for every $x \in X$ the (unique) image, $f(x) \in Y$, of that element x.
- the guarantee that if $f : X \to Y$ and $g : Y \to Z$ are mappings in my theory, I can form the **composition** $g \cdot f : X \to Z$ by the rule that for any element $x \in X$ the value $(g \cdot f)(x) \in Z$ is just $g(f(x))$.

We neither lose nor gain anything by adding the requirement that, for any object X, the identity mapping $1_X : X \to X$ is a bona fide mapping in our set theory, so for convenience let us do that. Also, we see that our composition rule is *associative* in the standard sense of multiplication, i.e., $(h \cdot g) \cdot f = h \cdot (g \cdot f)$, when these compositions can be made, and that our identity mappings play the role of "unit."

Much has been omitted from this synopsis–all traces of quantifications, for example— and certain things have been hidden. The repeated use of the word *repertoire* is already a hint that something big is being hidden. It would be downright embarrassing, for example, to have replaced the words "repertoire" in the above description by "set," for besides the blatant circularity, we would worry, with Russell, about what arcane restrictions we would have then to make regarding our universal quantifier, once that is thrown into the picture. "Repertoire" is my personal neologism; the standard word is *class* and the notion behind it deserves much discussion; we will have some things to say about it in the next section. You may notice that I refrained from using the word "repertoire" when talking about mappings. A subtle issue, but an important one, is that we *may* boldly require that all the mappings from a given set X to a given set Y form a bona fide *set* in our theory, and not merely an airy repertoire. This is a source of power, and we adopt it as a requirement. Let us refer, then, to a theory such as we have just sketched as a **bare set theory**.

A *bare set theory* can be stripped down even further by forgetting about the *elements* and a fortiori the *containment relations*. What is left?

We still have the *objects* of the theory, i.e., the repertoire (synonymously: class) of its sets. For any two sets X and Y we have the *set* of mappings from X to Y; and for any three sets X, Y, Z and two mappings $f : X \to Y$ and $g : Y \to Z$ we have the mapping that is the composition of the two, $g \cdot f : X \to Z$, this composition rule admitting "units" and satisfying the associative law.

This *further-stripped-down bare set theory* is our first example of a category: it is the underlying **category** of the bare set theory.

The concept of *class* which will occur in the definition of category, and has already occurred in our proto-example, now deserves some discussion.

6 Class *as a Library With Strict Rules for Taking Out Books*

I'm certain that there are quite precise formulations of the notion of *class*, but here is a ridiculously informal user's-eye-view of it. Imagine a library with lots of books, administered by a somewhat stern librarian. You are allowed to take out *certain* subcollections of books in the library, but not all. You know, for example, that you are forbidden to take out, at one go, *all* the books of the library. You assume, then, that there are other subcollections of books that would be similarly restricted. But the full bylaws of this library are never to be made completely explicit. This doesn't bother you overly because, after all, you are interested in reading, and not the legalisms of libraries.

In observing how mathematicians tend to use the notion *class*, it has occurred to me that this notion seems really never to be put into play without some *background* version of set theory understood already. In short by a *class*, we mean a collection of objects, with some restrictions on which subcollections we, as mathematicians, can deem *sets* and thereby operate on with the resources of our set theory. I'm perfectly confident that this seeming circularity can be–and probably has been–ironed out. But there it is.

7 Category

A category C is intrinsically a relative notion for it depends upon having a set theory in mind; a *bare set theory* such as sketched above will do.

Fixing on a "bare set theory," a **category** C (*modeled* on this bare set theory) is given by the following[4]:

- a *class* of things called **the objects of** C and denoted $\mathrm{Ob}(C)$;
- given any two objects X, Y of C, a *set* denoted $\mathrm{Mor}_C(X, Y)$, which we think of as the set of *transformations* from the object X to the object Y; we refer to these transformations as **morphisms from** X **to** Y and usually denote such a morphism f as a labelled arrow $f : X \to Y$;
- given any three objects X, Y, Z of C and morphisms $f : X \to Y$ and $g : Y \to Z$ we are provided with a law that tells us how to "compose" these morphisms to get a morphisms

$$g \cdot f : X \to Z.$$

Intuitively, we are thinking of f and g as "transformations," and *composition* of them means that we imagine "first" applying f to get us from X to Y and "then" applying g to get us from Y to Z. The rule that associates to such a pair (f, g) the composition,

$$(f, g) \mapsto g \cdot f$$

we think of as a sort of "multiplication law."

One also requires that morphisms playing the role of "identity elements" 1_X in $Mor_C(X, X)$ with respect to this composition law exist; that is, for any morphism $f : X \to Y$ we have $f \cdot 1_X = f$; and similarly, for any morphism $e : V \to X$ we have $1_X \cdot e = e$. Finally the composition law is assumed to be associative, in the evident sense.

As for the word *class* that enters into the definition, we will, at the very least, want, for any object X in our category, that the singleton set consisting of that object,$\{X\}$, be viewed as a bona fide set of our set theory.

---

[4] Category theorists will note that I am restricting my attention to what are called *locally small categories*.

This concept of *category* is an omni-purpose affair: we have our categories of sets, where the objects are sets, the morphisms are mappings of sets; we have the category of *topological spaces* whose objects are the eponymous ones, and whose morphisms are continuous maps. We have the algebraic categories: the category of *groups* where the morphisms are homomorphisms, the category of *rings with unit element* where the morphisms are ring homomorphisms (that preserve the unit element), etc. Every branch and sub-branch of mathematics can package their entities in this format. In fact, at this point in its career it is hard to say whether the role of category in the context of mathematical work is more descriptive, or more prescriptive. It frames a possible template for any mathematical theory: the theory should have *nouns* and *verbs*, i.e., objects, and morphisms, and there should be an explicit notion of composition related to the morphisms; the theory should, in brief, be packaged by a category. There is hardly any species of mathematical object that doesn't fit into this convenient, and often enlightening, template.

Template is a crucial feature of categories, for in its daily use, a category avoids any really detailed discussion of its underlying set theory. This clever manner in which category theory engages with set theory shows, in effect, that it has learned the Augustinian lesson. Category theory doesn't legislate *which* set theory we are to use, nor does it even give ground-rules for what "a" *set theory* should be. As I have already hinted, one of the beautiful aspects of category theory is that it is up to you, the category-theory-user to supply "a" set theory, a bare category of sets S, for example. A category is a B.Y.O.S.T. party, i.e., you *bring your own set theory* to it. Or, you can adopt an even more curious stance: you can view S as something of a free variable, and consequently, end up by making no specific choice!

So, for example, "the" category of rings with unit element is, more exactly, a mold that you can impress on any bare set theory. To be sure, you want your set theory to be sufficiently rich so as to hold this impression: if there were no sets at all in your *set theory*, you wouldn't get much.

You might wonder why the framers of the notion of category bothered to use two difficult words *class* and *set* rather than only one, in their definition. One could, after all, simply require that there be a *set* of objects of the category, rather than bring in the airy word *class*. In fact, people do that, at times: it is standard to call a category whose objects form a *set* a **small category**, and these small categories do have their uses. Indeed, if you are worried about foundational issues, it is hardly a burden to restrict attention to small categories. But I think the reason that the notion of class is invoked has to do with the high ambition we have for categories: categories are meant to offer a fluid vocabulary for whole 'fields of mathematics' like group theory or topology, with a Fregean desire for freedom from the contingency implicit in subjective choices.

8 Equality versus Isomorphism

The major concept that replaces *equality* in the context of categories is *isomorphism*. An **isomorphism** $f : A \to B$ between two objects A, B of the category C is a morphism in the category C that can be "undone," in the sense that there is another morphism $g : B \to A$ playing the role of the inverse of f; that is, the composition $gf : A \to A$ is the identity morphism 1_A and the composition $fg : B \to B$ is the identity morphism 1_B. The essential lesson taught by the categorical viewpoint is that it is usually either quixotic, or irrelevant, to ask if a certain object X in a category C is *equal* to an object Y. The query that is usually pertinent is to ask for a *specific isomorphism* from X to Y.

Note the insistence, though, on a *specific* isomorphism; although it may be useful to be merely assured of the *existence* of isomorphisms between X and Y, we are often in a much better position if we can pinpoint a specific isomorphism $f : X \to Y$ *characterized by an explicitly formulated property, or list of properties.* In some contexts, of course, we simply have to make do without being able to pinpoint a specific isomorphism. If, for example, I manage to construct an algebraic closure of the finite field $\mathbf{F}_2$ (i.e., of the field consisting of two elements), and am told that someone halfway around the world has also constructed such an algebraic closure, I *know* that there exists an isomorphism between the two algebraic closures but–without any further knowledge—I have no way of pinpointing a specific isomorphism. In contrast, desipte my ignorance of the manner in which my colleague at the opposite end of the world went about constructing her algebraic closure, I can, with utter confidence, put my finger on a *specific* isomorphism between the group of automorphisms of my algebraic closure and the group of automorphisms of the other algebraic closure[5]. The fact that the algebraic closures are not yoked together by a *specified* isomorphism is the source of some theoretical complications at times, while the fact that their automorphism groups are seen to be isomorphic via a cleanly specified isomorphism is the source of great theoretical clarity, and some profound number theory.

A uniquely specified isomorphism from some object X to an object Y characterized by a list of explicitly formulated properties—this list being sometimes, the truth be told, only implicitly understood—is usually dubbed a "canonical isomorphism." The "canonicality" here depends, of course, on the list. It is this brand of *equivalence*, then, that in category theory replaces *equality*: we wish to determine objects, as people say, *"up to canonical isomorphism."*

9 *An Example of Categorical Vocabulary: Initial Objects*

We also have at our immediate disposal, a broad range of concepts that can be defined purely in terms of the structure that we have already elucidated. For example, if we are given a category $\mathcal{C}$, an **initial object** Z of $\mathcal{C}$ is an object Z that has the property that given *any* object X of $\mathcal{C}$ there is a unique morphism of the category $i_X : Z \to X$ from Z to X; that is, the set $Mor_{\mathcal{C}}(Z, X)$ consists of the single element $\{i_X\}$.

Suppose that a category $\mathcal{C}$ has an *initial object* Z. There may, very well, be quite a number of objects vying for the role of *initial object of this category* $\mathcal{C}$. But given another contender, call it Z', there is a unique morphism $i_{Z'} : Z \to Z'$ since Z is an initial object, and a unique morphism $i'_Z : Z' \to Z$ since Z' is. Also, again since Z is an initial object, there can only be one morphism from Z to Z, and the identity morphism $\mathbf{1}_Z : Z \to Z$ fills this role just fine, so we must have that $i'_Z \cdot i_{Z'} = \mathbf{1}_Z$ and, for similar reasons, $i_{Z'} \cdot i'_Z = \mathbf{1}_{Z'}$. In summary, $i_{Z'}$ and i'_Z are (inverse) isomorphisms, and provide us with *canonical, in fact the only, isomorphisms* between Z and Z'. One way of citing this is to say, as people do, that the initial object of a category—if it exists—is *unique up to unique isomorphism.* To be sure it is not unique as "object," but rather, as "something else." It is this difference, what the "something else" consists in, that we are exploring.

---

[5] for these automorphism groups are both topologically generated by the field automorphism consisting of squaring every element.

10 *Defining the Natural Numbers as an "Initial Object"*

For this discussion, let us start by considering "the" initial object in the category of rings with unit. As we shall see, such an initial object does exist, given that the underlying bare set theory is not ridiculously impoverished. Such an initial object is "unique up to unique isomorphism," as all initial objects are. What is it?

Well, by the definition of initial object, it must be a ring **Z** (with a unit element) that admits a unique ring homomorphism (preserving unit elements) to any ring with unit. Since *the ring of ordinary integers* **Z** has precisely this property (there being one and only one ring homomorphism from **Z** to any ring with unit, the one that sends $1 \in Z$ to the unit of the range ring) "the" initial object in the category of rings with unit is nothing more nor less than **Z** but, of course, only "up to unique isomorphism."

The previous paragraph situated **Z** among its fellow rings with unit element. Let us fashion a similar discussion for the Natural Numbers, highlighting the type of structure that Peano focussed on, when formulating his famous axioms.

For this, I want to define a category denoted $\mathcal{P}$ that I will call **the Peano category**.

The objects, $Ob(\mathcal{P})$, of the Peano category consists of triples (X, x, s) where X is a set, $x \in X$ is an element (call it a **base point**), and $s : X \to X$ is a mapping of X to itself (a "self-map" which we might call the *successor map*).

Given two objects $\mathcal{X} = (X, x, s)$ and $\mathcal{Y} = (Y, y, t)$ of $\mathcal{P}$, a morphism

$$F : (X, x, s) \to (Y, y, t)$$

in the category $\mathcal{P}$ is a mapping of sets $f : X \to Y$ with the property that

- f preserves base points; i.e., $f(x) = y$, and
- f respects the self-maps s and t, in the sense that $f \cdot s = t \cdot f$, i.e., we have for all elements $z \in X$, $f(s(z)) = t(f(z))$.

We will denote by $\mathrm{Mor}_\mathcal{P}(X, Y)$ the set of morphisms of the Peano category from X to Y, i.e., the set of such F's.

For any choice of bare set theory, we have thereby formed the category which we will call $\mathcal{P}$. If our bare set theory, on which the category is modeled, is at all decent—e.g., is one of the standard set theories containing the set of natural numbers $\mathcal{N} = \{1, 2, 3, \dots\}$, then $\mathcal{N}$ may be viewed as an object of $\mathcal{P}$, its *base point* being given by $1 \in \mathcal{N}$, and the self-map $s : \mathcal{N} \to \mathcal{N}$ being given by the rule that sends a natural number to its successor, i.e., $n \mapsto n + 1$.

Given any object $\mathcal{X} = (X, x, s)$ in Ob $\mathcal{P}$ *there is one and only one* morphism from $\mathcal{N}$ to $\mathcal{X}$ in the category $\mathcal{P}$; it is given by the mapping of sets sending $1 \in \mathcal{N}$ to the base point $x \in X$ (for, indeed, any morphism in $\mathcal{P}$ is required to send base point to base point) and the mapping is then "forced," from then on, by the formula $f(n + 1) = sf(n)$.

In summary, there is a unique morphism in $\mathcal{P}$ from the natural numbers to any object in the category. That is, the natural numbers, $\mathcal{N}$, is an initial object of $\mathcal{P}$.

Moreover, as any initial object in any category is uniquely characterized, up to unique isomorphism, by its role *as initial object*, the natural numbers when viewed as *initial object of* $\mathcal{P}$ is similarly pinned down.

This strategy of defining the *Natural Numbers* as "an" initial object in a category of (what amounts to) discrete dynamical systems, as we have just done, is revealing, I think; it isolates, as

Peano himself had done, the fundamental role of mere *succession* in the formulation of the natural numbers. It also follows the third of the three formats we listed for defining natural numbers; it is, in a sense that deserves to be understood, a compromise strategy between a bureau-of-standards kind of definition, and a Fregean universal quantification approach. Notice, though, its elegant shifting sands. At the very least, we have a definition that depends upon a selection of *a* set theory, as well as an agreement to deal with the object Z pinned down "up to unique isomorphism." We have even further to go, but first let us discuss how our definition differs in approach from the standard way of expressing Peano's axioms.

11 Light, Shadow, Dark

In elementary mathematics classes, we usually describe Peano's axioms that characterize the natural numbers roughly as follows.

> The natural numbers $\mathcal{N}$ is a set with a chosen element $1 \in \mathcal{N}$ and an injective ("successor") function $s : \mathcal{N} \to \mathcal{N}$ such that $1 \notin s(\mathcal{N})$ and such that **mathematical induction** holds, in the sense that if $P(n)$ is any proposition which may be formulated for all $n \in \mathcal{N}$, and for which $P(1)$ is true, and which has the further property that whenever $P(n)$ is true then $P(s(n))$ is true, then $P(n)$ is true for all $n \in \mathcal{N}$.

This, of course, has shock-value: it recruits the entire apparatus of propositional verification to its particular end. The fact that, especially when taken broadly, mathematical induction has extraordinary consequences, is amply illustrated by the ingenious work of Gentzen.[6] To formalize things, we tame these axioms by explicitly providing a setting in which the words *proposition* and *true* make sense.

The easiest way of comparing the *Peano axioms* with the *Peano category* as modes of defining natural Numbers, is to ask what each of these formats

- shines a *spotlight* on?
- keeps in the *shadows*?

  and

- keeps in the *dark*?

---

[6] Gentzen developed a *normal form* for propositions and deductions in (Peano) arithmetic, and he noted that if it were permitted to employ—in one's demonstrations— a version of *mathematical induction* that ranges over *all* demonstrations in arithmetic (these demonstrations being organized according to their natural partial ordering) one can actually *prove* the *consistency of arithmetic*; see [Gentzen 1936] and [Gentzen 1938].

To be sure, there is an inherent circularity issue here, beyond the fact that one is calling forth an unusually powerful version of mathematical induction, but Gentzen's ideas are not the less interesting for all this. His tactic was to assume that the line

$$\text{“}0 = 1\text{”}$$

occurred in a demonstration expressed in normal form, and then to examine what the line immediately preceding $0 = 1$ in this putative demonstration could possibly be. From the structure of *normal form demonstrations* one sees that there could be no such line, and as a consequence, one could never deduce a contradiction in arithmetic by a demonstration that has been expressed in *normal form*.

Both ways of pinpointing *natural numbers* are fastidiously explicit about the fact that a certain discrete dynamical system is involved: each shine their spotlight on the essence of *iteration*, the successor function. The Peano axioms do this by focussing in somewhat more detail on the elementary properties of this successor function s, requiring as those axioms do, that 1 not be in the image of s, and that s be injective. The Peano category approach does this by simply considering the entire species of discrete dynamical systems with chosen base point.

Both modes of definition need a way of insisting on a certain "minimality" for the structure of natural numbers that they are developing. The Peano axioms formulate this "minimality" by dependence upon the domino effect of truth in a mathematically inductive context. The Peano category approach formulates "minimality" by considering the position of the natural numbers as a discrete dynamical system, among all discrete dynamical systems.

The Peano axiom approach calls up the full propositional apparatus of mathematics. But the details of the apparatus are kept in the shadows: you are required to "bring your own" propositional vocabulary if you wish to even begin to flesh out those axioms. The Peano category approach keeps all this in the dark: no mention whatsoever is made of propositional language.

The Peano axiom approach requires—at least explicitly—hardly any investment in some specific brand of set theory. At most one set is on the scene, the set of natural numbers itself. In contrast, the Peano category approach forces you to "bring your own set theory" to make sense of it.

When we gauge the differences in various mathematical viewpoints, it is a good thing to contrast them not only by what equipment these viewpoints ultimately invoke to establish their stance, for *ultimately* they may very well require exactly the same things, but also to pay attention to the order in which each piece of equipment is introduced and to the level of explicitness required for it to play its role.

12 Representing One Theory in Another

If categories package entire mathematical theories, it is natural to look for the shadow of one mathematical theory (as packaged by a category C) in another mathematical theory (as packaged by a category D). We might do this by establishing a "mapping" of the entire category C to the category D. Such a "mapping" should, of course, send basic features (i.e., objects, morphisms) of C to corresponding features of the category D, and moreover, it must relate the composition law of morphisms in C to the corresponding law for morphisms of D ; we call such a "mapping" a **functor** from C to D.

To give a functor F from C to D, then, we must stipulate how we associate to any object X of C a well-defined object $F(X)$ of D, and to any morphism between objects $f : X \to Y$ of C a well-defined morphism $F(f) : F(X) \to F(Y)$ between corresponding objects of D; and, as mentioned, this relationship between objects and morphisms in C to objects and morphisms in D must respect identity morphisms, and the composition laws of these categories[7]. Let us denote

---

[7] By definition, then, a functor F from C to D associates to each object U of C an object of D, call it $F(U)$; and to each morphism $h : U \to V$ of C, a morphism of D, call it $F(h) : F(U) \to F(V)$. If the morphism $\mathbf{1}_U : U \to U$ is the identity, then we require that $F(\mathbf{1}_U) = \mathbf{1}_{F(U)}$, i.e., that it be the identity morphism of the object $F(U)$ in D. When we say

such a functor F from C to D by a broken arrow

$$F : C \; -- \rightarrow \; D.$$

In this way, we have a vocabulary for establishing bridges between whole disciplines of mathematics; we have a way of representing grand aspects of, say, topology in algebra (or conversely) by establishing functors from the category of topological spaces to the category of groups (or conversely): construct the pertinent functors from the one category to the other!

The easiest thing to do, at least in mathematics, is to forget, and the forgetting process offers us some elementary functors, such as the functor from *topological spaces* to *sets* that passes from a topological space to its underlying set, thereby forgetting its topology. Of course one should also pay one's respects to the simplest of functors, the *identity functor*

$$C \overset{1_C}{--\rightarrow} C,$$

which when presented with any object U of C it gives it back to you intact, as it does each morphism.

The more profound bridges between fields of mathematics are achieved by more interesting constructions. But there is a ubiquitous type of functor, as easy to construct as one can imagine, and yet extraordinarily revealing. Given any object X in any category C we will construct (in section 14) an important functor (we will denote it F_X) from C to S, the category of sets upon which C was modeled. This functor F_X will be enough to "reconstruct" X, but—as you might guess—only "up to canonical isomorphism."

But before we do this, we need to say what we mean by a morphism from one functor to another.

13 Mapping One Functor to Another

If we are given two functors,

$$F, G : C \; -- \rightarrow \; D,$$

by a **morphism of functors**

$$\mu : F \longrightarrow G$$

we mean that we are given, for each object U of C, a morphism

$$\mu_U : F(U) \rightarrow G(U)$$

in the category D which respects the structures involved. [8]

---

that the functor F *respects* composition laws we mean that if $g : V \rightarrow W$ is a morphism of C (so that we can form the composition $g \cdot h : U \rightarrow W$ in the category C) we have the law

$$F(g \cdot h) = F(g) \cdot F(h) : F(U) \rightarrow F(W).$$

[8] in the sense that for every pair of objects U, V of C, and morphism $h : U \rightarrow V$ in $\mathrm{Mor}_C(U, V)$ we have the equality

$$G(h) \cdot \mu_U = \mu_V \cdot F(h),$$

both left- and right-hand side of this equation being morphisms $F(U) \rightarrow G(V)$ in the category D.

To offer a humble example, for any functor

$$F : \mathcal{C} \; -- \to \mathcal{D},$$

we have the *identity morphism of the functor F to itself,*

$$F \xrightarrow{1_F} F$$

which associates, to an object U of $\mathcal{C}$, the identity morphism $1_{F(U)} : F(U) \to F(U)$ (this being the identity morphism of the object $F(U)$ in the category $\mathcal{D}$). You might think that this example is not very enlightening, but it already holds its surprises; in any event, in the next section we shall visit an important large repertoire of morphisms of functors.

Once we have settled on the definition of *morphism* of functors, our way is clear to define **isomorphism of functors** for the definition of this notion follows the natural format for the definition of isomorphism related to absolutely *any* species of mathematical object. Namely, an *isomorphism of the functor $F : \mathcal{C} -- \to \mathcal{D}$ to the functor $G : \mathcal{C} -- \to \mathcal{D}$* is a morphism of functors $\mu : F \to G$ for which there is a morphism of functors going the other way, $\nu : G \to F$ such that $\nu \cdot \mu : F \to F$ and $\mu \cdot \nu : G \to G$ are equal to the respective identity morphisms (of functors).

To understand the notion of *isomorphism of functors* I find it particularly illuminating to consider, for the various categories of interest, what the *automorphisms of the identity functor* consist of. Note, to take a random example, that if $\mathcal{V}$ is the category of vector spaces over a field k, then multiplication by any nonzero scalar (i.e., element of k^*) is an automorphism of the identity functor. That is, let $1_\mathcal{V} : \mathcal{V} -- \to \mathcal{V}$ denote the identity functor; for any fixed nonzero scalar $\lambda \in k^*$ we can form (for all vector spaces U over k) the morphism in $\mathcal{V}$,

$$\lambda_U : U \to U$$

defined by $x \mapsto \lambda \cdot x$, and this data can be thought of as giving an isomorphism of functors

$$\lambda : 1_\mathcal{V} \cong 1_\mathcal{V}.$$

14 An Object "as" a Functor from the Theory-in-Which-it-Lives to Set Theory

Given an object X of a category $\mathcal{C}$, we shall define a specific functor (that we will denote F_X) that encodes the essence of the object X. The functor F_X will, in fact, determine X up to canonical isomorphism.

This functor F_X maps the category $\mathcal{C}$ to the category $\mathcal{S}$ of sets (the same category of sets on which $\mathcal{C}$ is "modeled," as we've described in section 7 above).

Here is how it is defined. The functor F_X assigns to any object Y of $\mathcal{C}$ the set of morphisms from X to Y; that is,

$$F_X(Y) := \mathrm{Mor}_\mathcal{C}(X, Y).$$

Now, $\mathrm{Mor}_\mathcal{C}(X, Y)$ is indeed a *set,* i.e., an object of $\mathcal{S}$, so we have described a mapping from *objects of $\mathcal{C}$* to *sets,*

$$Y \mapsto F_X(Y) = \mathrm{Mor}_\mathcal{C}(X, Y).$$

Moreover, to every morphism $g : Y \to Z$ of $\mathcal{C}$, our functor F_X assigns the mapping of sets

$$F_X(g) : F_X(Y) = \text{Mor}_{\mathcal{C}}(X, Y) \longrightarrow F_X(Z) = \text{Mor}_{\mathcal{C}}(X, Z)$$

given simply by composition with g; i.e., if $f \in F_X(Y)$, the mapping $F_X(g)$ sends f to $g \cdot f \in F_X(Y)$:

$$f \mapsto g \cdot f.$$

In this way, every object X of any category $\mathcal{C}$ gives us a functor, F_X from $\mathcal{C}$ to $\mathcal{S}$.

Also any morphism $h : X' \to X$ in $\mathcal{C}$ gives rise to a morphism of functors $\eta : F_X \to F_{X'}$ by this simple formula: for an element Y of $\mathcal{C}$, the morphism h gives us a mapping of sets $\eta_Y : F_X(Y) \to F_{X'}(Y)$ by sending any $f : X \to Y$ in $F_X(Y)$ to $f \cdot h : X' \to Y$ in $F_{X'}(Y)$. The rule associating to an object Y the mapping of sets η_Y produces our morphism of functors $\eta : F_X \to F_{X'}$.

The fundamental, but miraculously easy to establish, fact is that the object X is entirely retrievable (however, only up to canonical isomorphism, of course) from knowledge of this functor F_X. This fact, a consequence of a result known as Yoneda's Lemma, can be expressed this way:

Theorem: *Let X, X' be objects in a category $\mathcal{C}$. Suppose we are given an isomorphism of their associated functors $\eta : F_X \cong F_{X'}$. Then there is a unique isomorphism of the objects themselves,*

$$h : X' \cong X$$

that gives rise—as in the process described above—to this isomorphism of functors.

The beauty of this result is that it has the following decidedly structuralist, or Wittgensteinian language-game, interpretation:

> *an object X of a category $\mathcal{C}$ is determined (always only up to canonical isomorphism, the recurrent theme of this article!) by the network of relationships that the object X has with all the other objects in $\mathcal{C}$.*

Yoneda's lemma, in its fuller expression, tells us that the set of morphisms (of the category $\mathcal{C}$) from an object X to an object Y is naturally in one-one correspondence with the set of morphisms of the functor F_Y to the functor F_X.

In brief, we have (or rather, Yoneda has) reconstructed the category $\mathcal{C}$, objects and morphisms alike, purely in terms of functors to sets; i.e., in terms of networks of relationships that deal with the entire category at once.[9]

With all this, Yoneda's Lemma is one of the many examples of a mathematical result that is both extraordinarily consequential, and also extraordinarily easy to prove.[10]

---

[9] The connection between Yoneda's lemma and structuralist and/or Wittgensteinian attitudes towards *meaning* was discussed in Michael Harris's review of *Mathematics and the Roots of Postmodern Theory* [Harris 2003]

[10] A full proof, for example, is given neatly and immediately via a single diagram in the "wikipedia entry" (en.wikipedia.org/wiki/Yoneda's_lemma). For an accessible introductory reference to the ideas of category theory, see the article by Daniel K. Biss [Biss 2003]. For a more technical, but still relatively gentle, account of category theory, see Saunders MacLane's *Categories for the working mathematician* [Mac Lane 1971].

15 Representable Functors

The following definition (especially as it pervades the mathematical work of Alexander Grothendieck) marked the beginning of a significantly new viewpoint in our subject.

> A functor $F : \mathcal{C} \; - - \rightarrow \mathcal{S}$, from a category $\mathcal{C}$ to the category of sets $\mathcal{S}$ on which it is modeled, is said to be **represented** by an object X of $\mathcal{C}$ if an isomorphism of functors $F \cong F_X$ is given. The functor F is said, simply, to be **representable** if it can be represented by some object X.

If you consult the theorem quoted at the end of the last section you see that Yoneda's lemma, then, guarantees that *if* a functor F is representable, then F determines the object X that represents it up to unique isomorphism.

One of the noteworthy lessons coming from subjects such as algebraic geometry is that often, when it is important for a theory to make a construction of a particular object that performs an important function, we have a ready description of the functor F that it would represent, if it exists. Often, indeed, the basic utility of the object X that represents this functor F comes exactly from that: that X represents the functor![11] Although a specific construction of X may tell us more about the particularities of X, there is no guarantee that all the added information a construction provides—or any of it—furthers our insight beyond guaranteeing representability of F.

Some of the important turning points in the history of mathematics can be thought of as moments when we achieve a fuller understanding of what it means for one "thing" to *represent* another "thing." The issue of *representation* is already implicit in the act of *counting,* as when we say that *these two mathematical units "represent" those two cows.* Leibniz dreamed of a scheme for a universal language that would reduce ideas "to a kind of alphabet of human thought" and the ciphers in his universal language would be manipulable *representations* of ideas.

Kant reserved the term *representation* (Vorstellung) for quite a different role. Here is the astonishing way in which this concept makes its first appearance in the *Critique of Pure Reason:*[12]

> There are only two possible ways in which synthetic representations and their objects ... can meet one another. Either the object (*Gegenstand*) alone must make the representation possible, or the representation alone must make the object possible.

It is this *either-or*, this dance between *object* and *representation*, that animates lots of what follows in Kant's *Critique of Pure Reason.* With meanings quite remote from Kant's, the same two terms, *object* and *representation*, each provide grounding for the other, in our present discussion.

Nowadays, whole subjects of mathematics are seen as represented in other subjects, the "represented" subject thereby becoming a powerful tool for the study of the "representing" subject, and vice versa.

It sometimes happens that the introduction of a term in a mathematical discussion is the signal that an important shift of viewpoint is taking place, or is about to take place. An emphasis

---

[11] Students of algebra encounter this very early in their studies: the *tensor product* is (happily) nowadays first taught in terms of its functorial characterization, with its construction only coming afterwards; this is also the case for *fiber products*, for *localization in commutative algebra*; indeed this is the pattern of exposition for lots of notions in elementary mathematics, as it is for many of the grand constructions in modern algebraic geometry.

[12] [Kant 1961], p. 125.

on "representability" of functors in a branch of mathematics suggests an ever so slight, but ever so important, shift. The lights are dimmed on *mathematical objects* and beamed rather on the corresponding functors; that is, on the networks of relationships entailed by the objects. The functor has center stage, the object that it represents appears almost as an afterthought. The lights are dimmed on on *equality of mathematical objects* as well, and focussed, rather, on canonical isomorphisms, and equivalence.

16 The Natural Numbers as Functor

Allow me to define, for *any* category, a particularly humble functor. If C is a category with underlying set theory S define a functor

$$I : C - - \to S$$

as follows:

If X is an object of C, let

$$I(X) := \{X\};$$

that is, the set $I(X)$ is the singleton consisting in the set containing only one element: the object X. If $f : X \to Y$ is any morphism in C, $I(f) : \{X\} \to \{Y\}$ is the unique mapping of singleton sets. We may think of our functor I as a *singleton functor*: it is a functor from C to the category of set that assigns to each object of the category C a singleton set. Any two "singleton functors" are (uniquely) isomorphic as functors. Is our functor I representable?

The answer here is clean. The functor I is representable if and only if our category C has an initial object. For if Z is an initial object, then F_Z, by the very meaning of *initial object*, is a singleton functor (there is a unique morphism from Z to any object X of the category). Therefore F_Z is isomorphic as a functor to I. Conversely, any object that represents I has the feature that it needs for us to deem it an initial object of C.

This viewpoint gives us a way of pinning down the natural numbers from a different angle, which at first glance may seem quite strange.

> *The natural numbers are defined uniquely, up to unique isomorphism, as an object of the Peano category P that represents the singleton functor I.*

There is aspect to this definition that Frege might have liked: nothing "bureau-of-standards-like," nothing that smacks of a subjective choice of some particular exemplar, has entered this description. But where, in this definition, are the tangible, familiar, natural numbers? You may well ask this question; for—despite the crispness of the above definition—the concept embodied by the good old symbols $1, 2, 3, \ldots$ appears to have holographically smeared itself over the panoply of little "discrete dynamical systems" given by the objects of P. And the category P itself, remember, is but a template, dependent upon an underlying set theory. But we have even further to go.

17 Equivalence of Categories

If the grand lesson is that *equivalence* has some claim to priority over *equality* in the mathematical theories packaged by categories, why are *categories* themselves untouched by this insight? The answer is that they are not. With this brief Q & A, to say nothing of the title of this section, you will not be surprised to find that what is next on the agenda is

Definition. *A functor* $F : \mathcal{C} -- \to \mathcal{D}$ *from the category* $\mathcal{C}$ *to* $\mathcal{D}$ *is called an* **equivalence of categories** *if there is a functor going the other way,* $G : \mathcal{D} -- \to \mathcal{C}$ *such that* $G \cdot F$ *is isomorphic to the identity functor from* $\mathcal{C}$ *to* $\mathcal{C}$, *and* $F \cdot G$ *is isomorphic to the identity functor from* $\mathcal{D}$ *to* $\mathcal{D}$.

and that we are specifically interested in the nature of many of our categories, *only up to equivalence*. So with this elementary vocabulary, entire theories are allowed to shift—up to equivalence.

18 *Object and Problem*

Following Kronecker, we sometimes allow ourselves to think, say, of $\sqrt{2}$ as nothing more than a cipher that obeys the standard rules of arithmetic and about which *all we know* is that its square is 2. This characterization, to be sure, doesn't pin it down, for $-\sqrt{2}$ has precisely the same description. Nevertheless, there is no contradiction here, for having named our cipher $\sqrt{2}$ we have given birth to a specific creature of mathematics, and $-\sqrt{2}$ is just another creature with (evidently!) a different name. It is a clarifying move (in fact, the essence of algebra) to usher into the mathematical arena, and to name, certain mathematical objects that are unspecified beyond the sole fact that they are a *solution to a certain explicit problem*; in this case: a solution to the polynomial equation $X^2 = 2$.

When we do such a thing, what is sharply delineated is the *problem,* the object being a tag for (a solution to) the problem.

In the same spirit, any functor, explicitly given, from a category $\mathcal{C}$ to the category of sets $\mathcal{S}$ that the category is modeled on,

$$F : \mathcal{C} -- \to \mathcal{S}$$

may be construed as formulating an *explicit problem:*

Problem: *Find "an" object X of the category C together with an isomorphism of functors*

$$\iota : F_X \cong F.$$

In a word, solve the above problem for the unknown X. To be sure, if we find two solutions,

$$\iota : F_X \cong F \text{ and } \iota' : F_{X'} \cong F,$$

then

$$\iota^{-1} \cdot \iota' : F_{X'} \cong F_X$$

is an isomorphism of representable functors and so, by Yoneda's Lemma, is induced from a unique isomorphism

$$X' \cong X;$$

i.e., the solution is unique, up to unique isomorphism.

The moral here, is that it is the *problem* that is explicit, while the *object* (that represents the solution of the problem) follows the theme of this essay: it is *unique up to unique isomorphism.*

19 Object and Equality

The habitual format for discussions regarding the grounding of mathematics shines a bright light on modes of formulating assertions, organizing and justifying proofs of those assertions, and on setting up the substrate for it all—which is invariably a specific set theory. In doing this a battery of choices will be made. These choices smack of contingency, of viewing the clarity of mathematics through some *subjective lens* or other.

I imagine that all of us want to ignore—when possible—the contingent, and seize the essential, aspect of any idea. If we are of the make-up of Frege, who relentlessly strove to rid mathematical foundations of subjectivism (Frege excoriated the writings of Husserl—incorrectly, in my opinion—for ushering *psychologism* into mathematics), we look to *universal quantification* as a possible method of effacing the contingent— drowning it, one might say, in the sea of *all* contingencies. But this doesn't work.

A stark alternative—the viewpoint of categories— is precisely to *dim* the lights where standard mathematical foundations shines them brightest. Instead of focussing on the question of modes of justification, and instead of making any explicit choice of set theory, the genius of categories is to provide a vocabulary that keeps these issues at bay. It is a vocabulary that can say nothing whatsoever about proofs, and that works with any—even the barest–choice of a set theoretic language, and that captures the essential template nature of the mathematical concepts it studies, showing these concepts to be—indeed—separable from modes of justification, and from the substrate of ever-problematic set theory. *Separable* but not forever separated, effecting the kind of *aphairesis* that Aristotle might have wanted, for, as we have said, *you* must bring your own set theory, and *your* own mode of proof, to this party. With the other lights low, the mathematical concepts shine out in this new beam, as pinned down by the web of relations they have with all the other objects of their species. What has receded are set theoretic language and logical apparatus. What is now fully incorporated, center stage under bright lights, is the curious *class* of objects of the category, a template for the various manners in which a mathematical object of interest might be presented to us. The basic touchstone is that, in appropriate deference to the manifold ways an object can be presented to us, objects need only be given *up to unique isomorphism*, this being an enlightened view of what it means for one thing to be equal to some other thing.

References

[Biss 2003] D. Biss, "Which Functor is the Projective Line?" *The American Mathematical Monthly* (2003).

[Gentzen 1936] G. Gentzen, "Die Widerspruchfreiheit der reinen Zahlentheorie," *Mathematische Annalen*, **112** (1936) pp. 493–565. Translated as "The consistency of arithmetic," in [Szabo 1969].

[Gentzen 1938] ——, "Neue Fassung des Widerspruchsfreiheitsbeweises für die reine Zahlentheorie," *Forschungen zur Logik und zur Grundlegung der exakten Wissenschaften*, **4** (1938) pp. 19–44. Translated as "New version of the consistency proof for elementary number theory," in [Szabo 1969].

[Harris 2003] M. Harris, "Postmodern at an Early Age, a review of *Mathematics and the Roots of Postmodern Theory* by Vladimir Tasi'c," *Notices of the AMS*, **50** (August 2003), pp. 790–799.

[Kant 1961] Immanuel Kant, *Critique of Pure Reason,* translated by Norman Kemp Smith, Macmillan, London, (1961).

[Mac Lane 1971] S. Mac Lane, *Categories for the working mathematician*, Springer-Verlag, New York-Berlin (1971).

[Plato 1997] Plato, *Republic, transl.* G.M.A. Grube *rev* C.D.C Reeve, in *Plato, Complete Works*, ed. John Cooper, Hackett (1997).

[Szabo 1969] M. E. Szabo (ed.), *The collected works of Gerhard Gentzen*, North-Holland, Amsterdam (1969).

IV

The Nature of Mathematics and its Applications

In this final section we consider general questions about the nature of mathematics and about its applicability to the world. The question "What is mathematics?" can be looked at in a number of ways: what the objects of mathematics are, what topics do mathematicians study, what kinds of methods mathematics uses, whether mathematics belongs with the humanities or the sciences, and so on. Robert Thomas proposes an answer to the question by considering mathematics in relationship to the spectrum of the sciences. Guershon Harel proposes an answer from the viewpoint of a researcher in mathematics education. Keith Devlin discusses how the answers to this question have changed over time, and in what direction he sees the answer likely to change over this current century.

Mathematics appears to be abstract and independent of the physical world. Given this, the question of why it turns out to be so useful in scientific investigation of that physical world has been a topic of discussion for many years. The discussion has been carried on more by physicists (who are making use of that mathematics) and by philosophers, than by mathematicians. Mark Steiner, in his chapter, shows that some of the discussion has simply been due to philosophers and physicists meaning different things by the question. However, he also gives an extended example to suggest that there is still a mystery to be investigated. Alan Hájek looks at a particular topic, probability, which has had many interactions with problems external to mathematics throughout its development. He also discusses some of the philosophical confusions that are still being sorted out in this field.

12

Extreme Science: Mathematics as the Science of Relations as Such

R. S. D. Thomas
Professor of Mathematics
University of Manitoba

From the Editors

The question of what mathematics is has never received a satisfactory answer, we feel, although "mathematics is the science of patterns" may come close. Devlin's chapter (which takes that as its definition) discusses briefly some answers that have been tried. This chapter by Robert Thomas is a new contribution to the question, and, we feel, one worth serious consideration. It certainly helps with questions such as the relationship between mathematics and the (other) sciences, and has something to say about the applicability of mathematics.

As this chapter is exploring where mathematics fits into our overall understanding of the world, it is not likely to specifically influence how to teach mathematics. However, both in our courses for mathematics majors and in our service courses (for students who will use mathematics in the service of their majors, and for students taking mathematics to enhance their general education) there is some value to reflecting from on the nature of what we are teaching. Students often appreciate this reflection on where the whole enterprise is going and how it might fit into their world-view. This chapter may be of use as you consider how to talk with your students about mathematics and its role.

Robert Thomas is a Professor of Mathematics at the University of Manitoba in Canada (www.umanitoba.ca/science/mathematics/new/faculty/html/thomas.html). He is the current editor of Philosophia Mathematica, *the one journal devoted exclusively to the philosophy of mathematics, and was a founding member of the Canadian Society for the History and Philosophy of Mathematics. His research interests include applications of geometry and philosophy of mathematics, more specifically in the application of descriptions to the world. To study these applications to the world, he felt it essential to do some: classical applied mathematics from a geometric perspective, primarily applying differential geometry to continuum mechanics in collaboration with H. Cohen, studying linear elastic waves in shells. He has pursued an interest in the philosophy of mathematics, with a historical focus. This historical focus resulted in a book with J.L. Berrgren,* Euclid's Phænomena: A translation and study of a Hellenistic treatise in spherical astronomy, *reprinted as* History of Mathematics Sources, *Volume 29, by the American Mathematical Society and London Mathematical Society (2006). Beyond mathematics, he is interested in natural language as a rational structure.*

Among his philosophical articles that are likely to be of interest to readers of this volume are "Mathematicians and mathematical objects," *in* One Hundred Years of Russell's Paradox. Papers from the 2001 Munich Russell Conference, *Godehard Link, ed. (2004), which discusses similar issues as this chapter;* "Idea Analysis of Groups," Philosophical Psychology *(2002) (a response to George Lakoff and Rafael Núñez's,* Where Mathematics Comes From: How the Embodied Mind Brings Mathematics into Being*);* "Proto-Mathematics and/or Real Mathematics," For the learning of mathematics *(1996), and* "Meanings in ordinary language and in mathematics," Philosophia Mathematica *(1991). Also likely to be of interest are a series of articles comparing mathematics to literature:* "Mathematics and Narrative," The Mathematical Intelligencer *(2002),* "Mathematics and Fiction I: Identification," Logique et Analyse *(2000),* "Mathematics and Fiction II: Analogy," Logique et Analyse *(2002), and* "The comparison of mathematics with narrative," *in* Perspectives on mathematical practices: Bringing together philosophy of mathematics, sociology of mathematics, and mathematics education, *Bart Van Kerkhove and J. P. Van Bendegem, eds. (2006).*

1 Introduction

Consideration of any mathematical model, whether from science or operations research, can lead to consideration of the effectiveness of mathematical models for understanding and prediction of the non-mathematical world. This effectiveness was famously called 'unreasonable' by Eugene Wigner [1960] but 'reasonable' by Saunders Mac Lane [1990]. Whether reasonable or unreasonable, its effectiveness does require explanation—actually two explanations. One explanation is of why the *world* is the way it is that allows our rationality to function dependably. This explanation is probably religious even when it does not set out to be so (see [McGrath 2004]). Another explanation is required of how or why *mathematics* is the way it is as a successful vehicle for our rationality. This explanation is probably philosophical, and it is a virtue of the view of mathematics presented here that it makes the mathematical side of the effectiveness seem natural.

Mathematics works so well in the sciences, I say, because it is one of the sciences but not in the simple-minded way of being empirical (based on pattern observation) that is associated with John Stuart Mill. Having lived with this idea for a long time, I find it obvious, obvious but not necessarily correct. Certainly it is not the only way to look at mathematics. Another way to see mathematics is as an art, and I have nothing whatever to say against that view. Mathematics is a complex business, and it would be remarkable if there were only one informative way to see it. The science view has a certain simple plausibility that ought, it seems to me, to require at least a wave of dismissal for those presenting other views, especially those that make mathematics *sui generis*.[1] While it has its own character, as both an art and science, it is not all on its own. But even a wave of dismissal of the science view is often not forthcoming. With the exception of Saunders Mac Lane, whose very similar view led him to regard its effectiveness in the world as reasonable, the view is hardly even available to be dismissed. The intention of this essay is to present a sketch of the view in this context so that it is available to be dismissed, argued against, or even further developed.

Seeing mathematics as a science (though not exclusively) does not solve a lot of philosophical problems. In presenting a context of other sciences for mathematics, philosophy of mathematics is set into philosophy of science, leaving most of its problems intact. The effect on a couple of philosophical problems will be mentioned at the end. Mostly what the remainder of this essay will do is context setting.[2] In order to see mathematics as a science, one needs to see it in two of its contexts: historical and then scientific. Accordingly, I begin in the next section with the historical beginnings of mathematics. Next, I describe the sciences in a way that allows mathematics to fit into them. Finally, having fitted mathematics into a picture of the sciences, I say something about a couple of philosophical problems.

2 Historical Context

One needs to push back a long way in order to include the whole development of mathematics, since mathematical records go back a very long way indeed even if you don't count notches on sticks found in stone-age sites, and I don't. We have tablets several thousand years old—how many hardly matters—recording the solution of mathematical problems. If we think about the learned landscape of such a period, we note that most current scholarly disciplines had not been invented. History was a long way from birth. Of the sciences, astronomy, botany, and zoology were beginning their pre-scientific period of data collection. The social sciences and humanities, the latter based on what we call the classics, did not exist. Our classics had not yet been written! All there was on the learned landscape was myth, of which we still have records. Is the mathematics the same as or different from these myths? Obviously different in both manner and subject matter. Or are the manners so different? In previous writing [2002], in which I have compared mathematical proof to narrative, I have dismissed algorithms as being so obviously narrative in form as to require no comment. I thought that the interesting comparison was of the things that were not so obviously similar. But what of the things that *are* obviously similar? What we have from the most distant past that is mathematical is algorithmic rather than theoretical. Quite differently from contemporary presentation of algorithms, specific problems are solved using their peculiar data because, lacking algebraic notation with which to represent either general numbers or arithmetic operations, verbal description of how to solve a particular problem was all that could be written down. More could be learned, however. The apparent intention was that, by learning a few examples, the algorithm could be mastered and applied elsewhere. This is, as

---

[1] I have in mind those, like the realists of Charles Chihara's essay in this volume, that give mathematics a subject matter and way of knowing it totally different from other subjects.

[2] What is offered is, while not necessarily the 'big picture' of Charles Chihara's essay, a bigger picture than a portrait of mathematics alone.

one easily recognizes, a learning technique still used by students—with the same attendant pitfall that the range of usefulness of the algorithm is not learned. I draw attention to this because it is somewhat similar to how myths work.

A narrative can perform a variety of functions. Fiction informs us of possibilities, and history informs us about what happened in the past. A function of myth, according to a recent theory [Peterson 1999] is that myth informs us how things ought to be. Myth is one of the few ways in which value is communicated. Mythical stories are applied to present reality allegorically to indicate the distinction between good outcomes and bad outcomes. An example of how a story can suggest how things ought to be is the stories we heard of from the path of the 2004 Boxing Day tsunami. Some societies had stories that motivated them to seek high ground when the earth shook. Whether those stories were history or something like *Chicken Licken,*[3] they worked, and those that had no such stories suffered the avoidable consequence of lacking an appropriate myth. The same process is applied whenever we apply a proverb to assure ourselves that what seems to be happening or what has happened is in accordance with the expectations we ought to have had, even if we did not have them. The process of applying a numerical example to another numerical example by way of an unspecified algorithm is surprisingly similar to the application of a story allegorically to a situation. It appears then that myths, which existed long before history or literary fiction, both are important and compose the literary context in which the earliest serious mathematics was written down. And the modes of interpretation of a mythical narrative and a special-case algorithm are surprisingly parallel.

The way of communicating that I am attributing to ancient mathematicians involves writing about something without worrying what that something is going to turn out to be in the situation of the reader. This is like knowing that interpretation will be allegorical. Particulars stand in for other particulars. In the recounting on a clay tablet of how to solve a particular mathematical problem by an algorithm that could not itself be written down, a whole class of solutions for a whole class of inputs were being recorded in the only way available.

Long before the invention of what we recognize as natural science at the Renaissance, the mathematics from which ours descended had become theoretical—quite different from telling little numerical stories. Recipe mathematics can be extremely useful; ask most engineers. It was recipe mathematics that helped build the pyramids. By Euclid's time, whether in Indian or Hellenistic civilization, mathematics had ceased to be primarily algorithmic, a transformation that did not occur in China. Another shift took place along with that from recipes to theorems. The language used shifted from the particulars that were used to represent the ancient algorithms to attempts at generality. The ubiquitous triangle PQR in Euclid's *Elements* is just any (non-degenerate) triangle for the reader. Syntactically, it is a particular, but functionally it stands for any triangle at all. Any choice of a particular number, used to represent an arbitrary number in an algorithm, still has its own special properties, perhaps divisible by various factors and perhaps prime, but the generic triangle used by Euclid has no particular features. It can be right or isosceles or both, and so on. This indefiniteness became much more explicit when algebra was invented and a symbol not a number was specifically set to have the value of an unknown. Talking about things without knowing *which* things (if any) seems to be an easily observed fundamental

---

[3] Also known as *The sky is falling*.

mathematical technique. I am suggesting that it predates algebra and is more pervasive than is often thought.

Interpretation is so obviously a feature of accounts of the past that I cannot imagine that historical works have ever been straightforwardly and commonly accepted as telling it like it is in the way that mathematics does. Mathematics is the very paradigm of dependable knowledge and I think has been since it became theoretical with the axiom-proof format.[4] So used are we in the present day to accepting accounts as accurate that we forget that financial accounts, like historical accounts, are made to convince someone of something, even if the something is not quite true. People may believe what they read in newspapers because they believed what they read in textbooks at school. And television is believable because seeing is believing. In the world of two thousand years ago, in which deductive geometry flourished, one of the most important studies was rhetoric. Convincing folks may even be more important now, but we are less frank about needing to do it. Mathematics, in contrast, is mercifully free of that sort of thing. Logic, not rhetoric, rules. Mathematics is unusual in this. It is not faulty mathematics that is used to convince shareholders of Enron that their shares will increase in value; it is assumptions on which calculations are based. I mention this to emphasize that, in being logical and dependable, mathematics has for two thousand years—a long time in human affairs—been what thinking has aspired to if it is to be regarded as above reproach.

What I have been concerned to indicate with this sketch of mathematics before and after the invention of the theorem-proof way of expression is that, as written documents, it began similar to myths in form and interpretation and was transformed from narrative (always about particulars) to theoretical form with language intended to be general, assuming a unique place as what any serious intellectual enterprise would be if it could.

3 Scientific Context

Plainly there are many ways to look at mathematics. As my title indicates, I am putting one forward. Some would call it a view of the nature of mathematics.[5] The main point I want to make in this essay is that one way to see mathematics is as sitting at the extreme of a spectrum of sciences. Since such boundaries don't matter, I don't see it as important whether it is just beyond the extreme end of the spectrum and so not a science or just inside the end and so is the most abstract science. I do see it as sufficiently important to want to make two subsidiary points: The sciences do compose a spectrum, and something important can be learned about mathematics by seeing it in its place on (or beyond) that spectrum, with one end being chemistry, physics, mathematics and with the other end containing the subject matters that are interesting in their own right like psychology. Whether one finds mathematics, physics, chemistry, sociology, or psychology interesting is not what I am concerned with. That is a question about prior interests, modes of presentation, and inclinations of various sorts. I am drawing attention to the gradual difference in what these subjects *study*—what is there before the study begins. With psychology there are folks with their varied minds. With sociology there are whole groups of folks with their

---

[4] Does it matter that not all of mathematics is axiomatized? Not at all; axioms are just a remarkably effective coping strategy, as they are in science.

[5] I am sceptical about mathematics' having a nature. It has been pointed out to me by Carlo Cellucci that I suggest variously fictionalism, modal structuralism, and deductivism. All three have things to say worth hearing.

common mind and differences. But with chemistry there are just reagents; even a chemist would have a hard time working up interest in a jar of Glauber's salts. With physics there is anything at all, so long as it is flying through the air, sliding along a surface, flowing in a channel, or doing any of the other physical things that objects and substances do, but such contemplation abstracts from all of the aspects of the objects that make them interesting in themselves. The psychologist's person with a mental life has become simply a rigid body or a point mass with friction. The subject matters of chemistry and physics, near the one extreme, are sufficiently undifferentiated not to be of intrinsic interest, since chemistry considers relations among all substances that interact chemically and physics considers relations among all things that are physical. As I am going to spell out in greater detail in the next section, a physicist in considering a person as a point mass with friction is not negating personality but simply considering what physics considers, physical relations. Physics is about how things interact physically, chemistry about how they interact, as we say, chemically, biology about the new relations added by being alive, and psychology about the new relations added by thinking. Even a thinking thing has merely physical relations.

The contradictory view of what scientific subjects are about, which I consider myself to be combatting because it misrepresents mathematics, is that the different sciences simply have different stuffs as subject matters, and that the stuff of mathematics is the things philosophers call mathematical objects or even what a self-confessedly ill-informed poster to the POMSIGMAA listserve has too often said, numbers. I do not see that widening the focus from numbers to more than just numbers improves this view. At its narrowest, it is plain wrong, but even broadened I find it misleadingly uninformative. The argument against mathematics' having a stuff has been carried through some way by Charles Chihara in his essay in this volume.

I am putting forward a picture, the frame of which I have now described. In order to see mathematics as I see it, one needs to see chemistry and physics—and for that matter psychology—as I see them. I take the defining feature of science as we understand it now, as distinct from natural philosophy before Galileo, to be its study of relations rather than of the things themselves. The sciences are ways of understanding not minds, chemicals, and things in general but the ways minds interact with one another and the world, the way chemicals relate—chiefly react, and the way things in general behave physically: statics, kinematics, dynamics, thermodynamics, fluid mechanics. Psychologists do not pontificate on the nature of mind. You will not find a chemist say anything about the essence of antimony. And even when they profess to be thinking the thoughts of God, physicists do not tell us what gravity *is* any more than Newton did. Physicists tell us the magnitude and direction of gravitational acceleration, whether Newtonian or relativistic, but that is as far as they go. And clearly the science that opposed Galileo, based on Aristotelian common sense and observation, was essentialist to the core. It is an interesting historical question how conscious Galileo was of modelling his new science on mathematics in anything like the way I have suggested. But whether it was conscious or not, that is what he did. It is a possible interpretation of his famous statement that the laws of physics are written in mathematical language.

Mathematics, in its shift from algorithms to theory, moved from what we could do with numbers to the study of relations among numbers, relations among points, lines, and planes. That shift to the rigorous determination of relational consequences of relations had been consolidated nearly two millennia before Galileo. He had little choice. As I tried to indicate in the previous section, mathematics and its approach were all there was to imitate. Myth was hardly appropriate, history is similar to the natural philosophy he was replacing, and Europe was innocent of serious prose fiction. The systematic study of relations and their consequences, which mathematics had been doing apparently forever, was the right target.[6]

We all know that the definitions of points, lines, etc., at the beginning of Euclid's *Elements* are not used in the sequel. What are used in ruler-and-compasses geometry are the positions of lines and points and planes and the construction of new ones in specific positions and proofs of their relations. What a line *is* does not matter. What a magnitude *is* does not matter. *All that matters is how they are related.* When the *Elements* was written, it was thought to be about relations in physical space. But because the enterprise was to prove what followed from the assumptions, it did not matter whether Euclidean space was indeed an accurate model of physical space. The subject, modulo some implicit assumptions, was coherent in itself, and it showed that such study was possible. When it came time to do heliocentric astronomy, it was not only the conic sections that could be taken over from geometry but also the method of making assumptions about relations and seeing what those relations led to without regard to the obvious enough fact that the relations were among physical objects with their varied inner constitutions. The only thing that was required was that they have mass, and even that is a matter of inertial and gravitational behaviour not of essence. And so essences quietly became irrelevant. They turned out to be no loss, since no one had known what they were anyway. We no longer worry about the difference between magnitudes; they each have their dimensions, but aside from keeping track of those dimensions, we treat all magnitudes as numbers. They all behave identically as to arithmetic operations, and so from our structuralist mathematical point of view there is nothing other than their dimensions to distinguish them.

4 Mathematics as a Science

Evidence for the view that mathematics can be regarded as a science is the recent turn to experimental mathematics. Not just to experimenting to look for counterexamples or patterns that might turn out to be universal, but the search for evidence in the absence of proofs. Those writing about these matters in books [Borwein/Bailey 2004], [Borwein/Bailey/Girgensohn 2004], and the recent papers [Bailey/Borwein 2005], [Borwein 2008], are mathematicians interested in proving what they have evidence for, but it is easy to see that they need not have that interest or even be mathematicians. Moreover, there are the proofs up to a pre-set level of probability advocated by Doron Zeilberger [1993], for instance that a given number is prime with 99.9% probability. An old-fashioned mathematician will probably say that proofs help us to understand the results and do not *just* assure us of their correctness. You cannot say that of a 'proof' to a certain level of probability, no matter how high; it is just limited assurance of a scientific-like fact.

An aspect of this notion of scientific-like facts, of things that happen to be true, is that they do not have the interest of justified facts whose justification is much of why they are interesting. Gregory Chaitin has often said, e.g., [Chaitin 1998], that there are so very many such facts that

---

[6] More precisely, Galileo chose *the right way*, with primary properties and no essences, to imitate mathematics, Aristotle having attempted a more superficial and less successful imitation. The difference was noted by Kant in the preface of his [Kant 1992], according to [Cassirer 1923].

have no old-fashioned reason to be true. It remains to be seen whether such facts have any permanent place in mathematics, where, after all, interest is a joint value with correctness.

The reader that has come this far can now see that the case I am making can be put so simply that it looks as though I am saying nothing at all. Mathematics is like the sciences because the sciences have been constructed to imitate mathematics. This is easily said, but to mean anything—or to mean what I mean—some content has to be given to the words.

If one sees science in the way I do, then one can easily see mathematics as the next step, where it simply does not matter—is not taken into account—what the things are that are being talked about so long as they behave in the way that we are interested in studying. Mathematics is the extreme of the sciences because they are themselves approximations to varying degrees to its method and matter. If one calls the characterless objects of mathematics point masses, then suddenly one is doing physics, but if one does not, one is doing mathematics—perhaps, depending on motivation, applied mathematics.

Reviel Netz claims [1999, p. 197] that the relational view was present even in ancient Greek mathematics, taking his book on it to vindicate such a claim on the part of just one of the mathematicians that has held the view. Newton viewed numbers as relations (ratios) between magnitudes[7] but I don't know about geometry. The relational view is easier to see in pure mathematics than in mathematics before the invention of pure mathematics, perhaps by Riemann but which for our purposes we can date between Newton and Frege. But Newton may possibly have had a clearer idea of what he was doing than the average mathematician. Another clearer than average thinker was Gauss, to whom Bourbakiste Jean Dieudonné [1977] attributed the same view. After the invention of pure mathematics, the view becomes more common. It was Russell who made the claim Netz vindicated, that mathematics considers 'types of relation' [Russell 1956, p. 3]. Poincaré [1902, p. 20] states the view very clearly, 'Mathematicians do not study objects, but the relation between objects.' Hilbert and von Neumann seem to have agreed, although the latter, like Saunders Mac Lane, preferred functions, which are interdefinable with relations in general. Carnap's philosophy of physics was so formalistic that he was accused of turning physics into mathematics.[8]

Another philosopher that seems to have taken this view was Ernst Cassirer, quoting [1953, vol. 3, p. 293] approvingly more of the above quotation from Russell. [9] Gödel at one time at least appears to have embraced the scientific-style basis for axiom choice advocated by Russell ('their justification lies (exactly as in physics) in the fact that they make it possible for [what one wants] to be deduced' [Gödel 1944, p. 121]). Sir Michael Atiyah's presidential address to the Royal Society of London [Atiyah 1995] contained the following statement, speaking of abstraction, which he had been saying was used in science: 'Mathematics takes the process to its ultimate conclusion: the identity of the players is ignored, only their mutual relations are studied. It is this abstraction that makes mathematics such a universal language: it is not tied to any particular interpretation.' Note that this ignoring does not empty mathematical language of all meaning;

---

[7] Austin, translator of [Frege 1980], gives the reference *Arithmetica Universalis,* Vol. I, cap. ii, 3 at §19.

[8] [Carnap 1967, §15, p. 27], cited in [H. Wang 1974, p. 40] and n. 2 to that page.

[9] A contemporary philosopher that assimilates mathematics to science and in particular deduction in mathematics to deduction in science is Carlo Cellucci in his book [Cellucci 2002], of which only the introduction is available in English as [Cellucci 2005].

rather it allows its various patches to be filled with a variety of meanings including new meanings. My most up-to-date indication that this is a common view of reflective mathematicians is the recent book [Widdows 2004], in which the mathematician author expresses a similar view both at the beginning and at the end.

I have not been concerned to combat the blinkered view that mathematics is just about numbers, although my examples tend to be geometrical. That is an unnecessary limitation. If one finds something in the world that is an application of the four-group, that is not because the elements of that group bear any resemblance to the elements of Klein's four-group or of the general abstract four-group. It is because their squares are the identity and they multiply one another to give one another. It is group *behaviour* that we find in the world, not the elements of abstract groups. If one models something with a graph in the sense of graph theory, one may be modelling activities by the edges of a graph whose vertices just represent the termination of the tasks, as in the critical-path method, or one may be modelling places by vertices and the routes among them by edges, as in the travelling-salesman problem. But places and times are not significantly *like* the vertices, which have no character to be like. And activities and distances are not *like* edges, which are just pairs of vertices. In these and other cases, graphs represent naked relations with any quantitative elements as add-ons. Perhaps my exposure to graph theory as a beginning graduate student is what sensitized me to this aspect of mathematics.

One may reasonably ask what relations are studied in pure mathematics—those that are later applied elsewhere. Not to give arithmetic priority over geometry, one of the earliest relations has to be sameness. In arithmetic sameness is equality, in geometry congruence. I think it is fair to say that there is little to study about these two fundamental relations, but one does need them in order to consider others. Arithmetic appears to be based on the successor relation that allows us to assign both ordinal and cardinal numbers, whichever came first historically. Both kinds of number, that is, the objects invented to carry the successor relation as far as we wish, have long been studied. The relation that connects pairs of numbers to their sums can be thought of as connecting the number of things in the combination of two counted clusters to their counts. This can be thought of as an operation on two numbers to produce their sum or as a function on the cartesian product of the numbers with themselves. The cartesian product itself is based on the relation between two things and the single thing that is the pair of them. The relation expressed by the operation of addition gives rise to the relation expressed by the operation of subtraction. In order to make subtraction work more of the time, we consider what the numbers would have to be that would allow us to subtract always. In this way, repeated *mutatis mutandis* to form fractions, real numbers, and complex numbers, the raw material of the study of the relations is expanded in a way that the raw material of the other sciences cannot be. (And yet, the extension of physics to infrared, ultraviolet, and X-radiation is not utterly divorced from what had happened earlier in mathematics.) The study of addition logically produces two quite different offspring: the further scientific consideration of addition and the technology of addition. It is not enough to be able to add two numbers a and b by counting b numbers beginning with $a + 1$. We need to be able to take a and b, written in the customary notation for numbers and produce algorithmically the customary notation for $a + b$. As notation (to think only of Europe) changed from Greek to Roman to the contemporary decimal system, the technology of addition had to be changed and improved with much thought; the addition of Greek and of Roman numbers is not so easily performed. On the other hand, if one is adding a number of cases of the same number, then one stumbles on a new relation, that expressed by the operation of multiplication—again an operation on two numbers or a function from the cartesian product to the numbers. Multiplication gives rise to the new relation of two numbers to their quotient in much the same way as addition gives rise to subtraction. It is discovered that zero misbehaves seriously in division as it did not in subtraction. Beginning with the integers, one quickly needs a new sort of number to deal with practical examples of division. And again there is the technology of performing divisions and multiplications as well as the coming to understand how they work. The relation of equality is basic to this because one is often looking for the succinct and standard number that is equal to the one for which one has a cumbersome expression: one wants the standard expression for the integer $a + b$ or $a \div b$ rather than the expressions '$a + b$' or '$a \div b$', that is, the standard expression for the integer *equal to* $a + b$ or $a \div b$.

Similar basic notions are involved in geometry. Beyond congruence, there are several basic relations. Coxeter's *Introduction to Geometry* [Coxeter 1961] elaborates an axiom system for the ternary relation 'between' on a line. Euclid does not concern himself with such things, presumably taking their behaviour to be obvious. Collinearity seems to me to be the basic ternary relation in Euclidean geometry. If one has three points, no two identical, then either they are collinear or not. If they are, then fine, the line defined by each of the three pairs is the same. (The line segments may have very different lengths, however—with relations like those in arithmetic.) If they are not collinear, then they define a triangle with sides the line segments defined by each of the pairs. As soon as one has triangles, then one can consider their congruence and then the slightly more sophisticated relations of similarity and equality of area. Congruence will obviously not apply to figures of different shapes, but equality of area may. Differently shaped polygons can be studied for their equality properties, and eventually one reaches the possible relation of areas of polygons and circles and other figures with curved boundaries (eventually leading to integral calculus). But even to say this is to call upon the relations 'boundary of' and 'same number of sides' for polygons. A small increase in the level of sophistication of the relations involved brings one from the geometry well studied in the fourth century B.C.E. to the topology only brought to light in the nineteenth century C.E. This is not to say that the relations are more complicated; they may be simpler. But simple in mathematics does not always mean easy.

I conclude this indication of what sort of relation has been studied mathematically—an indication that is just the start, since all of classical mathematics can be looked at in this way— with a word on what lies beyond arithmetic and geometry. Multiplication and addition give us linear functions, which are just the simplest examples of functions as they were historically viewed. It has taken a long time to change thinking about functions from the process view with which functional relationships originally began to the set-of-ordered-pair view that allows any functional relationship to be specified. No matter how they are specified, functions are a specific sort of relation. Their study developed into analysis with particular attention to those of use in physics. The relation between a function as values and the rates of change of those values was particularly important in this developmental process. Differentiation was not originally thought of as a relation between functions, but that view has become standard. I have already alluded to modern algebra, which is a development of the study of operations (as in addition and multiplication) applied to what are not necessarily numbers. With the development of category theory, the relation view in its function form, which I attributed above to von Neumann and Mac Lane, appears to have triumphed as the practical way to organize mathematical ideas, while the mathematical-object view of set theory still holds centre stage in foundational discussions (a competition hotly disputed).

5 *Ontological Consequences*

One might reasonably ask what philosophical problem the relational view offers a solution to that the object view does not. My concern is not to solve philosophical problems but rather to have philosophical problems that purport to be about mathematics actually be about what I can recognize as mathematics. Since the relational or scientific view says nothing specific about metaphysics, it does not attempt to solve metaphysical problems, although it does try to avoid unnecessary ones. Ontological arguments concern what exists, often whether mathematical objects exist[10] or even whether abstract objects in general exist with mathematical objects taken as typical abstract objects. It is not clear to me that mathematical objects are typical examples of abstract objects, but I am trying to avoid a concentration on objects.[11] A switch to relations, if taken seriously, would I am sure produce different and more relevant problems. I think that argument about the existence of relations is harder to mount than analogous argument about objects. It is not clear what it means for a relation to exist. It does not seem to require that the relata exist, since a great deal of fictional literature takes its meaning from the fact that relations in fiction are intended to be of the same types as occur in the real world. If that were not so, it would be meaningless to have a fictional child of a fictional parent without explaining what parenthood meant in the world of the fiction. While parenthood *could* be differently defined in a science-fiction world, all such relations, when verbally described, are parasitic on the ordinary relations from which we derive our vocabulary. This view is subversive of ontology as well as trying to avoid it, but it does not seem to favour either answer to the existence question for mathematical objects.

While we cannot easily and convincingly say what we mean by a relation's existing or not, something we can say about some relations is that they *cannot* relate anything; they are impossible. The obvious example is the relation describable as being not self-identical, sometimes used to define the empty set as consisting of those things that are not self-identical. The empty set, by the way, is the only mathematical object that I am almost sure exists, in some sense of that slippery word. While there are some persons that find contradiction (logical contradiction, not argumentative contradiction) interesting, mostly mathematicians prefer to avoid it at almost any cost. Russell's notorious ridicule of Meinong was based on the latter's ontological espousal of impossible objects, not just non-existent objects. Just what espousal consisted of we need not go into, my point being that mathematicians have no time for them when doing mathematics—however amusing they may find Escher's impossible drawings. Given that we wish to avoid

---

[10] Such arguments are discussed by Stewart Shapiro and others in this volume.

[11] Jeremy Gray captures the connection to objects briefly in discussing implicit definitions by axioms, 'There was no attempt to show that the new, implicit, definitions somehow captured the essence of the real object, because the real object was only incidentally what it was about.' [Gray 2006, p. 390] Jean Dieudonné puts it more clearly as follows. To solve eighteenth-century problems in the nineteenth century, it was necessary to abandon 'the semi-"concrete" character of classical mathematical objects; it has to be understood that what is essential about these objects is not the particular features which they seem to have but the *relations* between them. These relations are often the same for objects which appear very different, and therefore they must be expressed in ways which do not take these appearances into account; for example; if we wish to specify a relation which can be defined either between numbers or between functions, it can only be done by introducing objects which are neither numbers nor functions, but which can be *specialized* at will as either numbers or functions, or indeed other kinds of mathematical objects. It is these "abstract" objects which are studied in what have come to be called mathematical *structures* . . . ' [Dieudonné 1992, pp. 2 f.].

impossible relations in particular and contradictions in general, what does the relational view of mathematics suggest to us? It seems to me that it suggests only that, in our study of relations by attributing them to objects made up for the purpose, we need to be careful to avoid the possibility of deducing any logical contradiction. We would like not only to avoid logical contradictions, which we might do by being careful, but also to avoid the possibility of them.[12] And to operate in the realm of no logical contradictions is to operate in the realm of the logically possible. To say that mathematics studies logical possibilities is, while certainly true for most of mathematics, no more informative than to say that we want our deductions to be logically valid because, without the relational subject matter to which I say the subject is devoted, that description would describe logic or logicism. Mathematicians explore, in a scientific spirit, relations that interest us to see how those relations are related to one another regardless of the objects that grammatically are their subject matter, any choice of which, as I have said, is an extra-mathematical enterprise called interpretation or application. A philosopher that studies the consequences of approximately this point of view is Geoffrey Hellman, who calls it [Hellman 1989] modal structuralism, a structuralism without structures. He genuinely attempts to avoid the basis in things that I regard as artificial and misleading. A different attempt to avoid things and structures is made by Charles Chihara [2008].

I have mentioned structuralism, both modal and otherwise, because non-modal structuralism is as close as most philosophy of mathematics gets to the relational view of the subject. Ordinary structuralism, which one can easily attribute to Bourbaki, has been philosophically elaborated chiefly by Michael Resnik [1997] and Stewart Shapiro [1997]. In order, it seems to me, to have objects to talk about, structuralism considers usually sets of objects and their relations as forming structures, which are then said to be the subject matter of mathematics. Mathematics is then about those structures rather than about the somehow lesser objects that compose them; how the objects are inferior to the structures other than by inclusion is unclear. Because structures are objects themselves, the usual discussion of their definition and existence is easy to launch. Structuralism does appear to be inspired by mathematicians' interest in structure rather than in a uniquely mathematical subject matter, whether objects or structures. It is close but, as is widely thought, not quite right. In criticizing structuralism, Fraser MacBride [2005] chooses as its weak link what is called the incompleteness of mathematical objects. Note the return to the objects that philosophers are happy talking about from the structures or positions or relations that structuralism tries to replace them with. I have not found in MacBride's article what seems to me the obvious criticism, namely that it is not positions in structures that have the relations we want but the hypothetical holders of those positions that have them. A set of eggs is what goes into an egg carton, not the egg carton. I do not see talk of the positions instead of their contents as satisfactory. And both of the main advocates of positions structuralism, Michael Resnik and Stewart Shapiro, use essentially this approach. I am concerned, however, to discuss very briefly the actual criticism made of both of them, which is that, as things having only the relations assumed or deduced, these structures are incomplete in the way that Hamlet is incomplete because we do not know the length of his nose or many other things about him that Shakespeare did not tell us. This incompleteness

---

[12] So far as I know, how to do this dependably is not known. Considering relations rooted in physical relations is one attempt. Intuitionism is another.

of fictional characters is an inevitable result of the way in which they are specified. In the next of a series of novels, the author is free to specify some feature of, say, a serial detective, that was not previously specified—even in the case of Dr Watson to specify features previously *differently* specified.[13] The freedom of fiction does not carry over to mathematics. The incompleteness of mathematical objects, the fact that we do not know everything about them, is I think a direct and harmless consequence of their role in mathematics.

When we use the definition of bachelor as unmarried adult human male, and then deduce that whatever is a bachelor has no husband, say, or wife, we are not using an incomplete object denoted by 'whatever.' We are using a pronoun, the non-fictional uses of which are all objects of whatever kind they may be, none of them incomplete. If we apply the term whatever/bachelor to Hamlet, then that bachelor is incomplete but not because we used the word we used. The incompleteness is not in the pronoun or term but in the antecedent. If we apply the pronoun or term to a real person, say, Sam Smith, then there is no incompleteness. Similarly, because our mathematical expressions are all ultimately of the form 'whatever satisfies the axioms of our system and the hypotheses of our theorem satisfies also the conclusion of our theorem', it is too hasty to claim that our whatevers are incomplete. The time for incompleteness claims is in application not in hypothetical pure mathematics. This incompleteness, by the way, is entirely distinct from the incompleteness (of mathematical systems not objects) of Gödel's famous theorems.

To summarize, the relational view does not settle any ontological problems, but it does suggest two conjectures. Ontology is less important to mathematics than contemporary philosophers often think. (I have in mind those that require an existent subject matter for worthwhile talk or even reference; one can only denote what is not blessed with existence, not refer to it.) Ontology that considers only objects and ignores relations (or regards them as non-existent or unimportant) is too simple-minded to cope with mathematics.

6 Epistemological Consequences

The philosophical aspect that the relation view is particularly useful for is epistemological—for explaining how it is that we can gain and apply mathematical knowledge. Object views have notorious difficulty with gaining mathematical knowledge because they make the subject matter of mathematics even more remote than the mathematical knowledge itself. Since we do not interact with mathematical objects, which are supposed to exist on an altogether different plane from us, we have no way to get information about them—no way to form reasonable hypotheses, and no way to disprove them if they are wrong—not even any way to see that our proofs about them are relevant to them. This is referred to as the problem of access because we have no access to timeless objects outside space. Aristotle's solution to this platonic problem was to locate mathematical objects, likewise mysteriously, 'in' physical objects. So the problem has been recognized as a problem for over two thousand years. It seems to me that the way in which the problem is to be solved is to notice what we mean by access in this context. Obviously it cannot mean sensory access to objects that are not sensible. The relational view allows a solution like Aristotle's but without the metaphorical 'in.' We have access to a relation whenever we can

---

[13] Conan Doyle is inconsistent in what he says about Watson.

consider objectively what stands in that relation, where 'in' is not metaphorical but is just the standard way in which we speak of x related to y by relation R. We say that x stands *in* relation R to y. Mathematically, we say xRy if the relation happens to be binary, as it need not be. And so, to consider relation R, all we need to be able to do is to talk objectively about any things at all that stand in relation R. To begin to consider marriage, all that we need is some persons that are married, not access to all married couples, past, present, and future. Because marriage is a relation among real persons, it will be difficult to say anything definitive about it, but it is not difficult to see examples of it. Its study would have to be empirical. Because mathematical relations are among mathematical objects, we are free to define them for ourselves—by consensus if we are going to communicate successfully.

We have access to the relation of successor, as used in arithmetic, both as Brouwer claimed with the passage of time and in many physical arrangements of one thing after another, where 'after' is not temporal but can be physical. To have access to this relation it is not necessary that the set of things standing in the relation all be accessible or on the other hand that the set be infinite even though in mathematics we extend the domain of the relation to infinity. Our puny physical access to this relation inspires us to imagine an infinite domain on which it is defined— by us. This is handy even though our physical means cannot even represent what is in such a domain, either on paper or electronically. In the natural sciences we likewise have access only to limited examples of the relations that are studied, some of which we assume to have infinite domains. To the infinite domain of arithmetic we have no access at all, and so our examples and counterexamples in arithmetic are often finite where we can specify and understand them. Once we have this ideal domain and become comfortable with it, we can consider relations within it other than the relation of successor. I have indicated above how we define ternary relations like sum and product, difference and quotient. Once we have these working well, in order to make them work better we further idealize to negative numbers and to rational numbers. Debit balances and portions of a pizza help us out with the *relations* involved, but they do not give us access to the negative numbers and fractions of which any mathematical discussion of such things makes use. Because we have access to ordinary things that stand in the relations among the mathematical objects, even if only approximately, we do not need access to the mathematical objects themselves. They are idealizations or reifications that it is extremely convenient to talk about, but we can get along perfectly well without access to them and—a nominalist will say— without their being in any sense real. But their reality is more irrelevant than relevantly false. Most mathematical relations arise among mathematical objects themselves. In physics we have no access to the perfect systems that physics books talk about either. I see mathematical objects as similar idealizations for the sake of how they are related, which is what the idealizations in the physics books are there for too. But it is essential that we have very clear ideas about their relations widely agreed upon so that our reasoning can be correct and be seen to be correct. We do not need to agree on their non-mathematical existence or location or when or how they were created or discovered (or which it was). These aspects are mathematically irrelevant and so mathematically neutral.

The above paragraph concerns the big leap from ordinary objects and their relations to the mathematical objects that bear the idealizations of those relations. We have access to the former and not to the latter. But we have something more powerful than mere access to the mathematical objects and relations; we decide what they shall be and stipulate that to suit ourselves (collectively, not individually). It is a smaller step to move from one set of mathematical objects and the relations among them to another set of mathematical objects and relations based on them. For example, studies in analysis from the seventeenth century to the nineteenth had produced many functions, and it was observed that functions could be added and multiplied like the values in their common ranges. This allowed the creation of altogether new structures, function spaces, based on the relations among the functions composing those spaces. Such creation is a smaller step than the move from combining sets of physical objects to the addition of their cardinal numbers. It appears to be much easier to make such intra-mathematical steps than to find altogether new physical relations to mathematize. That would appear to be why graph theory, topology, and modern algebra are so much more recent inventions than arithmetic and geometry. No one can specify in advance what might be mathematized or what cannot be. That may be because of our comparative ignorance of how it is done or because it is so creative an action that we shall never understand it.

While the above considerations go some way, I think, to clarifying the problem of access—largely by recasting it, it deals not at all with the somewhat different philosophical problem of reference. One can ask, if one takes the view that mathematical statements are made about mathematical objects, how we refer to them. Since I think that mathematical statements do not refer to specifically mathematical objects but to whatever might satisfy our axioms, the problem of reference in what might be called its platonic form does not arise. We have genuine reference to objects only in cases of applied mathematics, and there it is usually only approximate, and in informal discussion, where we are free to talk in ways that no doubt defy philosophical analysis. The same is true of the relations involved in our mathematical statements. Just as the nouns in our theorems are really pronouns standing for whatever satisfies the axioms or conventions of our theory, the relating words stand for whatever relations work in the ways specified by the theory. An example of this apparent indefiniteness that is particularly clear is plane projective geometry. One can state the axioms in terms of points lying on lines (in the plane, which itself need not be mentioned), for example, every pair of points lies on one and only one line. The obvious and intuitive content of lying on is something much like set membership because one thinks of a line as composed of the points that lie on it. But, because of the point-line duality of plane projective geometry, the word 'point' can be taken to refer to the lines in the plane, and the word 'line' can be taken to refer to the points in the plane. Then *lying on* is transferred to what intuitively one would think of as passing through, with the result that the content of the above axiom is that every pair of lines passes through one and only one point—perfectly correct. Coxeter [1955] shows that this can be done with the whole theory.[14] This example shows that there is a *whatever* aspect to the relations in mathematical theories that parallels the whatever aspect to the objects. This whatever aspect is why Russell in the quotation above[15] said 'types of relation', not just 'relations'. It was a mistake for Carnap to think that mathematics is logical syntax of language, but the mistake was not totally lacking in excuse; there is in mathematics an element of how it is possible to talk objectively and reason correctly.

---

[14] This example, where it is unclear which are the sets and which are the elements, seems to me more dramatic than the mere automorphism problem of distinguishing between i and $-i$ in the complex numbers.

[15] Section 4, paragraph 5.

When we ask whether there exists a rational square root of two, we are not asking a metaphysical question. The question can be thought of as being a metaphysical one, but that is a mistake. The relational view suggests a non-metaphysical attitude that completely avoids any question about existence of rational numbers. The question is rather whether a rational number—whatever it may be—can stand in the relation that squared it equals two. Since none of the rational numbers—whatever they may be—can stand in that relation, we say that the root of two does not exist among the rationals. Only when we are happy with infinite sequences can we extend the arithmetic relations to the limits of Cauchy sequences and find that, among the Cauchy sequences, there *are* those whose limits have squares equalling two. So the root of two exists among the real numbers, but nothing was either created or found in a metaphysical sense. We simply extended our domain of discourse by making our relations apply to new materials.[16] We did not even create the Cauchy sequences; we just decided to talk about unending sequences of numbers, something we had got used to after we decided that the positive integers would be better thought of as an unending sequence. And why did we think that? Because the last positive integer would have embarrassing relations to the others.

I remarked earlier that the relational view helps with understanding the application of mathematics. This has always seemed to me one of its fundamental and obvious advantages, but the problem itself (discussed in this volume by Mark Steiner) is so little spoken of that I ought to elaborate. The simplest application of arithmetic, to make the example as simple as possible, to everyday objects requires only that they be discrete so that they can be counted rather than continuous like water. Even when what is counted *is* continuous, like time, we can agree on chunks, days for instance, that can be counted rather than measured. As soon as we recognize an order, as in the case of days, or impose an order, as in the case of pebbles, we can count them and perform arithmetic meaningfully because the relevant relations among the things are those based on order, which gave us the integers with which to do arithmetic. The relations among measurements rather than counts are more complicated, and we have created (epistemically not metaphysically) rational numbers to bear those relations to one another. This allows us to apply the arithmetic of rational numbers meaningfully to measurements. However amusing it may be to think that the hypotenuse of a 1-1-$\sqrt{2}$ triangle has an irrational side in mathematics, when one measures distances outside mathematics one always deals in rational numbers. Note that on this view the analogy is between relations among counts and relations among integers, and between relations among measurements and relations among rationals rather than directly between objects counted and integers or between material measured and rational numbers. One sees occasionally the questions, why should platonic mathematical objects 'apply' to ordinary physical objects, and what does it mean that they 'apply?' If one is concerned with those mathematical objects as primary, that's a very puzzling question. If one views them as being the bearers of important relations that we have decided to think about, then their application—through the relations in which they stand—is not at all mysterious.

I end with a return to disclaiming any exclusivity for the view of mathematics I have presented and discussed. In particular, I have much respect for mathematics as art. Art too is concerned with

---

[16] I am alluding to the standard introduction of arithmetic on equivalence classes of Cauchy sequences in which it is shown that the same arithmetic applies to ordinary numbers so that formally we replace the rationals with equivalence classes of Cauchy sequences so as to have a uniform theory.

relations but with their presentation rather than with their scientific study. You will note that I have not argued for the scientific view, lacking any premises from which to do so. I have just presented evidence and consequences. The aim is to offer a 'big picture,' as Charles Chihara puts it, a picture big enough to contain both mathematics and a context, in this case the scientific context. As a picture, its virtue is meant to be representational rather than merely aesthetic. It ought to correspond to what it represents. Like any representation, it omits much of what it represents. And non-correspondences mar it. It is not a myth, not being intended to motivate action. Let me end with a speculation. I have suggested that mathematics does not have a subject matter of things that could exercise a directing influence on our study of them by drawing themselves to our attention as matters of survival. Perhaps this is a reason why aesthetic influence is enhanced in mathematics beyond its considerable function in the natural sciences. Something has to ground our choices of topics, our choices of results to prove, our choices of proofs. What better than how attractive they are?

Acknowledgements I acknowledge with gratitude comments on earlier drafts of this essay by Carlo Cellucci, Jim Franklin, Hardy Grant, Gianluigi Oliveri, Grant Woods, and the editors, conversation with Jonathan Kirby, and the stimulus and hospitality of Dan Isaacson's seminar at the University of Oxford. A preliminary version of part of this essay was presented to the annual meeting of the Canadian Society for History and Philosophy of Mathematics in 2005 and appears in its privately printed proceedings.

References

[Atiyah 1995] Atiyah, Michael, Anniversary Address by the President of the Royal Society, *Bull. Inst. Maths App.* **31** (1995), pp. 82–84.

[Bailey/Borwein 2005] Bailey, David H., and Jonathan M. Borwein, "Experimental mathematics: Examples, methods and implications", *Notices of the Amer. Math. Soc.* **52** (2005), pp. 502–514.

[Borwein/Bailey 2004] Borwein, Jonathan, and David Bailey, *Mathematics by experiment: Plausible reasoning in the 21st century.* Wellesley, Mass.: A K Peters, 2004.

[Borwein/Bailey/Girgensohn 2004] Borwein, Jonathan, David Bailey, and Roland Girgensohn, *Experimentation in mathematics: Computational paths to discovery.* Wellesley, Mass.: A K Peters, 2004.

[Borwein 2008] Borwein, Jonathan, "Implications of Experimental Mathematics for the Philosophy of Mathematics", this volume.

[Carnap 1967] Carnap, Rudolf, *The logical structure of the world: Pseudoproblems in philosophy.* Translated by Rolf A. George from *Logische Aufbau der Welt.* Berkeley: University of California Press, 1967.

[Cassirer 1923] Cassirer, Ernst, *Einstein's theory of relativity considered from the epistemological standpoint* in *Substance and function* and *Einstein's theory of relativity considered from the epistemological standpoint.* W.C. Swabey and M.C. Swabey, trans. Chicago: Open Court. Reprint New York: Dover, 1953.

[Cassirer 1953] ——, *The philosophy of symbolic forms.* Ralph Manheim, trans. New Haven: Yale University Press, 1953.

[Cellucci 2002] Cellucci, Carlo, *Filosofia e matematica.* Bari: Laterza, 2002.

[Cellucci 2005] ——, "Introduction to *Filosofia e matematica*", pp. 17–36 in Reuben Hersh, ed. *18 unconvenentional essays on the nature of mathematics.* New York: Springer, 2005.

[Chaitin 1998] Chaitin, Gregory J. , *The limits of mathematics: a course on information theory and the limits of formal reasoning*. Singapore: Springer-Verlag, 1998.

[Chihara 2008] Chihara, Charles, "The existence of mathematical objects", this volume.

[Coxeter 1955] Coxeter, H.S.M., *The real projective plane*. 2nd ed. Cambridge: Cambridge University Press, 1955.

[Coxeter 1961] ——, *Introduction to Geometry*. New York: Wiley, 1961.

[Dieudonné 1977] Dieudonné, Jean, Lecture to the Gauss Bicentennial Symposium of the Royal Society of Canada, Toronto, June 3, 1977.

[Dieudonné 1992] ——, *Mathematics—The music of reason*. H.G. and J.C. Dales, trans. Berlin: Springer, 1992.

[Frege 1980] Frege, Gottlob, *The Foundations of Arithmetic*. J.L. Austin, trans. Evanston, Illinois: Northwestern University Press, 1980.

[Gray 2006] Gray, Jeremy J., "Modern mathematics as a cultural phenomenon", in José Ferreirós and Jeremy J. Gray, eds. *The architecture of modern mathematics: Essays in history and philosophy*. Oxford: Oxford University Press, 2006.

[Hellman 1989] Hellman, Geoffrey, *Mathematics without numbers*. Oxford: Oxford University Press, 1989.

[Kant 1992] Kant, Immanuel, *Attempt to introduce the concept of negative magnitudes into philosophy*, in Immanuel Kant, *Theoretical Philosophy, 1755–1770*, pp. 203–242. Edited and translated by David Walford in collaboration with Ralf Meerbote. Cambridge: Cambridge University Press, 1992.

[MacBride 2005] MacBride, Fraser, "Structuralism reconsidered", pp. 563–589 in [Shapiro 2005].

[Mac Lane 1986] Mac Lane, Saunders, *Mathematics: Form and Function*. Berlin: Springer-Verlag, 1986.

[Mac Lane 1990] ——, "The reasonable effectiveness of mathematical reasoning", pp. 115–121 in [Mickens 1990].

[McGrath 2004] McGrath, Alister E., *Dawkins' God*. Oxford: Blackwell Publishing 2004.

[Mickens 1990] Mickens, R., ed. , *Mathematics and Science*. Singapore: World Scientific Publishing, 1990.

[Netz 1999] Netz, Reviel, *The Shaping of Deduction in Greek Mathematics: A Study in Cognitive History*. Cambridge: Cambridge University Press, 1999.

[Peterson 1999] Peterson, Jordan, *Maps of meaning: The architecture of belief*. New York: Routledge, 1999.

[Poincaré 1902] Poincaré, Henri, "La grandeur mathématique et l'experience" in *La Science et l'Hypothèse*. Paris: Flammarion, 1902. English translation by W. J. G., "Mathematical magnitude and experiment", pp. 17–34 in *Science and Hypothesis*. Walter Scott Publishing Company. Reprinted New York: Dover, 1952.

[Resnik 1997] Resnik, Michael, *Mathematics as a Science of Patterns*. Oxford: Clarendon, 1997.

[Russell 1956] Russell, Bertrand, *Logic and Knowledge*, R. C. Marsh, ed. London: Allen & Unwin, 1956.

[Shapiro 1997] Shapiro, Stewart, *Philosophy of mathematics: Structure and ontology*. New York: Oxford University Press, 1997.

[Shapiro 2005] ——, ed., *The Oxford handbook of philosophy of mathematics and logic*. New York: Oxford University Press, 2005.

[Shapiro 2008] ——, "Mathematical objects", this volume.

[Thomas 2002] Thomas, R.S.D., "Mathematics and Narrative", *The Mathematical Intelligencer* **24**, No. 3 (2002), pp. 43–46.

[Wang 1974] Wang Hao, *From Mathematics to Philosophy*. London: Routledge & Kegan Paul 1974.

[Widdows 2004] Widdows, Dominic, *Geometry and meaning*. Stanford: [Stanford University] Center for the Study of Language and Information, 2004.

[Wigner 1960] Wigner, E., "The unreasonable effectiveness of mathematics in the natural sciences", *Communications on Pure and Applied Mathematics* **13** (1960), pp. 1–14. Reprinted in E. Wigner, *Symmetries and Reflections*. Bloomington and London: Indiana University Press, 1967, pp. 232–237, and in [Mickens 1990].

[Zeilberger 1993] Zeilberger, Doron "Theorems for a price: Tomorrow's semi-rigorous mathematical culture", *Notices of the Amer. Math. Soc.* **40** (1993), pp. 978–981.

13

What is Mathematics? A Pedagogical Answer to a Philosophical Question

Guershon Harel
Professor of Mathematics
University of California, San Diego

From the Editors

We had hoped to have several contributions to this book from people doing research in undergraduate mathematics education, since we believe that one's philosophy of mathematics often has a significant impact on how one teaches mathematics. Unfortunately, several of those we invited either declined or initially accepted and then found themselves too involved in other projects. Fortunately, Guershon Harel does some of the most interesting work in undergraduate mathematics education, although of course he does not represent the whole community. He includes the conceptual tools of mathematics as part of mathematics and hence part of what we must teach along with definitions, theorems and proofs. Those who are involved in teaching undergraduate mathematics are likely to find his chapter worthwhile.

We have one suggestion that we think may help you as you read this chapter. There is a nice parallelism in the language, "ways of thinking," "ways of understanding," but the first several times we read the chapter we had difficulty separating in our minds how a way of thinking differs from a way of understanding. We suggested to the author that he replace "ways of understanding" simply by "understandings" (by which we mean how each of us understands the objects and facts of mathematics). But he was reluctant to do this for reasons he explains in his chapter. As you read the chapter, you may want to mentally drop "ways of" from "ways of understanding"—at least, this helped us keep the distinction clear between his notions of "ways of thinking" and what we prefer to call "understandings."

Guershon Harel is a Professor of Mathematics at the University of California, San Diego (www.math.ucsd.edu/~harel/). He has research interests in cognition and epistemology of mathematics and their applications in mathematics curricula and teacher education. He has focused on the concept of mathematical proof, the learning and teaching of linear algebra, and the development of proportional reasoning of the multiplicative conceptual field. Dr. Harel's current work focuses on the concept of mathematical proof and the development of a theoretical framework, called DNR-based instruction in mathematics, *which stipulates conditions for achieving critical goals such as provoking students' intellectual need to learn mathematics, helping them acquire mathematical ideas and practices, and assuring that they internalize, organize, and retain the mathematics they learn. He developed (with L. Sowder) a taxonomy of students cognitive mathematical-proof schemes—based, in part, on parallels between historical and individual epistemologies—and a system of foundational pedagogical principles for instructional treatments that facilitate the development of deductive reasoning among students. Harel worked on several projects, funded, in part, by the National Science Foundation, the US Department of Education, and the State of California. These include the Rational Number Project (a research project investigating the learning and teaching of multiplicative concepts, such as fractions and linearity); the Algebraic Thinking Institute (a professional development project for secondary mathematics teachers); Proof Understanding, Production, and Appreciation (a research project aimed at mapping the development of the concept of proof with students); and Development of Mathematics Teachers Knowledge Base Through DNR-Based Instruction (a research project aimed at investigating the development of teachers' knowledge of mathematics and of pedagogy). He is the author of over 80 articles (many joint) in mathematics education at all levels from elementary school through collegiate mathematics. Some that are especially likely to interest readers of this chapter include "Learning and Teaching Linear Algebra: Difficulties and an Alternative Approach to Visualizing Concepts and Processes,"* Focus on Learning Problems in Mathematics *(1989); "The General, the Abstract, and the Generic,"* For the Learning of Mathematics *(1991, with D. Tall); "The Role of Conceptual Entities in Building Advanced Mathematical Concepts and Their Symbols," in D. Tall (ed.),* Advanced Mathematical Thinking *(1991, with J. Kaput); "The Process Conception of Function," in G. Harel & E. Dubinsky,* The Concept of Function: Aspects of epistemology and pedagogy *(1992, with E. Dubinsky); "Two Dual Assertions: The First on Learning and the Second on Teaching (Or Vice Versa),"* The American Mathematical Monthly *(1998); "The Development of Mathematical Induction as a Proof Scheme: A Model for DNR-Based Instruction," in S. Campbell & R. Zaskis (eds.).* Learning and Teaching Number Theory *(2001); "Advanced Mathematical-Thinking at Any Age: Its Nature and Its Development,"* Mathematical Thinking and Learning *(2005, with L. Sowder); and "Toward a Comprehensive Perspective on Proof," in F. Lester (ed.),* Second Handbook of Research on Mathematics Teaching and Learning *(2007, with L. Sowder).*

0 Introduction

Why do we teach the long division algorithm, the quadratic formula, techniques of integration, and so on when one can perform arithmetic operations, solve many complicated equations, and integrate complex functions quickly and accurately using electronic technologies? Typical answers teachers give to these questions include "these materials appear on standardized tests," "one should be able to solve problems independently in case a suitable calculator is not present," "such topics are needed to solve real-world problems and to learn more advanced topics." From a social point of view, there is nothing inadequate about these answers. Teachers must prepare students for tests mandated by their educational system, they must educate students to carry out elementary calculations independent of computer technologies, especially calculations one might encounter in daily life, and they must prepare students to take advanced courses where certain computational skills might be assumed by the instructors of these courses. These answers, however, are external to mathematics as a discipline, in that they offer justifications that are neither cognitive (about thought processes) nor epistemological (regarding the philosophical theory of knowledge) but mainly social. For example, nothing in these answers suggests the role of computational skills in one's conceptual development of mathematics; nor do these answers reflect the role of computations in the development of mathematics. A related question is: why teach proofs? The most typical answer given by teachers to this question was, "so that students can be certain that the theorems we present to them are true." While this is an adequate answer—both cognitively and (by inference) epistemologically—it is incomplete. The teachers who were asked this question had little to say when skeptically confronted about their answers by being asked: Do you or your students doubt the truth of theorems that appear in textbooks? Is certainty the only goal of proofs? The theorems in Euclidean geometry, for example, have been proven and re-proven for millennia. We are certain of their truth, so why do we continue to prove them again and again?

Overall, these teachers' answers do not address the question of what intellectual tools one should acquire when learning a particular mathematical topic. Such tools, I argue, define the nature of mathematical practice. Judging from current textbooks and teaching practices, teachers at all grade levels, including college instructors, tend to view mathematics in terms of subject matter, such as definitions, theorems, proofs, problems and their solutions, and so on, not in terms of the conceptual tools that are necessary to construct such mathematical objects. While knowledge of and focus on subject matter is indispensable for quality teaching, I argue it is not sufficient. Teachers should also concentrate on conceptual tools such as problem-solving approaches, which, I argue, constitute an important category of knowledge different from the subject matter category, as I will explain shortly.

What exactly are these two categories of knowledge, subject matter and conceptual tools? And what is the basis for the argument that both categories are needed? Initially, pedagogical considerations, not philosophical ones, engendered the two questions. However, my inquiries into these questions, especially in relation to students' conceptions of proof, have led me into historical and philosophical analyses not initially intended. These analyses have shed considerable light on my understanding of cognitive processes of learning. For example, the philosophical debate during the Renaissance as to whether mathematics conforms to the Aristotelian definition of science helped me understand certain difficulties able students have with a particular kind of proof (see [Harel 1999]). The juxtaposition of such epistemological and cognitive analyses compelled me to look deeply into the nature of mathematical knowledge and its implications for curriculum development and instruction. Thus, my answers to the above two questions—the main concern of this paper—draw upon epistemological, cognitive, and pedagogical considerations. These answers are situated within a broader theoretical framework called *DNR-based instruction in mathematics* (*DNR* for short).The initials, D, N, and R, stand for three leading principles in the framework—*duality*, *necessity*, and *repeated-reasoning*—to be presented in Section 4. *DNR* stipulates conditions for achieving critical goals such as provoking students' intellectual need to learn mathematics, helping them acquire mathematical ideas and practices, and assuring that they internalize, organize, and retain the mathematics they learn.

The paper consists of five sections: Section 1 discusses a triad of key *DNR* constructs: "mental act," "way of understanding," and "way of thinking." On the basis of these constructs, a definition according to which mathematics consists of two categories of knowledge is offered in Section 2. Epistemological considerations and pedagogical consequences of this definition are discussed in Section 2 and Section 3. Section 3 focuses mainly on long term curricular and research goals, with particular attention to lessons from history. The three foundational principles of *DNR* along with examples of other *DNR constructs* are briefly presented in Section 4. However, *DNR* concepts and themes are on every page of the paper. The paper concludes with a summary in Section 5.

1 Mental Act, Way of Understanding, and Way of Thinking

1.1 Mental Act

Humans' reasoning involves numerous mental acts such as interpreting, conjecturing, inferring, proving, explaining, structuring, generalizing, applying, predicting, classifying, searching, and problem solving.[1]   These are examples of mental acts as opposed to *physical acts*. "Lifting" and "pulling" an object are examples of the latter. However, many terms may refer to either physical acts or mental acts. For example, *searching* may refer to the act of physically looking for a missing material object—such as when one searches for missing keys—or to the act of mentally looking for an abstract object—such as when one searches for the value of an equation's unknown. The distinction between "mental act" and "physical act" is not without difficulty, as one can learn from the work of Lakoff and Johnson ([Lakoff/Johnson 2003]) and Johnson ([1987]), who argue that meaning, imagination, and reason have a bodily basis. This debate, however, is beyond the scope and goals of this paper.

Humans perform mental acts, and they perform them in every domain of life, not just in science and mathematics. Although all the aforementioned examples of mental acts are important in the learning and creation of mathematics, they are not unique to mathematics—people interpret, conjecture, justify, abstract, solve problems, etc. in every area of their everyday and professional life. Professionals from different disciplines are likely to differ in the extent to which they carry out certain mental acts; for example, a painter is likely to abstract more often than a carpenter, a chemist to model more often than a pure mathematician, and the latter to conjecture and justify more often than a pianist. But a more interesting and critical difference among these professionals is in the nature, the characteristics, of the mental acts they perform. A biologist, chemist, physicist, and mathematician all carry out problem-solving acts in every step in their professional activities, and they may even produce similar solutions to problems their fields have in common. The four, however, are likely to differ in the nature of the problem-solving act and other related mental acts they perform while solving problems. Mental acts are basic elements of human cognition. To describe, analyze, and communicate about humans' intellectual activities, one must attend to their mental acts.

1.2 Way of Understanding Versus Way of Thinking

It was suggested earlier that teachers at all grade levels tend to view mathematics in terms of "subject matter" (e.g., definitions, theorems, proofs, problems and their solutions), not in

---

[1] The notion of "mental act" is taken as undefined in this paper.

terms of the conceptual tools that are necessary to construct such mathematical objects, and that cognitively, pedagogically, and epistemologically both categories are needed. In this section, these two categories of knowledge will be defined more precisely in terms of the triad, *mental act, way of understanding* and *way of thinking*.[2]

Mental acts can be studied by observing peoples' statements and actions. A person's statements and actions may signify cognitive products of a mental act carried out by the person. Such a product is the person's way of understanding associated with that mental act. Repeated observations of one's ways of understanding associated with a given mental act may reveal certain cognitive characteristics of the act. Such a characteristic is referred to as a way of thinking associated with that act. In the rest of this section, these definitions will be explained and illustrated.

Again, *a way of understanding is a particular cognitive product of a mental act carried out by an individual*. For example, upon seeing the symbol 3/4 one may carry out the interpreting act to produce a meaning for this symbol. The interpretation the person produces is her or his way of understanding the symbol. Such a way of understanding may vary with context, and when judged by an observer, it can be deemed right or wrong. For example, in one context a person may produce the meaning "3 objects out of 4 objects," and in another the meaning "the sum $1/4 + 1/4 + 1/4$." One person may produce a mathematically sophisticated way of understanding, such as "the equivalence class $\{3n/4n \mid n$ is an integer different from zero$\}$," and another a naive way of understanding, such as "two numbers with a bar between them." Likewise, a particular solution to a problem and a particular proof of an assertion are products of the problem-solving act and proving act, respectively; hence, each is a way of understanding.

A way of thinking, on the other hand, is *a cognitive characteristic of a mental act*. Such a characteristic is always inferred from observations of ways of understanding—cognitive products of a mental act. For example, a teacher following her student's mathematical behavior may infer that the student's interpretation of mathematical symbols is characteristically inflexible, devoid of quantitative referents, or, alternatively, flexible and connected to other concepts. Likewise, the teacher may infer that a student's justifications of mathematical assertions are typically based on empirical evidence, or, alternatively, based on rules of deduction.

Methodologically, when analyzing students' mathematical behavior in terms of ways of understanding and ways of thinking, one begins with, and fixes, a mental act under consideration, looks at a class of its products (i.e., ways of understanding associated with it), and attempts to determine common cognitive properties among these ways of understanding. Any property found is a way of thinking associated with the mental act.

To further illustrate the distinction between ways of understanding and ways of thinking, consider the three mental acts of "interpreting," "problem solving" and "proving."

1.2.1 The Interpreting Act

The actual interpretation one gives to a term or a string of symbols is a way of understanding because it is a particular cognitive product of her or his act of interpreting. For example, one may

---

[2] Until the terms, *way of understanding* and *way of thinking*, are defined and illustrated, the reader might find it helpful to associate this pair of terms with the pair, *subject matter* and *conceptual tools*, respectively. The two pairs of terms, however, are not synonymous (see also Footnote 5).

interpret the string of symbols $y = \sqrt{6x - 5}$ in different ways: as an equation (a condition on the variables x and y), as a number-valued function (for each number x, there corresponds the number $\sqrt{6x - 5}$), or as a proposition-valued function (for each ordered pair (x, y) there corresponds the value "true" or the value "false."). These ways of understanding manifest certain characteristics of the interpreting act—for example, that "symbols in mathematics represent quantities and quantitative relationships." A person who holds more than one such way of understanding is likely to possess, in addition, the way of thinking that "mathematical symbols can have multiple interpretations." And a person who is able to vary the interpretation of symbols according to the problem at hand is likely to possess the way of thinking that "it is advantageous to attribute different interpretations to a mathematical symbol in the process of solving problems." These are examples of mature ways of understanding and ways of thinking, which are absent for many high school and college students. For example, when a class of calculus students was asked what $y = \sqrt{6x - 5}$ meant to them, many were unable to say more than what one of their classmates said: "It is a thing where what you do on the left you do on the right." For many students the act of interpreting algebraic symbols can be characterized as being free of quantitative meaning.

It is not uncommon that students manipulate symbols without meaningful basis in the context (as in $(\log a + \log b)/\log c = (a + b)/c)$. Matz ([1980]) connects this (erroneous) way of understanding and a wide range of algebra errors to an overgeneralization of the distributive property. Students factor out the symbol log from the numerator and cancel it, without attending to the quantitative meaning of their action. The behavior of operating on symbols as if they possess a life of their own, not as representations of entities in a coherent reality, is referred to as the *non-referential symbolic* way of thinking. With this way of thinking, one does not attempt to attend to meaning. For example, one does not ask questions such as "What is the definition of $\log a$?", "Does $\log \cdot a$ (multiplication) have a quantitative meaning?", "Is $\log a + \log b = \log(a + b)$?" and so on, for symbols are not conceived as representations of a coherent mathematical reality. Of course, one may produce correct results and still operate with the non-referential symbolic way of thinking. For example, we have observed students correctly solve systems of equations without attaching meaning to the operations they apply or to the solution they obtain. Our current work ([Harel et al. forthcoming]) points to an expected source of the phenomenon of non-attendance to meaning by students: the way algebra is taught in school. We demonstrate how current teaching practices of algebra teachers tend to deemphasize, ignore, or misuse mathematical meaning.

Before proceeding with the other two mental acts, a remark on the non-referential symbolic way of thinking, which I have just mentioned, is needed. The characterization of this way of thinking may have evoked with the reader a different image from the one portrayed here since relative to the reader's practice of mathematics it is not uncommon that symbols are treated as if they possess a life of their own, and, accordingly, are manipulated without (necessarily) examining their meaning. I will return to discuss this point in Section 3.2.

1.2.2 The Problem Solving Act

As to the mental act of problem solving, the actual solution—correct or erroneous—one provides to a problem is a way of understanding because it is a particular cognitive product of the person's problem-solving act. A problem-solving approach, on the other hand, is a way of thinking. For example, problem-solving approaches such as "look for a simpler problem," "consider alternative possibilities while attempting to solve a problem," and "just look for key words in the problem statement" characterize, at least partially, the problem-solving act; hence, they are instances of ways of thinking.

The problem-solving act is not of the same status as the other mental acts listed above, in that any of these acts is, in essence, a problem-solving act. The acts of interpreting, generalizing, and proving, for example, are essentially acts of problem solving. Despite this, the distinction among the different mental acts is cognitively and pedagogically important, for it enables us to better understand the nature of mathematical practice by individuals and communities throughout history, and, accordingly, set explicit instructional objectives for instruction. This viewpoint will be demonstrated in Sections 2 and 3.

1.2.3 The Proving Act

While problem-solving approaches are instances of ways of thinking associated with the problem-solving act, *proof schemes* are ways of thinking associated with the *proving* act. *Proving* is defined in [Harel/Sowder 1998] as the act employed by a person to remove or instill doubts about the truth of an assertion. Any assertion can be self-conceived either as a *conjecture* or as a *fact*. A *conjecture* is an assertion made by a person who has doubts about its truth. A person ceases to consider an assertion to be a conjecture and views it to be a *fact* once the person becomes certain of its truth. In [Harel/Sowder 1998], a distinction was made between two variations of the proving act, *ascertaining* and *persuading*. *Ascertaining* is the act one employs to remove one's own doubts about the truth of an assertion (or its negation), whereas *persuading* is the act one employs to remove others' doubts about the truth of an assertion (or its negation). A *proof scheme* characterizes one's collective acts of ascertaining and persuading; hence, it is a way of thinking.

A common proof scheme among students is the *authoritative proof scheme*, a scheme by which proving depends mainly on the authority of the teacher or textbook. Another common proof scheme among students is the *empirical proof scheme*—a scheme marked by its reliance on evidence from examples or visual perceptions. Against these proof schemes stands the *deductive proof scheme*, a scheme by which one proves an assertion with a finite sequence of steps, where each step consists of a conclusion which follows from premises (and previous conclusions) through the application of rules of inference.[3] Note that while a proof scheme is a way of thinking, a *proof*—a particular statement one offers to ascertain for oneself or convince others— is, by definition, a way of understanding.

Mathematical reasoning centers on the deductive proof scheme. In contrast, the authoritative proof scheme and the empirical proof scheme are examples of undesirable ways of thinking. While undesirable, a dash of the authoritarian proof scheme is not completely harmful and is unavoidable; people may use this scheme to some extent when they are sampling an area outside their specialties. In two of its worst forms, however, either the student is helpless without an authority at hand, or the student regards a justification of a result as valueless and unnecessary. As with the authoritarian proof scheme, the empirical proof scheme does have value. Examples and nonexamples can help to generate ideas or to give insights. The problem arises in contexts in which a deductive proof is expected, and yet all that is necessary or desirable in the eyes of the student is verification by one or more examples.

---

[3] For an extensive taxonomy of proof schemes drawn from students' mathematical behaviors and the historical development of proof, see [Harel/Sowder 1998].

1.2.4 Terminology

Two remarks on terminology are in order. The first remark concerns the adjective "cognitive" in the definitions of "way of understanding" and "way of thinking." A way of understanding is a *cognitive* product of a mental act, and a way of thinking is a *cognitive* characteristic of a mental act. This is to indicate that the focus here is on cognition rather than affect or physiology. For example, the product of feeling confusion or frustration as one attempts to interpret a statement, prove an assertion, or solve a problem, is not dealt with in the conceptual framework offered here. Nor does this framework deal with physiological characteristics of mental acts—those that include, for example, certain neurological activities in the brain. Thus, the adjective "cognitive" in the above definitions intends to single out one type of products and characteristics—that which signifies *cognition.* The focus on cognition rather than affect and physiology is also evident in the examples discussed to illustrate the definitions.

The second remark concerns ease of terminology. It may not be easy to get accustomed to the technical distinction between the terms "way of understanding" and "way of thinking" as is made here. This is partly because in communication among educators and in the literature on learning and teaching the two terms are often used interchangeably (and without exact definitions). Also, the phrase "way of" seems to connote a sort of a process and, hence, a dynamic image, whereas the definition of "way of understanding" as a *product* of a mental act may connote an outcome, a static image. My intention in using the phrase "way of" is to insinuate "one of several possible ways," which suggests that a mental act in mathematics can, and should, have multiple products and characteristics—an implied view in the *DNR* perspective, as we will see. The verbs "to understand" and "to think" are used in this paper in accordance with the definitions of the corresponding terms: "to understand" means to "have a way of understanding," and "to think" means to "apply a way of thinking." In DNR, and throughout this paper, "ways of understanding" and "ways of thinking" are distinguished from their values. For example, one's way of understanding can be judged as correct or wrong, useful or impractical in a given context, etc.

2 A Definition of Mathematics: Epistemological Considerations and Pedagogical Implications

The notions of "ways of understanding" and "ways of thinking" as defined here are key constructs in the definition of mathematics I will now state. Mathematicians, the practitioners of the discipline of mathematics, practice mathematics by carrying out mental acts with particular characteristics (ways of thinking) to produce particular constructs (ways of understanding). Accordingly, mathematics consists of these two categories of knowledge. Specifically:

Definition: Mathematics consists of two complementary subsets:

- The first subset is a collection, or structure, of structures consisting of particular axioms, definitions, theorems, proofs, problems, and solutions. This subset consists of all the institutionalized[4] ways of understanding in mathematics throughout history. It is denoted by WoU.

---

[4] *Institutionalized* ways of understanding are those the mathematics community at large accepts as correct and useful in solving mathematical and scientific problems. A subject matter of particular field may be viewed as a structure of institutionalized ways of understanding.

- The second subset consists of all the ways of thinking, which are characteristics of the mental acts whose products comprise the first set. It is denoted by WoT.

By this definition, mathematics is like a living organism. It grows continually as mathematicians carry out mental acts and their mathematical communities assimilate the ways of understanding and ways of thinking associated with the mathematicians' mental acts. The assimilation is attained when new ways of understandings are integrated into an existing mathematical edifice and ways of thinking are adopted in subsequent mathematical practices. As one can learn from the history of mathematics, the assimilation process is gradual and often not without conceptual struggle. Some ways of understanding and ways of thinking are regarded as inaccurate or faulty—sometimes long after they have been institutionalized. They, too, are part of mathematics according to this definition, as I explain later in this paper. In the rest of this section, I shall discuss several epistemological issues concerning this definition and examine their pedagogical consequences.

2.1 Listability

Mathematics as a union of WoU and WoT is not listable—capable of being completely listed. WoU contains more than the collection of all the statements appearing in mathematical publications, and the members of WoT are largely unidentified. I explain.

Consider a statement—say, a new theorem—that has appeared in a mathematical publication, such as a book or research paper. Its publication indicates recognition by a community that a new way of understanding has been accepted. Individual mathematicians might believe and act as if the published theorem represents a way of understanding *shared* by the community at large, whereas, in fact, each individual mathematician possesses an idiosyncratic way of understanding the theorem. Of all the latter "private" ways of understanding, consider only the subset of those that are consistent with the former "public" ways of understanding. These, too, are considered institutionalized, since it is assumed by the mathematics community that any way of understanding that is consistent with a "public" way of understanding is acceptable. Thus, WoU contains all the statements that have appeared in mathematical publications—which the mathematics community views as representations of shared ways of understanding. WoU also contains individual mathematicians' ideas which are consistent with the published assertions. While statements in the first set are listable, those in the latter are not. The reason they are not listable is this. Let S be a statement that has been published, and let S′ be a particular person's way of understanding S. Once this person has expressed S′ to the community, S′ moves to the domain of those ways of understanding assumed-to-be shared by the community. But the members of the community, including this person, possess idiosyncratic ways of understanding S′.

A pedagogical implication of this analysis is that a way of understanding, such as a definition, theorem, proof, or solution to a problem, cannot and should not be treated by teachers as an absolute universal entity shared by all students. Any statement a teacher (or a classmate) utters or puts on the board will be translated by each individual student into a way of understanding that depends on her or his experience and background. The goal of the teacher is then that these necessarily different individual mental constructs are compatible with each other. A classroom environment that promotes discussion and debate among students is both necessary for and instrumental in achieving this goal.

As to the WoT subset of mathematics, its members are not formally recognized by the mathematics community. They are neither explicitly targeted as instructional objectives by mathematicians nor investigated and reported in formal publications. Occasionally, however, they are informal parts of communications between collaborators. Pólya's book "How to Solve It" ([Pólya 1957]) is a rare attempt by a professional mathematician to explicate desirable problem-solving approaches, which, as was explained earlier, are ways of thinking. (For a discussion on Pólya's pedagogical and epistemological assumptions on mathematical heuristics, see [Schoenfeld 1992].) It is much more difficult to reflect on and express in precise words ways of thinking than ways of understanding. In *DNR*-based instruction, considerations of ways of thinking are central; they are an essential part of curriculum development and instruction, as we will see.

2.2 Boundaries

A consequence of my definition of mathematics is that mathematics must include ways of understanding and ways of thinking that from the vantage point of contemporary mathematicians are imperfect or erroneous; Euclid's *Elements* is an example. This leads to the following question: should ways of understanding and ways of thinking used or produced by individuals (students, for example) while they are engaged in a mathematical activity be considered mathematical even if they are narrow or faulty? My answer to this question is affirmative *in so far as* the individual has utilized—with or without the help of an expert—such ways of understanding and ways of thinking for the construction of institutionalized knowledge—knowledge accepted at the time by the mathematics community at large.

This position is consistent with the definitions of "way of understanding" and "way of thinking." As can be seen from the examples discussed in the previous section, these terms do not imply correct knowledge. The terms only indicate the knowledge currently held by a person, which may be correct or erroneous, useful or impractical. Having said this, it must be emphasized that *the ultimate goal of instruction must be unambiguous: to help students develop ways of understanding and ways of thinking that are compatible with those that are currently accepted by the mathematics community at large.* From a pedagogical point of view this goal is meaningless without realizing that the process of learning necessarily involves the construction of imperfect and even erroneous ways of understanding and deficient, or even faulty, ways of thinking. Teachers must be aware of this phenomenon when working toward an instructional goal, and their teaching actions must be consonant with this awareness. In particular, they must attempt to identify students' current ways of understanding and ways of thinking, regardless of their quality, and help students gradually refine and modify them toward those that have been institutionalized—those the mathematics community at large accepts as correct and useful in solving mathematical and scientific problems.

The repeated use of the term "institutionalized" here raises the question: what about creativity—the discovery of *new,* not necessarily institutionalized, ways of understanding and ways of thinking? Are such discoveries mathematical? By my definition of mathematics they are not. This position is based on the premise that mathematics is a human endeavor, not a predetermined reality. As such, it is the community of the creators of mathematics who makes decisions about the inclusion of new discoveries in the existing edifice of mathematics. Such decisions may never be made by the community at large, and the new discoveries may be forever lost as a result. The work of Ramanujan would have likely been lost had G. H. Hardy not recognized the precious mathematical discoveries in the letter Ramanujan sent to him around 1913. Other decisions may be delayed; the work of Grassman (19th century) and the work of Cantor (19th century) are examples. Grassman's work was ignored for many years but became later the basis for vector and tensor analysis and associative algebras. Cantor's set theory, too, was ignored or boycotted for some time, but was later recognized as one of the most important discoveries of twentieth century mathematics. At the time of their discoveries, prior to their institutionalization, these works did not belong to mathematics, according to the definition of mathematics I propose in this paper.

2.3 Relation to Ontology

There is a danger of confounding the above definition of mathematics with a particular philosophical stance with which I vehemently disagree. Mathematics, according to this definition, consists of ways of understanding and ways of thinking that have evolved throughout history. Inevitably, some of these constructs are narrow and even faulty if judged from a contemporary perspective. *This does not entail that particular mathematical statements could be true for some people and false for others*—a view that is implied by an extreme form of post-modernism, which asserts that mathematical truth depends on the culture or bias of the mathematician (see [Buss 2005] for a discussion against this view). Such disputed statements cannot be part of mathematics according to my definition, for they have never been institutionalized by any mathematics community in the history of mathematics. That is, no mathematics community, as far as I know, has ever accepted that a statement A *and* its negation, $\sim$A, can both be true within the same system of rules of inference. A statement such as "Every function on the real numbers is continuous" is true for intuitionists but false for the rest of us because the two communities are considering the statement within different systems of rules of inference. Also the term "function" has different meanings for the two communities. Thus, a statement must not be considered in isolation but within a context that constitutes its meaning.

What is disputed among philosophers, and to a lesser extent among mathematicians, is the answer to an ontological question: what is the nature of the *being and existence* of mathematics? For example, is mathematical practice an act of discovery of eternal objects and ideas that are independent of human existence, an intuition-free game in which symbols are manipulated according to a fixed set of rules, or a product of constructions from primitive intuitive objects, most notably the integers? The three positions expressed in this question correspond, respectively, to the three major schools of thought, Platonism, Formalism, and Constructivism. Since Constructivism insists that mathematical objects must be computable in a finite number of steps, it does not admit many results accepted by the other schools as true. The basis for this rejection is not "cultural difference" or "personal bias," as the extreme forms of post-modernism imply; rather, the basis for the rejection is philosophical: nothing can be asserted unless there is a proof—a constructivist proof—for it.

It is an open, empirical question whether mathematicians' ontological stances on the nature of mathematical practice have any bearing on their views of how mathematics is learned and, consequently, how it should be taught. I conjecture that teachers' approaches to the learning and teaching of mathematics are not determined by their ontological stance on the being and existence of mathematics. Dieudonné, a prominent member of the Bourbaki group, calls in the following statement for an uncompromising Formalist view:

> Hence the absolute necessity from now on for every mathematician concerned with intellectual probity to present his reasoning in *axiomatic* form, i.e., in a form where propositions are limited by *virtue of rules of logic only*, all intuitive "evidence" which may suggest expressions to the mind being deliberately disregarded.
>
> ([Dieudonné 1971], p. 253).

Yet, he cautions his reader:

> *We are saying that this is a form imposed on the presentation of the results; but this does not lessen in any way the role of intuition in their discovery.* Among the majority of researchers the role of intuition is considerable, and no matter how confused it may be, an intuition about the mathematical phenomena being studied often puts them on the track leading to their goal.   (Emphases added; [Dieudonné 1971], p. 253).

One can reasonably infer from these statements that Dieudonné's approach to teaching is to emphasize intuition despite his adherence to the Formalist school.

What, then, does determine one's approach to learning and teaching of mathematics?

2.4 Quality of Teacher's Knowledge Base

Quality of instruction is determined largely by what teachers know. Building on Shulman's ([1986], [1987]) work and consistent with current views ([Brousseau 1997] [Cohen/Ball 1999], [Cohen/Ball 2000]), *teacher's knowledge base* was defined in [Harel 1993] in terms of three components: *knowledge of mathematics, knowledge of student learning,* and *knowledge of pedagogy*. Here I present a refined definition of these components that is aligned with the definition of mathematics I have just discussed:

- *Knowledge of mathematics* refers to a teacher's *ways of understanding* and *ways of thinking*. It is the quality of this knowledge that is the cornerstone of teaching for it affects both what the teachers teach and how they teach it.
- *Knowledge of student learning* refers to the teacher's understanding of fundamental psychological principles of learning, such as how students learn and the impact of their previous and existing knowledge on the acquisition of new knowledge.
- *Knowledge of pedagogy* refers to teachers' understanding of how to teach in accordance with these principles. This includes an understanding of how to assess students' knowledge, how to utilize assessment to pose problems that stimulate students' intellectual curiosity, and how to help students solidify and retain knowledge they have acquired.

Thus, while mathematical knowledge is indispensable for quality teaching, it is not sufficient. Teachers must also know how to address students as learners. In *DNR*, however, teacher's knowledge of student learning and pedagogy rests on the teacher's knowledge of mathematics. That is to say, although each of the three components of knowledge is indispensable for quality teaching, they are not symmetric: the development of teachers' knowledge of student learning and of pedagogy depends on and is conditioned by their knowledge of mathematics. A brief example to illustrate this claim follows. The example is from an on-site professional development study, currently underway, aimed at investigating the evolution of teachers' knowledge base. One of the findings of this study is that teachers' appreciation for students' struggle with a particular concept is a function of the quality of the teachers' way of understanding that concept. For example, Lisa, one of the teacher participants in this study, developed and enthusiastically implemented an instructional activity where her tenth-grade class gradually discerned the formula for the sum of the interior angles in a convex polygon along with a mathematically acceptable justification for it. In one of the interviews with Lisa, she pointed out, with great satisfaction and a sense of accomplishment, that the class understood well the proof of the formula and some students even developed it on their own. On the other hand, Lisa, who had insufficient understanding of graphical representation of solutions to systems of linear inequalities, struggled to see the benefit of a multi-stage instructional activity that was designed to involve students in developing a solid understanding of how to solve and graph the solution of such systems. She inclined, instead, to provide the students with a prescribed procedure of how to solve these systems. Thus, Lisa's lack of a deep understanding of systems of linear inequality prevented her from pursuing good teaching of this topic. Overall, Lisa's knowledge of pedagogy and of student learning seems to evolve hand in hand with the growth of and self reflection on her knowledge of mathematics, not out of institutional demand to improve her students' mathematical performance.

3 Long-Term Curricular and Research Goals

In the opening of this paper, it was argued that the instructional objectives teachers set for their classes correspond merely to subject matter in terms of products of mental acts—ways of understanding, such as particular definitions, procedures, techniques, theorems, and proofs. Neither the actions of the teachers nor the justifications they provide for their objectives indicate attention to the characteristics of mental acts—to the ways of thinking that students are to develop by learning particular subject matter. Objectives formulated in terms of ways of understanding are essential, as it is asserted in one of the *DNR* principles, to be presented in Section 4, but without targeting ways of thinking, students are unlikely to become independent thinkers when doing mathematics. This brings up the question, when should we start targeting ways of thinking with students?

3.1 Elementary Mathematics

The formation of ways of thinking is extremely difficult and those that have been established are hard to alter. This is one of the main findings of our research (see for example, [Harel/Sowder 1998]). Hence, the development of desirable ways of thinking should not wait until students take advanced mathematics courses; rather, students must begin to construct them in elementary mathematics, which is rich in opportunities to help students begin acquiring crucial ways of thinking. Consider, for example, the concept of fraction. In current mathematics teaching, even when students learn mathematics symbolism in context, the context is usually limited. For example, the most common way of understanding the concept of fraction among elementary school students is what is known in the literature as the *part-whole* interpretation: m/n (where m and n are positive integers) means "m out of n objects." Many students never move beyond this limited way of understanding fraction and encounter, as a result, difficulties in developing meaningful knowledge of fraction arithmetic ([Lamon 2001]) and beyond ([Pustejovsky 1999]).

Seldom do students get accustomed to other alternative ways of understanding such as m/n means "the sum $1/n + \cdots + 1/n$, m times" or "the quantity that results from m units being divided into n equal parts" or "the measure of a segment m-inches long in terms of a ruler whose unit is n inches" or "the solution to the equation $nx = m$" or "the ratio $m : n$; namely, m objects for each n objects." This range of ways of understanding a fraction makes the area of fractions a powerful elementary mathematics topic—one that can offer young students a concrete context to construct desirable—indeed, crucial—ways of thinking, such as: mathematical concepts *can* be understood in different ways, mathematical concepts *should* be understood in different ways, and *it is advantageous* to change ways of understanding of a mathematical concept in the process of solving problems. These ways of thinking will be needed in the development of future ways of understanding. Indeed, without the above cluster of ways of thinking students are bound to encounter difficulties in other parts of mathematics. In calculus, for example, depending upon the problem at hand, one would need to interpret the phrase "derivative of a function at a," or the symbol $f'(a)$, as "the slope of a line tangent to the graph of a function at a" or "the $\lim_{h \to 0} (f(a + h) - f(a))/h$" or "the instantaneous rate of change at a" or "the slope of the best linear approximation to a function near a." Likewise, in solving linear algebra problems it is often necessary—or at least advantageous—to convert one way of understanding into another way of understanding by using the equivalence among problems on systems of linear equations, matrices, and linear transformations.

The history of mathematics can provide a guide to ways of thinking worth pursuing—in the classroom and in mathematics education research. In the rest of this section, I will illustrate this claim with examples from the history of algebra and proof.

3.2 Algebra

According to Klein ([1968]) the revival and assimilation of Greek mathematics during the 16th century resulted in a conceptual transformation that culminated in Vieta's development of symbolic algebra. Until then, mathematics had evolved for at least three millennia with hardly any symbols. The following is an example to illustrate the colossal role symbolic algebra played in defining modern mathematics. The work of Vieta that led to the creation of algebra and that of Descartes and Fermat that led to the creation of analytic geometry constituted the conceptual foundation for the critical shift from "results of operations" as the object of study to the operations themselves as the object of study. While the Greeks restricted their attention to attributes of spatial configurations and paid no attention to the operations underlying them, 19th century mathematics investigated the operations, their algebraic representations, and their structures. In particular, Euclidean constructions using only a compass and straightedge were translated into statements about the constructability of real numbers, which, in turn, led to observations about the structure of constructible numbers. A deeper investigation into the theory of fields led to the understanding of why certain constructions are possible whereas others are not. The Greeks had no means to build such an understanding, since they did not attend to the nature of the operations underlying Euclidean construction. Thus, by means of analytic geometry, mathematicians realized that all Euclidean geometry problems can be solved by a single approach, that of reducing the problems into equations and applying algebraic techniques to solve them. Euclidean straightedge-and-compass constructions were understood to be equivalent to equations, and hence the solvability of a Euclidean problem became equivalent to the solvability of the corresponding equation(s) in the constructible field.

The monumental role that symbolic algebra played in defining modern mathematics might be obvious to many, but it is worth pointing out in debates on the future direction of school mathematics, particularly when attempts are made to deemphasize symbolic manipulation skills. Often the rationale behind these attempts is the availability of electronic technologies equipped with computer algebra programs that can carry out complex computations of all kinds and in all areas of mathematics. While these technologies can have a positive role in the teaching of mathematics (see [Kaput/ Hegedus 2003]) they can, if not used wisely, deprive the students of the experience necessary for developing critical mathematical ways of thinking. In particular, they can deprive students of the opportunity to develop one of the most crucial mathematical ways of thinking, that of *algebraic invariance*.

Algebraic invariance refers to the way of thinking by which one recognizes that algebraic expressions are manipulated not haphazardly but with the purpose of arriving at a desired form and maintaining certain properties of the expression invariant. If this way of thinking were set as an instructional objective, elementary algebra—especially symbol manipulation skills— would be taught differently and more meaningfully. The method of completing the square, for example, would have an added value, not just as a method for solving quadratic equations but as an activity to advance students toward acquiring the algebraic invariance way of thinking. Assuming the students have already learned how to solve equations of the form $(x + T)^2 = L$, the teacher's action would be geared toward helping them manipulate the quadratic equation $ax^2 + bx + c = 0$ with a goal in mind—that of transforming the latter equation form into the former known equation form but maintaining the solution set unchanged. The intellectual gain is that students learn that algebraic expressions are re-formed for a reason and would, accordingly, develop a sense of the actions needed in order to reach a desired algebraic form. Without this ability, symbol manipulation is largely a mysterious activity for students—an activity they carry out according to prescribed rules but without a goal in sight. With this ability, on the other hand, symbol manipulation is not a matter of magic tricks performed by the teacher but goal-directed operations learnable by all students. Of course, one reason symbolic manipulation is being deemphasized is that this is not how it's being taught!

With the algebraic invariance way of thinking as an instructional objective, teaching techniques of integration, for example, will have an added value: would one teach such techniques not only so that students know how to determine antiderivatives of functions and values of integrals, but also to help students develop a critical way of thinking in mathematics—that of utilizing the power of mathematical symbolism to solve problems and make and prove conjectures. Techniques of integration provide an excellent context to advance students toward this goal, which is why I believe this topic should be maintained as part of the calculus curriculum. Take, for example, the simplest technique of Reduction to Standard Formulas. In solving an integral such as $\int \tan \theta \, d\theta$, students in freshman calculus learn to set a goal of transforming this unknown integral into an equivalent form that is familiar. Even if the students do not note that the symbolic representation $\tan \theta = \sin \theta / \cos \theta$ suggests the substitution $u = \cos \theta$, they would learn to appreciate such a representation when they see how it is utilized to *change the form, of the integral without changing its value* through a sequence of symbolic transformations, e.g.,

$$\int \tan \theta \, d\theta = - \int du/u = - \ln |u| + C = \ln \left|u^{-1}\right| + C = \ln |\sec \theta| + C.$$

Likewise, the algebraic invariance way of thinking is the basis for the concept of "equivalent systems;" that is, for manipulating a system of equations but maintaining its solution set.

Algebraic invariance is of course a special case of the more general way of thinking where one attempts to reduce a given problem into a familiar problem, represent a mathematical entity by another, more useful one, etc. The algebraic invariance way of thinking is not learned at once—one constructs it gradually by applying it in different contexts, such as techniques of integration, systems of linear equations, matrix factorization, etc. Students can start acquiring it in elementary mathematics, for example when transforming fractions into decimals, and vice versa. It is crucial, however, that such transformations are carried out meaningfully. Often students are taught to carry out symbolic transformations without adequate emphasis on their justification. For example, we have seen students learn to solve division problems involving decimal numbers (e.g., $0.14\sqrt{12.91}$ by transforming them into division problems involving whole numbers (e.g., $14\sqrt{1291}$) without ever attending to the mathematical basis for the transformation. Such exercises—devoid of meaning—have no value in advancing the algebraic invariance way of thinking among students, and they deprive the students of the opportunity to develop other critical ways of thinking. When justifying the equivalence of $0.14\sqrt{12.91}$ and $14\sqrt{1291}$, for example, students reason proportionally, e.g., when justifying that

$$\frac{0.14}{12.91} = \frac{0.14 \times 100}{12.91 \times 100},$$

attend to the nature of the number system (e.g., when justifying that $0.14 \times 100 = 14$ and $12.91 \times 100 = 1291$), and begin to develop algorithmic ways of thinking (when dividing 1291 into 14 by using the long-division algorithm). Obviously, such opportunities will not occur if the non-referential way of thinking dominates students' actions or if the students obtain the answer to $0.14\sqrt{12.91}$ by using a calculator.

It should be clear that in applying the algebraic invariance way of thinking, it is never the case that every single symbol is referential. It is only in critical stages—viewed as such by the person who carries the symbol manipulations—that one forms, or attempts to form, referential meanings. One does not usually attend to interpretation in the middle of symbol manipulations unless one encounters a barrier or recognizes a symbolic form that is of interest to the problem at hand; thus, for most of the process the symbols are treated as if they have a life of their own. It is in this sense that symbol manipulation skills should be understood and, accordingly, be taught.

One might ask, what is then the difference between the algebraic invariance way of thinking and the non-referential symbolic way of thinking? The answer is that the former includes the ability to pause at will to probe into a referential meaning for the symbols involved, whereas the latter does not. In applying the algebraic invariance way of thinking, the attempt to form a referential meaning does not have to occur, and even if it occurs it does not have to succeed. It is only that the person who carries out the manipulation has the ability to investigate the referential meaning of any symbol and transformation involved. In the non-referential symbolic way of thinking this ability is largely absent.

It is worth pointing out that the practice of manipulating symbols without *necessarily* examining their meaning played a significant role in the development of mathematics. For example, during the nineteenth century significant work was done in differential and difference calculus using a technique called the "operational method," a method whose results are obtained by symbol manipulations without understanding their meaning, and in many cases in violation of well-established mathematical rules. (See, for example, the derivation of the Euler-MacLaurin summation formula for approximating integrals by sums, in [Friedman 1991].) Mathematicians sought meaning for the operational method, and with the aid of functional analysis, which emerged early in the twentieth century, they were able to justify many of its techniques. Hence, the operational method technique is a manifestation of the algebraic invariance way of thinking, not the non-referential symbolic way of thinking.

In sum, with the algebraic invariance way of thinking, teachers would recognize that the goals of teaching manipulation skills include both learning how to compute solutions to particular problems and constructing conceptual tools that are an essential part of mathematical practice. The goal of teaching techniques of integration, for example, is not just to obtain an antiderivative for a given function, but also to help students acquire an important way of thinking—that of manipulating symbols with a goal of changing the form of an entity without changing a certain property of the entity, a way of thinking that is ubiquitous and essential in mathematical practice. The role of symbolic algebra in the reconceptualization of mathematics raises a critical question about the role of symbolic manipulation skills in students' conceptual development of mathematics. In response to increasing use of electronic technologies in schools, particularly computer algebra systems, educators should ask: might these tools deprive students of the opportunity to develop algebraic manipulation skills which are needed for the development of the algebraic invariance way of thinking?

3.3 Proof

Certain obstacles students encounter with the concept of proof seem to parallel obstacles in the development of mathematics. I discuss here two related observations. The first involves the transition from Greek mathematics to modern mathematics and the second the notion of Aristotelian causality.

3.3.1 Transition from Greek mathematics to Modern Mathematics

The deductive mode of thought was conceived by the Greeks more than 20 centuries ago and is still dominant in the mathematics of our day. The mathematicians of the civilizations that preceded the Greeks established their observations on the basis of empirical measurements; hence, they mainly possessed and employed empirical proof schemes—schemes marked by their reliance on evidence from examples or visual perceptions. In Greek mathematics, logical deduction is central in the reasoning process, and it was necessary for the geometric edifice they created. This edifice, however, represents a single model—that of idealized physical reality. This ultimate bond to a real-world context had an impact on the Greeks' deductive proof scheme, in that Euclid often uses arguments that are not logical consequences of his initial assumptions but are rooted in humans' intuitive physical experience. While Euclid's *Elements* is restricted to a single interpretation—namely that its content is a presumed description of human spatial realization—Hilbert's *Grundlagen* is open to different possible realizations, such as Euclidean space, the surface of a half-sphere, ordered pairs and triples of real numbers, etc., including the interpretation that the axioms are meaningless formulas. In other words, the *Grundlagen* characterizes a structure that fits different models. To reflect this fundamental conceptual difference, I refer to the Greeks' method of proving as the *Greek axiomatic proof scheme* and to the modern mathematics' method of proving as the *modern axiomatic proof scheme*. The transition between these two proof schemes is revolutionary: it marks a monumental conceptual change in humans' mathematical ways of thinking. Understanding this transition may shed light on epistemological obstacles students encounter upon moving from concrete models of their quantitative or spatial reality—such as the ones held by the Greeks—to a more abstract setting—such as that offered by Hilbert. As a historian might ask what events—social, cultural, and intellectual—necessitated the transition from one way of thinking to another (e.g. from pre-Greek mathematics to Greek mathematics to the mathematics of the Renaissance and to modern mathematics), a mathematics educator should ask what is the nature of the instructional interventions that can bring students to refine and alter an existing way of thinking to a more desirable one?

3.3.2 Causality

According to the definition of "proof scheme" presented in Section 1, certainty is achieved when an individual determines—by whatever means he or she deems appropriate—that an assertion is true. Truth alone, however, may not be the only aim of an individual, and he or she may also desire to know *why* the observation is true—the cause that makes it true. An individual may be certain of the truth of an observation and still strive to understand what in that truth liberates her or him from doubt. "Proofs really aren't there to convince you that something is true—they're there to show you why it is true," said Gleason, one of the solvers of Hilbert's Fifth Problem ([Yandell 2002], p. 150). Two millennia before him, Aristotle, in his Posterior Analytic, asserted "To grasp the why of a thing is to grasp its primary cause."

The 16–18th century conception of mathematics reflects global epistemological positions that can be traced back to this position of Aristotle, according to which explanations in science must be causal. According to the philosophers of the time, this position entails the rejection of proof by contradiction, for when a theorem "A implies B" is proved by showing how not B (under the assumption of A) leads logically to an absurdity, a person does not learn anything about the causality relationship between A and B, nor—one might add—does one gain insight into how the theorem was—or might have been—conjectured. Some students' behavior with proof can be explained in terms of this epistemological position, in that many able students search for causal relationships in proofs and dislike indirect proofs (see [Harel 1999]). Likewise, for the decisive majority of mathematicians the purpose of a proof is not only demonstrating that the assertion is true, but also explaining why it is true. Proofs by contradiction, while accepted in modern mathematics, usually lack the explanatory power direct proofs can have. As an example, it is worth mentioning the controversy that Hilbert's proof of Gordan's Conjecture[5] raised. Hilbert didn't find a basis that everyone had searched for but merely proved that if we accept Aristotle's law of the excluded middle ("Any statement is either true or its negation is true") then such a basis had to exist, whether we could produce it or not. Why was Hilbert's use of proofs-by-contradiction so controversial—after all, he was not the first to use this method of argument? According to Yandell ([2002]), previous uses had not dealt with a subject of such obvious calculational complexity. A pure existence proof does not produce a specific object that can be checked—one had to trust the logical consistency of the growing body of mathematics to trust the proof. The presence of an actual object that can be evaluated provides more than mere certainty; it can constitute a cause—in the Aristotelian sense—for the observed phenomenon. The philosophers of the Renaissance rejected proof by contradiction, and the practice of many

---

[5] The conjecture states: There is a finite basis from which all algebraic invariants of a given polynomial form could be constructed by applying a specified set of additions and multiplications.

mathematicians of that period, such as Cavalieri, Guldin, Descartes, and Wallis, reflected this position by explicitly avoiding proofs by contradiction in order to conform to the Aristotelian position on what constitutes perfect science ([Mancosu 1996]).

The implication of this history is not to avoid proofs by contradiction in mathematics curricula. On the contrary, proofs by contradiction represent an important, institutionalized way of thinking, which students should acquire. The point of this history is that modern proof schemes were born out of an intellectual struggle—a struggle in which Aristotelian causality seems to have played a significant role. It is an open question whether the development of students' proof schemes necessarily includes some of these epistemological obstacles. The fact that even able students encounter these obstacles makes this question even more relevant to the matter at hand. An answer to this question may shed light on some of the roots of the obstacles students encounter with certain kinds of proof, such as proof by contradiction. Accordingly, appropriate instructional interventions can be devised to help students develop desirable proof schemes as they encounter these obstacles, which, perhaps, are unavoidable.

4 DNR Based Instruction in Mathematics

DNR-based instruction in mathematics (DNR, for short) is a theoretical framework for the learning and teaching of mathematics—a framework that provides a language and tools to formulate and address critical curricular and instructional concerns. In this framework the *mathematical integrity* of the content taught and the *intellectual need* of the student are at the center of the instructional effort. *DNR* has been developed from a long series of teaching experiments in elementary, secondary, and undergraduate mathematics courses, as well as teaching experiments in professional development courses for teachers at each of these levels. Briefly, *DNR* can be thought of as a system consisting of three categories of constructs:

1. *Premises*—explicit assumptions underlying the *DNR* concepts and claims.
2. *Concepts*—referred to as *DNR determinants*.
3. *Instructional principles*—claims about the potential effect of teaching actions on student learning.

It goes beyond the scope of this paper to do more than present a brief outline of these constructs. For more about *DNR,* see [Harel 1989], [Harel 2001], [Harel et al. forthcoming] and [Harel forthcoming].

Premises. One of the *DNR* premises is the *conceptualization premise:*

Humans—all humans—possess the ability to develop a desire to be puzzled and to learn to carry out physical and mental acts to fulfill their desire to be puzzled and to solve the puzzles they create.

This premise, which follows from Aristotle, is one of eight *DNR* premises. Note that it assumes not only humans' desire *to solve puzzles* but also humans' desire *to be puzzled.* It serves as a basis for many themes in *DNR*—the *necessity principle,* to be stated shortly, is one of them. It is also the basis of *DNR'*s interpretation of *equity: all* students are capable of learning if they are given the opportunity to be puzzled, create puzzles, and solve puzzles.

Concepts. "Mental act," "way of understanding," and "way of thinking" are examples of *DNR determinants*.

Instructional Principles. Not every *DNR* instructional principle is explicitly labeled as such. The system states only three foundational principles: the *duality principle*, the *necessity principle*, and the *repeated-reasoning principle;* hence, the acronym *DNR*. The other principles in the system are derivable from and organized around these three principles.

Recall, according to my definition, mathematical knowledge consists of ways of understanding and ways of thinking. The *duality principle* concerns the developmental interdependency between these two constructs:

> **The *Duality Principle*:** Students develop ways of thinking only through the construction
> of ways of understanding, and the ways of understanding they produce are determined
> by the ways of thinking they possess.

The reciprocity between ways of understanding and ways of thinking claimed in the duality principle is of mutual effect: a change in ways of thinking brings about a change in ways of understanding, and vice versa. The claim intended is, in fact, stronger. Not only do these two categories of knowledge affect each other but a change in one cannot occur without a corresponding change in the other.

Implied from the duality principle is that preaching ways of thinking to students would have no effect on the quality of the ways of understanding they would produce. For example, talking to them about the nature of proof in mathematics or advising them to use particular heuristics would have minimal or no effect on the quality of the proofs and solutions they would produce. Only by producing desirable ways of understanding—by way of carrying out mental acts of, for example, solving mathematical problems and proving mathematical assertions—can students construct desirable ways of thinking. This seems obvious until one observes, for example, teachers teaching problem-solving heuristics explicitly and students following them as if they were general rules rather than rules of thumb for solving problems.

Attention to ways of thinking, on the other hand, is necessary—according to the duality principle—for they direct teachers as to which teaching actions to avoid and which to pursue. As we have discussed earlier, attention to desirable ways of thinking—such as algebraic invariance, proportional reasoning, and algorithmic reasoning—highlights the need to focus on particular ways of understanding certain concepts and processes (e.g., the solution process of quadratic functions, techniques of integration, and division of decimal numbers; see Section 3.2). In particular, teachers must take into consideration students' current ways of thinking. For example, a college instructor may start a course in geometry with finite geometries as a preparation for non-Euclidean geometries. We found ([Harel/Sowder 1998]) that most undergraduate students taking college geometry are not prepared for such an instructional treatment because they do not possess the modern axiomatic proof scheme—which includes the way of thinking that geometric properties are not limited to spatial imageries.[6] As was discussed earlier, this way of thinking was born at the turn of the 20th century with the publication of Hilbert's *Grundlagen* and is considered revolutionary in the development of mathematics.

---

[6] For example, students in our study encountered insurmountable difficulty interpreting the statement "Given a line and a point not on the line, there is a line which contains the given point and is parallel to the given line" in a finite geometry.

Of critical pedagogical importance is the question: what is the nature of instructional treatments that can help students construct desirable ways of understanding and ways of thinking? This is addressed by the other two *DNR* principles: the *necessity principle* and the *repeated reasoning principle*.

> The *Necessity Principle*: For students to learn what we intend to teach them, they must have a need for it, where 'need' refers to *intellectual need*, not social or economic need.

Most students, even those who are eager to succeed in school, feel intellectually aimless in mathematics classes because we—teachers—fail to help them realize an *intellectual need* for what we intend to teach them. The term *intellectual need* refers to a behavior that manifests itself internally with learners when they encounter an intrinsic problem—a problem they understand and appreciate. For example, students might encounter a situation that is incompatible with, or presents a problem that is unsolvable by, their existing knowledge. Such an encounter is intrinsic to the learners because it stimulates a desire within them to search for a resolution or a solution, whereby they might construct new knowledge. There is no guarantee that the learners construct the knowledge sought or any knowledge at all, but whatever knowledge they construct is meaningful to them since it is integrated within their existing cognitive schemes as a product of effort that stems from and is driven by their personal, intellectual need. While one should not underestimate the importance of students' social need (e.g., mathematical knowledge can endow me with a respectable social status in my society) and economic need (e.g., mathematical knowledge can help me obtain a comfortable means of living) as learning factors, teachers should not and cannot be expected to stimulate (let alone fulfill) these needs. Intellectual need, on the other hand, is a prime responsibility of teachers and curriculum developers.

Even if ways of understanding and ways of thinking are necessitated through students' *intellectual need* there is still the task of ensuring that students internalize, organize, and retain this knowledge. This concern is addressed by the *repeated-reasoning principle:*

> The *Repeated Reasoning Principle:* Students must *practice* reasoning in order to internalize, organize, and retain ways of understanding and ways of thinking.

Research has shown that repeated experience, or practice, is a critical factor in these cognitive processes (see, for example, [Cooper 1991]). *DNR-based instruction* emphasizes repeated reasoning that reinforces desirable ways of understanding and ways of thinking. Repeated reasoning, not mere drill and practice of routine problems, is essential to the process of internalization—a conceptual state where one is able to apply knowledge autonomously and spontaneously. The sequence of problems must continually call for reasoning through the situations and solutions, and they must respond to the students' changing intellectual needs.

These instructional principles are the basis for many of the pedagogical positions expressed in this paper, and they have been used to organize my instruction, in general, and teaching experiments, in particular. Consider the following unit taken from a recent teaching experiment with secondary mathematics teachers with limited mathematics background. The teachers worked on justifying the quadratic formula. Prior to this problem, they had repeatedly worked with many quadratic functions, finding their roots by essentially completing the square. They abstracted this process to develop the quadratic formula. In doing so they repeatedly transformed particular equations of the form $ax^2 + bx + c = 0$ into an equivalent equation of the form $(x + T)^2 = L$ for some terms T and L, in order for them to solve for x (as $-T + \sqrt{L}$ and $-T - \sqrt{L}$). To get to the desired equivalent form, they understood the reason and need for dividing through by a, bringing to the other side of the equation, and completing the square. For these teachers, the symbolic manipulation process stems from an intellectual need—the need to arrive at a particular form in order to determine the equation's unknown—and conditioned by quantitative considerations—to transform the algebraic expressions without altering their quantitative value. In these activities, the teachers practiced the algebraic invariance way of thinking, whose importance I have discussed in Section 3.2. We see here the simultaneous implementation of the duality principle, the necessity principle, and the repeated reasoning principle. In particular, the repeated application of the invariance way of thinking helped the participant teachers internalize it, whereby they become autonomous and spontaneous in applying it.

5 Summary

Current teaching practices tend to view mathematics in terms of subject matter, such as definitions, theorems, proofs, problems and their solutions, not in terms of the conceptual tools that are necessary to construct such mathematical objects. The tenet of this paper is that instruction should focus on both categories of knowledge: subject matter and conceptual tools. The paper defines these two categories and explains why both categories are needed. The definitions and explanations are oriented within the *DNR* framework. Central to *DNR* is the distinction between "way of understanding" and "way of thinking." "Way of understanding" refers to a cognitive product of a person's mental act, whereas "way of thinking" refers to its cognitive characteristic. Accordingly, mathematics is defined as the union of two sets: the set WoU, which consists of all the institutionalized ways of understanding in mathematics throughout history, and the set WoT, which consists of all the ways of thinking that characterize the mental acts whose products comprise the first set.

The members of WoT are largely unidentified in the literature, though some significant work was done on the problem-solving act (e.g., [Schoenfeld 1985]; [Silver 1985]) and the proving act (see an extensive literature review in [Harel/Sowder 2007]). The members of WoU include all the statements appearing in mathematical publications, such as books and research papers, but it is not listable because individuals (e.g., mathematicians) have their idiosyncratic ways of understanding. A pedagogical consequence of this fact is that a way of understanding should not be treated by teachers as an absolute universal entity shared by all students, for it is inevitable that each individual student is likely to possess an idiosyncratic way of understanding that depends on her or his experience and background. Together with helping students develop desirable ways of understanding, the goal of the teacher should be to promote interactions among students so that their necessarily different ways of understanding become compatible with each other and with that of the mathematical community.

Since mathematics, according to the definition offered in this paper, includes historical ways of understanding and ways of thinking, it must include ones that might be judged as imperfect or even erroneous by contemporary mathematicians. Also included in mathematics are imperfect ways of understanding and ways of thinking used or produced by individuals in the process of constructing institutionalized knowledge. The boundaries as to what is included in mathematics are in harmony with the nature of the process of learning, which necessarily involves the construction of imperfect and erroneous ways of understanding and deficient and faulty ways of thinking. These boundaries, however, are not to imply acceptance of the radical view that particular mathematical statements could be true for some people and false for others.

My definition of mathematics implies that an important goal of research in mathematics education is to identify desirable ways of understanding and ways of thinking, recognize their development in the history of mathematics, and, accordingly, develop and implement mathematics curricula that aim at helping students construct them. This claim was illustrated in the contexts of algebra and proof. The discussion on algebra highlights the need to promote the algebraic invariance way of thinking among students. With it, students learn to manipulate symbols with a goal in mind—that of changing the form of an entity without changing a certain property of the entity. It also points to the risk that the use of electronic technologies in schools, particularly computer algebra systems, can deprive students of the opportunity to develop this crucial way of thinking. The discussion on proof focuses on the transition from Greek mathematics to modern mathematics and the role of Aristotelian causality in the development of mathematics during the Renaissance. It raises the question of whether the development of students' proof schemes parallels those of the mathematicians and philosophers of these periods. An answer to this question would likely have important curricular and instructional implications.

Since the formation of desirable ways of thinking is difficult and undesirable ways of thinking that have been formed are hard to relinquish, an effort must be made in early grades to help students acquire desirable ways of thinking. The concept of fraction, for example, can be taught with multiple ways of understanding, and in a context where students can develop ways of thinking necessary for the acquisition of advanced mathematics. Similarly, arithmetic problems such as division of decimals can provide invaluable opportunities to engage in proportional reasoning and algorithmic reasoning and revisit the nature of the decimal-number system.

Pedagogically, the most critical question is how to achieve such a vital goal as helping students construct desirable ways of understanding and ways of thinking. DNR has been developed to achieve this very goal. As such, it is rooted in a perspective that positions the mathematical integrity of the content taught and the intellectual need of the student at the center of the instructional effort. The mathematical integrity of a curricular content is determined by the ways of understanding and ways of thinking that have evolved over many centuries of mathematical practice and continue to be the ground for scientific advances. To address the need of the student as a learner, a subjective approach to knowledge is necessary. For example, the definitions of the process of "proving" and "proof scheme" are deliberately student-centered (see Section 1). It is so because the construction of new knowledge does not take place in a vacuum but is shaped by one's current knowledge. What a learner knows now constitutes a basis for what he or she will know in the future. This fundamental, well-documented fact has far-reaching instructional implications. When applied to the concept of proof, for example, this fact requires that instruction takes into account students' current proof schemes, independent of their quality. Again, despite this subjective definition *the goal of instruction must be unambiguous—namely, to gradually refine current students' proof schemes toward the proof scheme shared and practiced by contemporary mathematicians.* This claim is based on the premise that such a shared scheme exists and is part of the grounds for advances in mathematics.

Instruction concerns *what* mathematics should be taught as well as *how* to teach it. While the definition of mathematics offered in this paper dictates the kind of knowledge to teach—ways of understanding *and* ways of thinking—the three DNR principles stipulate how to teach that knowledge:

The duality principle concerns the developmental interdependency between ways of under-standing and ways of thinking. Students would be able to construct a way of thinking associated with a certain mental act or refine or modify an existing one *only* if they are helped to construct suitable ways of understanding associated with that mental act. Conversely, students would be able to construct a way of understanding associated with a certain mental act or refine or modify an existing one *only* if they are helped to construct suitable ways of thinking associated with that mental act in the form of problem-solving approaches or proof schemes.

According to the necessity principle, problem solving is not just a goal but also the means—the only means—for learning mathematics. Learning grows only out of problems intrinsic to the students, those that pose an intellectual need for them. In general, an intellectually-based activity is one where students' actions are driven by a desire to solve intrinsic problems. In a socially-based activity, on the other hand, students' actions are carried out merely to satisfy a teacher's will.[7] In an intellectually-based teaching environment, students are continually challenged with new problems from which they elicit new concepts and ideas. Such an environment is necessary for learning, and is conducive to creativity.

The repeated reasoning principle is complementary to the duality principle and the necessity principle, in that its aim is for students to internalize what they have learned through the application of these two principles. Through repeated reasoning in solving intrinsic problems, the application of ways of understanding and ways of thinking become autonomous and spontaneous.

Acknowledgements I wish to acknowledge the helpful comments from John Baldwin, Evan Fuller, Boris Koichu, Alfred Manaster, and the Editors of this book. The framework presented here is part of the *DNR* Project, supported, in part, by the National Science Foundation (REC 0310128). Opinions expressed are those of the author and not necessarily those of the Foundation.

References

[Buss 2005] Buss, S., "Nelson's work on logic and foundations and other reflections on foundations of mathematics," in W. Faris, ed., *Quantum Theory and Radically Elementary Mathematics,* Princeton University Press, 2005.

[Brousseau 1997] Brousseau, G., *Theory of Didactical Situations in Mathematics*, Dordrecht: Kluwer Academic Publishers, 1997.

[Cohen/Ball 1999] Cohen, D. K. & Ball, D. L., *Instruction, Capacity, and Improvement* (CPRE Research Report No. RR-043), Philadelphia, PA: University of Pennsylvania, Consortium for Policy Research in Education, 1999.

[Cohen/Ball 2000] ——, *Instructional innovation: Reconsidering the story.* Paper presented at the Annual Meeting of the American Educational Research Association, New Orleans, 2000.

[Cooper 1991] Cooper, R., "The role of mathematical transformations and practice in mathematical de-velopment," in L. Steffe, ed., *Epistemological Foundations of Mathematical Experience,* New York, Springer-Verlag, 1991.

[Dieudonné 1971] Dieudonné, "Modern axiomatic methods and the foundations of mathematics," in F. Le Lionnais, ed., *Great currents of mathematical thought* (Vol.1), New York, Dover, 1971.

---

[7] The notion of intellectually, but not socially, based activity is similar to what Brousseau (1997) calls an *adidactical situation*.

[Friedman 1991] Friedman, B., *Lectures on Applications-Oriented Mathematics,* New York, John Wiley & Sons, 1991.

[Harel 1993] Harel, G., "On teacher education programs in mathematics," *International Journal for Mathematics Education in Science and Technology,* 25 (1993), pp. 113–119.

[Harel 1998] ——, "Two Dual Assertions: The First on Learning and the Second on Teaching (Or Vice Versa)," *The American Mathematical Monthly,* 105 (1998), pp. 497–507.

[Harel/Sowder 1998] Harel, G., & Sowder, L., "Students' proof schemes: Results from exploratory studies," pp. 234–283 in A. Schoenfeld, J. Kaput, & E. Dubinsky (Eds.), *Research in collegiate mathematics education III,* Providence, RI, American Mathematical Society, 1998.

[Harel 1999] Harel, G., "Students' understanding of proofs: A historical analysis and implications for the teaching of geometry and linear algebra," *Linear Algebra and Its Applications,* 302–303 (1999), pp. 601–613.

[Harel 2001] ——, "The development of mathematical induction as a proof scheme: A model for DNR-based instruction," pp. 185-212 in S. Campbell & R. Zazkis, eds., *The learning and teaching of number theory,* Dordrecht, The Netherlands, Kluwer, 2001.

[Harel 2007] ——, "The DNR System as a Conceptual Framework for Curriculum Development and Instruction," pp. 263–280 in R. Lesh, E. Hamilton., J. Kaput, eds., *Foundations for the Future in Mathematics Education,* Erlbaum, 2007.

[Harel et al. forthcoming] Harel, G., Fuller, E., & Rabin, J. (submitted), "Attention to meaning by algebra teachers."

[Harel forthcoming] Harel, G. (in preparation), *DNR-Based Instruction in Mathematics* (a book manuscript).

[Harel/Sowder 2007] Harel, G., & Sowder, L., "Toward comprehensive perspectives on the learning and teaching of proof," pp. 805–842 in F. Lester (Ed.), *Second Handbook of Research on Mathematics Teaching and Learning: a project of the National Council of Teachers of Mathematics,* Charlotte, NC: Information Age Pub., 2007.

[Kaput/Hegedus 2003] Kaput, J., & Hegedus, S., "The effect of SimCalc connected classrooms on students' algebraic thinking," pp. 47–54 in N. A. Pateman, B. J. Dougherty & J. Zilliox, eds., *Proceedings of the 27th Conference of the International Group for the Psychology of Mathematics Education* (Vol. 3), Honolulu, Hawaii, College of Education, University of Hawaii, 2003.

[Klein 1968] Klein, J., *Greek mathematical thought and the origin of algebra* (E. Brann, trans.), Cambridge, MA, MIT Press, 1968. (Original work published 1934).

[Kleiner 1991] Kleiner, I., "Rigor and proof in mathematics: A historical perspective," *Mathematics Magazine,* 64(5) (1991), pp. 291–314.

[Johnson 1987] Johnson, M., *The Body in the Mind,* The University of Chicago Press, Chicago, 1987.

[Lakoff/Johnson 2003] Lakoff, G., Johnson, M., *Metaphors we live by,* Chicago, The University of Chicago Press, 2003.

[Lamon 2001] Lamon, S., "Presenting and representing: From fractions to rational numbers," in Cuoco, A., ed., *The roles of representation in school mathematics* (2001 Yearbook of the National Council of Teachers of Mathematics). Reston, VA, NCTM, 2001.

[Mancosu 1996] Mancosu, P., *Philosophy of mathematical practice in the 17th century,* New York, Oxford University Press, 1996.

[Matz 1982] Matz, M., "Towards a process model for high school algebra errors," pp. 25–50 in D. Sleeman & J. S. Brown, eds., *Intelligent Tutoring Systems,* New York, Academic Press, 1982.

[Pustejovsky 1999] Pustejovsky, P., *Beginning Undergraduate Students' Understanding of Derivative: Three Case Studies,* Doctoral Dissertation, Marquette University, 1999.

[Polya 1957] Polya, G., *How to Solve It,* Princeton University Press, 1957.

[Schoenfeld 1985] Schoenfeld, A., *Mathematical Problem Solving*, Orlando, FL, Academic Press, 1985.

[Schoenfeld 1992] ——, "Learning to think mathematically: Problem solving, metacognition, and sense making in mathematics, pp. 334–370 in D. Grouws, ed., *Handbook for Research on Mathematics Teaching and Learning*, New York, Macmillan, 1992.

[Shulman 1986] Shulman, L., "Those who understand: Knowledge growth in teaching," *Educational Researcher, 15* (1986), pp. 4–14.

[Shulman 1987] ——, "Knowledge and teaching: Foundations of the new reform," *Harvard Educational Review, 57* (1987), pp. 1–22.

[Silver 1985] Silver, E., *Teaching and learning mathematical problem solving: Multiple research perspective*, Hillsdale, NJ: Lawrence Erlbaum, 1985.

[Weber 2001] Weber, K., "Student difficulty in constructing proof: The need for strategic knowledge," *Educational Studies in Mathematics, 48* (2001), pp. 101–119.

[Yandell 2003] Yandell, B., *The Honors Class: Hilbert's Problems and Their Solvers*, A K Peters Ltd., 2003.

14

What Will Count as Mathematics in 2100?

Keith Devlin
Senior Researcher and Executive Director
CSLI, Stanford

From the Editors

In this chapter, Keith Devlin examines how our concept of mathematics has changed over the centuries and suggests a direction it may grow over this next century. He anticipates the focus of mathematics continuing to shift from science and engineering towards the humanities and social sciences. As a result, students who are more people-oriented may be increasingly attracted to mathematics. Of course, as our idea of mathematics changes, our approach to mathematics courses will have to change correspondingly.

Keith Devlin is a scholar with broad interests, and a penchant for trying to communicate mathematics to a broad audience. He is a Senior Researcher at, and Executive Director of, the Center for the Study of Language and Information, as well as a Consulting Professor of Mathematics at Stanford University (www.stanford.edu/~kdevlin/). He is a co-founder of the Stanford Media X research network, which provides an industry portal to Stanford departments and centers which study people and technology—from engineering and computer science to psychology, linguistics, education, and art. He is also a co-founder of the university's H-STAR (Human Sciences and Technology Advanced Research) institute (a new Stanford interdisciplinary research center focusing on people and technology—how people use technology, how to better design technology to make it more usable, how technology affects people's lives, and the innovative use of technologies). He is a World Economic Forum Fellow and a Fellow of the American Association for the Advancement of Science. His current research is focused on the use of different media to teach and communicate mathematics to diverse audiences. He also works on the design of information/reasoning systems for intelligence analysis. Other research interests include theory of information, models of reasoning, applications of mathematical techniques in the study of communication, and mathematical cognition. He has written 27 books and over 75 published research articles. His books tend to either be in mathematical logic or aimed at a broad audience. A number of the latter, however, are likely to be of interest to readers of this volume, including The Math Gene: How Mathematical Thinking Evolved and Why Numbers Are Like Gossip *(2000),* The Language of Mathematics: Making the Invisible Visible, *(1998),* Mathematics: The New Golden Age *(Second edition, 1999),* Goodbye Descartes: The End of Logic and the Search for a New Cosmology of the Mind *(1997), and* Mathematics: The Science of Patterns—The Search for Order in Life, Mind, and the Universe *(1994). Between 1983 and 1990, he wrote a twice-monthly column on mathematics and computing for the British newspaper* The Guardian. *He is "the Math Guy" on National Public Radio, and writes the Devlin's Angle column on MAA Online.*

1 What is Mathematics Today?

What *is* mathematics? That's one of the most basic questions in the philosophy of mathematics. The answer has changed several times throughout history.

Up to 500 B.C. or thereabouts, mathematics was—if it was anything to be given a name—the systematic use of numbers. This was the period of Egyptian, Babylonian, and early Chinese and Japanese mathematics. In those civilizations, mathematics consisted primarily of arithmetic. It was largely utilitarian, and very much of a cookbook variety. ("Do such and such to a number and you will get the answer.")

Modern mathematics, as an area of *study*, traces its lineage to the ancient Greeks of the period from around 500 B.C. to 300 A.D. From the perspective of what is classified as mathematics today, the ancient Greeks focused on properties of number and shape (geometry). (The word "mathematics" itself comes from the Greek for "that which is learnable." As always when interpreting one culture or age with another, it has to be acknowledged that things often appear quite different to those within a particular culture or age than when viewed from the other.)

It was with the Greeks that mathematics came into being as an identifiable discipline, and not just a collection of techniques for measuring, counting, and accounting. Greek interest in mathematics was not just utilitarian; they regarded mathematics as an intellectual pursuit having both aesthetic and religious elements. Around 500 B.C., Thales of Miletus (now part of Turkey) introduced the idea that the precisely stated assertions of mathematics could be logically proved by a formal argument. This innovation marked the birth of the theorem, the central focus of modern mathematics.

The next major change in the overall nature in mathematics (again from the perspective of looking back at the chain of development that led to today's subject) was when Isaac Newton (in England) and Gottfried Leibniz (in Germany) independently invented the calculus. Calculus is the study of continuous motion and change. Previous mathematics had been largely restricted to the static issues of counting, measuring, and describing shape. With the introduction of techniques to handle motion and change, mathematicians were able to study the motion of the planets and of falling bodies on earth, the workings of machinery, the flow of liquids, the expansion of gases, physical forces such as magnetism and electricity, flight, the growth of plants and animals, the spread of epidemics, the fluctuation of profits, and so on. After Newton and Leibniz, mathematics became the study of number, shape, motion, and change.

Most of the initial work involving calculus was directed toward the study of physics; indeed, many of the great mathematicians of the period are also regarded as physicists. But from about the middle of the 18th century there was an increasing interest in the mathematics itself, not just its applications, as mathematicians sought to understand what lay behind the enormous power that the calculus gave to humankind. Here the old Greek tradition of formal proof came back into ascendancy, as a large part of present-day pure mathematics was developed. By the end of the 19th century, mathematics had become the study of number, shape, motion, change, and of the mathematical tools that are used in this study, together with a number of other topics, such as formal logic and the theory of probabilities. With the growth and diversification of the subject, it became quite difficult to say what mathematics is without writing a short essay.

In the 1980s, however, a definition of mathematics emerged on which most mathematicians now agree, and which captured the broad and increasing range of different branches of the subject: *mathematics is the science of patterns*. This definition does, admittedly, require some elaboration as to what exactly constitutes a pattern, but that aside it captures very well what the subject is about.

According to this new definition, what the mathematician does is examine abstract patterns— numerical patterns, patterns of shape, patterns of motion, patterns of behavior, voting patterns in a population, patterns of repeating chance events, and so on. Those patterns can be either real or imagined, visual or mental, static or dynamic, qualitative or quantitative, purely utilitarian or of little more than recreational interest. They can arise from the world around us, from the depths of space and time, or from the inner workings of the human mind. Different kinds of patterns give rise to different branches of mathematics. For example:

- Arithmetic and number theory study the patterns of number and counting.
- Geometry studies the patterns of shape.
- Calculus allows us to handle patterns of motion (including issues such as velocity and acceleration, polynomial motion, exponential motion, etc.).
- Logic studies patterns of reasoning.
- Probability theory deals with patterns of chance.
- Topology studies patterns of closeness and position.

and so forth.

It is mathematics viewed in this way that I will attempt to project one hundred years into the future, to the start of the 22nd century. But before I do that, I should note that, around 150 years ago, in the middle of the period when mathematics was growing in scope, the subject also changed in nature.

2 The Last Revolution in Mathematics

For most of its history, mathematics was regarded as primarily about calculation or symbolic manipulation. Proficiency in mathematics was measured primarily in terms of an ability to carry out calculations or manipulate symbolic expressions to solve problems. In the middle of the 19th century, however, a revolution took place. Generally regarded as having its epicenter in the small university town of Göttingen in Germany, the revolution's leaders were the mathematicians Lejeune Dirichlet, Richard Dedekind, and Bernhard Riemann. In their new conception of the subject, the primary focus was not performing a calculation or computing an answer, but formulating and understanding abstract concepts and relationships. This represented a shift in emphasis from *doing* to *understanding*. For the Göttingen revolutionaries, mathematics was about "Thinking in concepts" (*Denken in Begriffen*). Mathematical objects, which had been thought of as given primarily by formulas, came to be viewed rather as carriers of conceptual properties. Proving was no longer a matter of transforming terms in accordance with rules, but a process of logical deduction from concepts.

For example, one post-Göttingen concept is the modern notion of a function. Prior to the 19th century, mathematicians were used to the fact that a formula such as $y = x^2 + 3x - 5$ specifies a rule that produces a new number (y) from any given number (x). Then along came Dirichlet who said, forget the formula and concentrate on what the function *does* in terms of input–output behavior. A *function*, according to Dirichlet, is any rule that produces new numbers from old. The rule does not have to be specified by an algebraic formula. In fact, there's no reason to restrict your attention to numbers. A function can be any rule that takes objects of one kind and produces new objects from them.

Mathematicians began to study the properties of abstract functions, specified not by some formula but by their behavior. For example, does the function have the property that when you present it with different starting values it always produces different answers? (Injectivity.)

This approach was particularly fruitful in the development of real analysis, where mathematicians studied the properties of continuity and differentiability of functions as abstract concepts in their own right.

Karl Weierstrass in Germany and Augustin Cauchy in France analyzed continuity and differentiability, finally coming up with today's famous epsilon-delta definitions. With Cauchy's contributions, in particular, mathematicians finally had a rigorous way to handle infinity, a concept that their predecessors had grappled with since the ancient Greek era. Riemann spoke of mathematics having reached "a turning point in the conception of the infinite."

In the 1850s, Riemann defined a complex function by *its property of differentiability*, rather than a formula, which he regarded as secondary. Karl Friedrich Gauss's residue classes were a forerunner of the approach—now standard—whereby a mathematical structure is defined as a set endowed with certain operations, whose behaviors are specified by axioms. Taking his lead from Gauss, Dedekind examined the new concepts of ring, field, and ideal—each of which was defined as a collection of objects endowed with certain operations.

Like most revolutions, the Göttingen one had its origins long before the main protagonists came on the scene. The Greeks had certainly shown an interest in mathematics as a conceptual endeavor, not just calculation, and in the 17th century, Gottfried Leibniz thought deeply about both approaches. But for the most part, until the Göttingen revolution, mathematics was viewed primarily as a collection of procedures for solving problems. To today's mathematicians, however, brought up entirely with the post-Göttingen conception of mathematics, what in the 19th century was a revolution is simply taken to be what mathematics is.

How will mathematicians one hundred years from now view their subject? Will it be more of the same? Will there be radically new branches of the subject? Or will there be another revolution, which changes the very nature of what is viewed as "mainstream mathematics"? These are the questions I want to address here.

3 How and Why Mathematics Changes

The first thing to note is that, for all its abstraction, much of mathematics has been developed in response to the needs of society. For example, the needs of commerce and trade led to the development of techniques of arithmetic, and navigation and architecture gave rise to geometry and trigonometry. A great deal of the mathematics developed since the 17th century was created with applications in the physical world in mind. In particular, the invention of calculus was motivated in large part by the need to bring precision to the study of the motion of the planets. In the physical domain for which it was developed, today's mathematics has been highly successful. That success depends in large part upon the deterministic nature of much of the physical universe, which makes it amenable to a mathematical approach.

While foretelling the future is always a perilous activity, most present-day scientists seem to agree that the next hundred years will see major advances in the life sciences—some of those sciences very new—such as biology, psychology, sociology, neuroscience, and the study of mind and consciousness. Some of these areas seem amenable to the application of current mathematical techniques. For example, many parts of biology already are highly mathematical. Other areas may yield to techniques that, while new, are not radically different from current mathematics. When it comes to the social and psychological world of people, however, we are in a (seemingly) highly nondeterministic realm that appears to rule out more than the occasional, fairly superficial use of mathematics *as we understand it today*. (The exception is where larger populations are concerned, when statistical techniques can capture the deterministic order that can emerge from the often nondeterministic actions of the individuals.) It is these areas that will, I believe, give rise to the development of new mathematics. But here's the rub. I suspect that this new mathematics will not look very much like today's mathematics. Whereas the Göttingen revolution changed the *nature* of mathematics, but left it *looking* on the surface much as it always had, I believe the next revolution will leave the fundamental *nature* of mathematics unchanged but will lead to something that *looks* very different on the surface.

The reason why this new mathematics will look different is precisely because it will be applied to domains having either a significant degree of nondeterminism or else such high complexity as to defy capture within a traditional mathematical framework in any form intelligible to the human mind. Handling such matters will require a new form of analysis that, *to present day eyes*, will look like a blend of mathematics with rigorous, logical—but *not mathematical*—reasoning (of the kind you can find today in, say, psychology or sociology—I'll give some examples presently).

This new form of reasoning—what I am suggesting will come to be viewed as a new form of *mathematics*—will result from the meeting of two approaches that can be witnessed today: top down (where attempts are made to apply current mathematical techniques to some human domain) and bottom up (where attempts are made to make human-science arguments more mathematical). At present, the gap between those two approaches is generally large. But a hundred years from now, it will, I believe, in many cases have been bridged.

Let me first illustrate my point with three examples from fields where both the bottom up and top down approaches have progressed quite far. All three involve quantifying uncertainties.

4 Bernoulli's Utility Concept

My first example is Daniel Bernoulli's work on risk assessment. The great 18th century Swiss mathematician, whose uncle Jacob did pioneering work in the mathematical theory of probability, set out to try to understand why people assess risk the way they do.

A modern-day example that shows how the human assessment of risk can differ from the mathematical analysis is the fear some people have about flying. Such individuals may know that the probability of being involved in a major airline accident is far less than being in a major auto accident. The issue is the nature of an airline crash and the importance they attach to such an event, however unlikely it may be. Fear caused by lightning is a similar phenomenon, where the tiny mathematical probability of being struck by a lightning bolt is far outweighed by the significance many individuals attach to the possibility of such an event.

It was this, essentially human, aspect of risk assessment that interested Daniel Bernoulli. To try to capture mathematically the way people actually assessed risk, he introduced the concept of *utility*.

Utility depends upon another notion of probability theory that preceded it: *expectation* (or *expected value*). Your *expectation* in, say, playing a certain game is a measure of what you can "expect" to win. It is the average amount you would win per game if you were to play repeatedly. To compute your expectation, you take the probability of each possible outcome and multiply it by the amount you would win in that case, and then add together all those amounts. By taking account of both the probabilities and the payoffs, the expectation measures the value to an individual of a particular risk or wager. The greater the expectation, the more attractive the risk.

For many examples, expectation seemed to work well enough. But there was a problem, and it was most dramatically illustrated by a tantalizing puzzle proposed by Daniel's cousin Nicolaus, commonly known as the Saint Petersburg Paradox. Here it is.

Suppose I challenge you to a game of repeated coin tosses. If you throw a head on the first toss, I pay you \$2 and the game is over. If you throw a tail on the first throw and a head on the second, I pay you \$4 and the game is over. If you throw two tails and then a head, I pay you \$8 and the game is over. We continue in this fashion until you throw a head. Each time you throw a tail and the game continues, I double the amount you will win if and when you throw a head.

Now imagine that a friend comes along and offers to pay you \$10 to take your place in the game. Would you accept or decline? What if he offered you \$50? Or \$100? In other words, how much do you judge the game to be worth to you?

The expectation of this game works out to be infinite, so in theory you should not give up your opportunity to play for any amount of money. But most people—even knowledgeable probability theorists—would be tempted to take a fairly low offer. Why is this?

This was the kind of problem with expectation that led Bernoulli to replace the highly mathematical concept of expectation (an example of what I am calling a top-down use of mathematics) by the far less formal and less precise concept of utility (which is very definitely bottom-up).

Utility is intended to measure the significance you attach to a particular outcome. As such, utility is very much an individual thing. It depends on the value a person puts on a particular event. Your utility and mine might differ.

At first glance, the move to replace the mathematically precise concept of expectation by the decidedly personal idea of utility might appear to render impossible any further mathematical analysis. Even for a single individual, it may well be impossible to assign specific numerical values to utility. Nevertheless, Bernoulli was able to make a meaningful, and definitely mathematical, observation about the concept. He wrote: "[The] utility resulting from any small increase in wealth will be inversely proportionate to the quantity of goods previously possessed."

Bernoulli's utility law explains why even moderately wealthy individuals will generally find it much more painful to lose half their fortune than the pleasure or benefit gained by doubling it. As a result, few of us are prepared to gamble half our wealth for the chance of doubling it. Only when we are truly able to declare "What have I got to lose?" are most of us prepared to take a big gamble.

For instance, suppose you and I each has a net worth of $10,000. I offer you a single toss of a coin. Heads and I give you $5,000; tails and you give me $5,000. The winner comes out with $15,000, the loser with $5,000. Since the payoffs are equal and the probability of each of us winning is 1/2, we each have an expectation of zero. In other words, according to expectation theory, it makes no difference to either of us whether we play or not. But few of us would play. We would almost certainly view it as taking an unacceptable risk. The 0.5 probability of losing $5,000 (half our wealth) far outweighs the 0.5 probability of winning the same amount.

Bernoulli's concept of utility likewise explains the Saint Petersburg paradox. According to Bernoulli's law, once you reach the stage where your minimum winning represents a measurable gain *in your terms*, the benefit to be gained by playing longer starts to decrease. That determines the amount for which you would be prepared to sell your place in the game.

So much for expectation. In fact, a similar fate was to befall Bernoulli's utility concept in due course, when mathematicians and economists of a later generation collaborated with psychologists to look more closely at human behavior. But the fact remains that it was Bernoulli who first insisted that if you wanted to apply mathematics to real world problems that involve chance, and if you want the results of that analysis to be of real use, then you had to take account of the human factor. In so doing, he was approaching the issue in a bottom-up fashion, and thereby making one of the first steps toward what I am suggesting will eventually be classified as a fully-fledged branch of mathematics—something that will be done in university departments of mathematics and taught (*as mathematics*) to mathematics students.

5 Bayesian Inference

My second example of what the 22nd century will view as mathematics addresses the question: How do you use inconclusive evidence to assess the probability that a certain event will occur? One method that has become increasingly popular in recent years depends on a mathematical theorem proved by an 18th century English Presbyterian minister by the name of Thomas Bayes. Curiously, Bayes' theorem languished largely ignored and unused for over two centuries before statisticians, lawyers, medical researchers, software developers, and others started to use it in earnest during the 1990s.

Bayesian inference, as the method using Bayes' theorem is called, is a step toward a new mathematics because it uses an honest-to-goodness mathematical formula (Bayes' theorem) in order to improve—on the basis of evidence—the best (human) estimate that a particular event will take place. In the words of some statisticians, it's "mathematics on top of common sense." You start with an initial estimate of the probability that the event will occur and an estimate of the reliability of the evidence. The method then tells you how to combine those two figures—in a precise, mathematical way—to give a new estimate of the event's probability in the light of the evidence.

In some highly constrained situations, both initial estimates may be entirely accurate, and in such cases Bayes' method will give you the correct exact answer. In a more typical real-life situation, you don't have exact figures, but as long as the initial estimates are reasonably good, then the method will give you a better estimate of the probability that the event of interest will occur. Thus, in the hands of an expert in the domain under consideration, someone who is able to assess all the available evidence reliably, Bayes' method can be a powerful tool.

Specifically, Bayes' theorem shows you how to calculate the probability of a certain hypothesis H, based on evidence E. Let $P(H)$ be the probability that the hypothesis H is correct in the absence of any evidence—the *prior probability*. Let $P(H|E)$ be the probability that H is correct given the evidence E. This is the revised estimate you want to calculate. Let $P(E|H)$ be the probability that E would be found if H were correct. This is called the *likelihood*. To compute the new estimate, you first have to calculate $P(H\text{-wrong})$, the probability that H is false, and you have to calculate $P(E|H\text{—wrong})$, the probability that the evidence E would be found in the event that H were false.

Bayes' theorem says that:

$$P(H|E) = \frac{P(H) \times P(E|H)}{P(H) \times P(E|H) + P(H - wrong) \times P(E|H - wrong)}.$$

A quantity such as $P(H|E)$ is known as a *conditional probability*—the conditional probability of H occurring, given the evidence E.

Unscrupulous lawyers have been known to take advantage of the lack of mathematical sophistication among judges and juries by deliberately confusing the two conditional probabilities $P(G|E)$, the probability that the defendant is guilty given the evidence, and $P(E|G)$, the conditional probability that the evidence would be found assuming the defendant were guilty. Such misuse of probabilities is a real possibility in cases where scientific evidence such as DNA testing is involved, such as paternity suits and rape and murder cases. Prosecuting attorneys in such cases have been known to provide the court with a figure for $P(E)$, whereas the figure relevant to deciding guilt is $P(G|E)$, which, as Bayes' formula shows, is generally much lower than $P(E)$. Unless there is other evidence that puts the defendant into the group of possible suspects, such use of $P(E)$ is highly suspect, and indeed should perhaps be prohibited. The reason is that it ignores the initial low prior probability that a person chosen at random is guilty of the crime in question.

In addition to its use—or misuse—in court cases, Bayesian inference methods lie behind a number of new products on the market. For example, chemists make use of Bayesian methods to improve the resolution of nuclear magnetic resonance (NMR) spectrum data. When NMR spectroscopy is used to determine three-dimensional molecular structures, problems remain in translating the data into atomic coordinates. The data is usually insufficient to uniquely define a structure, and subjective choices in data treatment and parameter settings make it difficult to judge the precision of NMR structures. To overcome this problem, probabilistic methods are used to calculate structures from NMR data. The idea is to view structure determination as an inference problem, and use a Bayesian approach to derive a probability distribution that represents the calculated structure and its precision. This approach can improve the resolution of the data by several orders of magnitude.

Other recent uses of Bayesian inference are in the evaluation of new drugs and medical treatments, the analysis of human DNA to identify particular genes, analyzing police arrest data to see if any officers have been targeting one particular ethnic group, and counter-terrorism intelligence analysis.

At the moment, the uses of Bayesian methods are viewed as a combination of mathematics (in the traditional sense) and other forms of reasoning. What will change, I believe, is that the entire reasoning process will come to be viewed as, simply, *mathematics*, as I shall argue toward the end of this article.

6 Black-Scholes Theory

My third example of what I think will be part of 22nd century mathematics is provided by the field of economics and finance.

The 1997 Nobel Prize for economics was awarded for the 1970 discovery of a mathematical formula. The prizewinners were Stanford University professor of finance (emeritus) Myron Scholes and economist Robert C. Merton of Harvard University. The prize would undoubtedly have been shared with a third person, Fischer Black, but for the latter's untimely death in 1995.

Discovered by Scholes and Black, and developed by Merton, the Black-Scholes formula tells investors what value to put on a financial derivative, such as a stock option. Use of the Black-Scholes formula is a clear example of how mathematics can be blended with other forms of reasoning. Human judgment is required both in providing numerical values to some of the formula's input variables and in deciding how much weight to attach to the derivative's value the formula provides.

When the Black-Scholes method was first introduced, the idea that you could use mathematics to (help) price derivatives was so revolutionary that Black and Scholes had difficulty publishing their work. When they first tried in 1970, Chicago University's *Journal of Political Economy* and Harvard's *Review of Economics and Statistics* both rejected the paper without even bothering to have it refereed. It was only in 1973, after some influential members of the Chicago faculty put pressure on the journal editors, that the *Journal of Political Economy* published the paper. [Black/Scholes 1973].

Industry was far less shortsighted than the ivory-towered editors at the University of Chicago and Harvard. Within six months of the publication of the Black-Scholes article, Texas Instruments had incorporated the new formula into their latest calculator, announcing the new feature with a half-page ad in *The Wall Street Journal*.

Modern risk management, including insurance, stock trading, and investment, rests upon the fact that you can use mathematics to predict the future well enough that you can make a wise decision as to where to put your money. When you take out insurance or purchase stock, the real commodity you are dealing with is risk. The underlying ethos in the financial markets is that the more risk you are prepared to take, the greater the potential rewards. Using mathematics can never remove the risk. But it can help to tell you just how much of a risk you are taking, and help you decide on a fair price.

What Black and Scholes did was find a way to determine the fair price to charge for a derivative such as a stock option. The idea with stock options is that you purchase an option to buy stock at an agreed price prior to some fixed later date. If the value of the stock rises above the agreed price before the option runs out, you buy the stock at the agreed lower price and thereby make a profit. If you want, you can simply sell the stock immediately and realize your profit. If the stock does not rise above the agreed price, then you don't have to buy it, but you lose the money you paid out to purchase the option in the first place.

What makes stock options attractive is that the purchaser knows in advance what the maximum loss is: the cost of the option. The potential profit is theoretically limitless: if the stock value rises dramatically before the option runs out, you stand to make a killing. Stock options are particularly attractive when they are for stock in a market which sees large, rapid fluctuations, such as the computer and software industries. Most of the many thousands of Silicon Valley millionaires became rich because they elected to take a portion of their initial salary in the form of stock options in their new company.

The question is, how do you decide a fair price to charge for an option on a particular stock? This is precisely the question that Scholes, Black, and Merton investigated back in the late 1960s. Black was a mathematical physicist with a recent doctorate from Harvard, who had left physics and was working for Arthur D. Little, the Boston-based management consulting firm. Scholes had just obtained a Ph.D. in finance from the University of Chicago. Merton had obtained a bachelor of science degree in mathematical engineering at New York's Columbia University, and had found a job as a teaching assistant in economics at MIT.

The three young researchers—all were still in their twenties—set about trying to find an answer using mathematics, exactly the way a physicist or an engineer approaches a problem. But would a mathematical approach work in the highly volatile world of options trading, which was just being developed at the time. (The Chicago Board Options Exchange opened in April 1973, just one month before the Black-Scholes paper appeared in print.) Many senior market traders thought such an approach could not possibly work, and that options trading was beyond mathematics. If that were the case, then options trading was an entirely wild gamble, strictly for the foolhardy.

The old guard were wrong. Mathematics could be applied. It was heavy duty mathematics at that, involving stochastic differential equations. The formula takes four input variables—duration of the option, prices, interest rates, and market volatility—and produces a price that should be charged for the option.

Not only did the new formula work, it transformed the market. When the Chicago Options Exchange first opened in 1973, less than 1,000 options were traded on the first day. By 1995, over a million options were changing hands each day.

So great was the role played by the Black-Scholes formula (and extensions due to Merton) in the growth of the new options market that, when the American stock market crashed in 1978, the influential business magazine *Forbes* put the blame squarely onto that one formula. Scholes himself has said that it was not so much the formula that was to blame, rather that market traders had not grown sufficiently sophisticated in how to use it.

At present the use of the Black–Scholes formula in assessing the value of stock options is viewed as a combination of (standard) mathematics and other forms of reasoning, but again I believe that in due course the entire reasoning process will be thought of as mathematics.

7 Mathematical Theories of Language

One feature of the examples I have presented so far is that the top down and bottom up approaches have to some extent already met. My next, and final set of examples are taken from a field where this has not yet happened: linguistics.

The first major attempt to develop a mathematical theory of ordinary language began with the publication of Noam Chomsky's seminal work *Syntactic Structures* in 1957 [Chomsky 1957].

Chomsky based his analysis on the axiomatic approach to mathematics, where the mathematician starts with an initial set of assumptions, or *axioms*, and then proceeds to deduce truths (*theorems*) from those axioms. He was inspired, in particular, by the dramatic advances that had been made in mathematical logic in the first half of the 20th century, and by the new branch of mathematics known as recursion theory.

Just as the logicians had been able to formulate axioms and rules that show how a mathematical proof is constructed as a chain of mathematical statements, so too Chomsky formulated rules that show how grammatical sentences are built up from words and phrases. For example, one such rule is that a sentence can be constructed by taking a determinate noun phrase (i.e., a noun phrase that starts with a determiner such as *the* or *a*) and following it by a verb phrase. For example, the sentence *The large black dog licked the tabby kitten* consists of the determinate noun phrase *The large black dog* followed by the verb phrase *licked the tabby kitten*. Using the letter S to stand for "sentence," DNP to denote "determinate noun phrase," and VP to denote "verb phrase," this rule may be written like this:

$$S \rightarrow DNP\,VP.$$

This expression is read as "S arrows DNP VP," or more colloquially, "a sentence results from taking a determinate noun phrase and following it by a verb phrase." The sentences (S), determinate noun phrases (DNP), and verb phrases (VP) are examples of what are called *syntactic categories*.

The rule for generating determinate noun phrases (DNP) from noun phrases (NP) is

$$DNP \rightarrow DET\,NP$$

where DET is given by the rule

$$DET \rightarrow the, a.$$

The first of the above two rules reads "a determinate noun phrase can be generated by taking a determiner and following it by a noun phrase." The second rule is an example of what is called a lexical rule, since it assigns particular words (i.e., items of the lexicon) to a syntactic category, namely the syntactic category DET of determiners. It says that either of the words *the* and *a* constitutes a determiner.

Thus, the determinate noun phrase *the large black dog* is generated by taking the determiner *the* and following it by the noun phrase *large black dog*.

Here are some further rules of syntax:

$$VP \rightarrow V\,DNP$$

"A verb phrase results from taking a verb and following it by a determinate noun phrase."

$$NP \rightarrow A\,NP$$

"A noun phrase results from taking an adjective and following it by a noun phrase." This rule has a circular property, in that you start with a noun phrase and the rule gives you another noun phrase. For example, the noun phrase *large black dog* results from combining the adjective *large* with the noun phrase *black dog*. The rule could then be applied again to give the noun phrase *old large black dog*, etc.

$$NP \rightarrow N$$

"A noun is (itself) a noun phrase."

$$S \rightarrow \textit{If } S \textit{ then } S$$

"A sentence can consist of the word *If* followed by a sentence followed by the word *then* followed by another sentence." For example,

> *If John comes home then we will play chess.*

On their own, these kinds of rules for combining phrases to give sentences produce stilted, machine-like, and often ungrammatical utterances of the kind produced by a robot or a space alien in a low budget science fiction movie. To obtain the correct verb forms, gender and plurality agreements, etc. to give a genuinely grammatical sentence, the strings of words generated by the initial composition rules have to be "massaged." Chomsky introduced *transformation rules* to perform this task.

Transformation rules take a word string produced by the generative grammar, such as *The large black dog lick the tabby kitten*, and turn it into a grammatical sentence such as *The large black dog licks the tabby kitten* or *The large black dog licked the tabby kitten*. Similarly, transformation rules generate other variants of the initial word string, such as the passive form *The tabby kitten is licked by the large black dog* or the question *Did the large black dog lick the tabby kitten?*

For example, the grammar-generated string
> *The large black dog lick the tabby kitten.*

is transformed to the sentence
> *The large black dog licks the tabby kitten.*

by the rule

$$\text{DNP}_{\text{sing}} \, \text{V}_{\text{stem}} \, \text{DNP} \rightarrow \text{DNP}_{\text{sing}} \, \text{V}_{\text{stem}} \text{s} \, \text{DNP}$$

In words, this reads: "Starting with a grammar-generated string consisting of a singular determinate noun phrase, add the letter *s* to the verb stem." (*Note that this kind of rule uses the arrow to mean something different from its meaning in the generative grammar or the lexicon.*)

The same grammar-generated string
> *The large black dog lick the tabby kitten.*

is transformed to the passive sentence
> *The tabby kitten is licked by the large black dog.*

by the transformation rule

$$\text{DNP}^1_{\text{sing}} \, \text{V}_{\text{stem}} \, \text{DNP}^2 \rightarrow \text{DNP}^2 \, \textit{is} \, \text{V}_{\text{stem}} \, \textit{ed by} \, \text{DNP}^1_{\text{sing}}$$

The superscripts 1 and 2 on the two DNPs are there simply to indicate that there are two different phrases involved, and to keep track of which goes where. This rule reads: "Given a grammar-generated string consisting of a singular determinate noun phrase, a verb stem, and a determinate noun phrase, to obtain the passive form, put the second noun phrase first, follow it by the word *is*, followed by the verb stem with the string *ed* appended, followed by the word *by*, followed in turn by the first noun phrase." This rule does not apply to irregular verbs, such as *break–broken*, which form their passives in a different manner.

Even with an extensive list of transformation rules, Chomsky's mathematical treatment of syntax provided at best a fairly crude *model* of syntactic structure, and in his subsequent work he adopted a different approach. The significance of his original work from my present perspective is that it was a first attempt to develop a mathematical description of linguistic structure in a top down fashion, by formulating formal, symbolic rules—a formal system—and then trying to modify the framework to provide a better fit to real language.

Chomsky's was not the only attempt to develop a mathematical model of language. The logician Richard Montague developed an elaborate mathematical model of linguistic meaning that came to be known as Montague semantics, inspired by the work on mathematical truth of his doctoral advisor Alfred Tarski. (See [Montague 1974] for details.)

8 Grice's Maxims

Both Chomsky and Montague took an approach that was predominantly top down. An example of a bottom-up approach to language was presented in a lecture given by the British philosopher and logician H. P. (Paul) Grice at Harvard University in 1967. Grice subsequently published his lecture under the title *Logic and Conversation*. (See [Grice 1975].) In his talk, he formulated a set of "maxims" that participants in a conversation implicitly follow. It was a bold attempt to apply a mathematical approach to the structure of conversation, very much in the spirit of Euclid's formulation of axioms for plane geometry.

Grice was trying to analyze the structure any conversation must have, regardless of its topic and purpose. He began by observing that a conversation is a cooperative act, which the two participants enter into with a purpose. He tried to encapsulate the cooperative nature of conversation by what he called the Cooperative Principle:

> Make your conversational contribution such as is required, at the stage at which it occurs,
> by the accepted purpose or direction of the talk exchange in which you are engaged.

In other words, be cooperative.

Grice went on to derive more specific principles—his maxims—from the Cooperative Principle, by examining its consequences under four different headings: quantity, quality, relation, and manner. He illustrated these four headings by means of non-linguistic analogies:

- *Quantity.* If you are assisting a friend to repair his car, your contribution should be neither more nor less than is required; for example, if your friend needs four screws at a particular moment, she expects you to hand her four, not two or six.
- *Quality.* If you and a friend are making a cake, your contributions to this joint activity should be genuine and not spurious. If your friend says he needs the sugar, he does not expect you to hand him the salt.
- *Relation.* Staying with the cake making scenario, your contribution at each stage should be appropriate to the immediate needs of the activity; for example, if your friend is mixing the ingredients, he does not expect to be handed a novel to read, even if it is a novel he would, at some other time, desire to read.
- *Manner.* Whatever joint activity you are engaged in with a friend, your partner will expect you to make it clear what contribution you are making, and to execute your contribution with reasonable dispatch.

In terms of conversation, the category of *quantity* relates to the amount of information the speaker should provide. In this category, Grice formulated two maxims:

1. Make your contribution as informative as is required.
2. Do not make your contribution more informative than is required.

Under the category of quality, Grice listed three maxims, the second two being refinements of the first:

1. Try to make your contribution one that is true.
2. Do not say what you believe to be false.
3. Do not say that for which you lack adequate evidence.

Under the category *relation*, Grice gave just one maxim:

Be relevant.

However, Grice observed that it would take a great deal more study to come up with more specific maxims that stipulate what is required to be relevant at any particular stage in a conversation.

Finally, under the category of *manner*, Grice listed five maxims, a general one followed by four refinements, although he remarked that the list of refinements might be incomplete:

1. Be perspicuous.
2. Avoid obscurity of expression.
3. Avoid ambiguity.
4. Be brief.
5. Be orderly.

As Grice observed, his maxims are not laws that have to be followed. In that respect, they are not like mathematical axioms. If you want to perform an arithmetical calculation in a proper manner, you have to obey the rules of arithmetic. But it is possible to engage in a genuine and meaningful conversation and yet fail to observe one or more of the maxims Grice listed. The maxims seem more a matter of an obligation of some kind. In Grice's own words: "I would like to be able to think of the standard type of conversational practice not merely as something which all or most do *in fact* follow, but as something which it is *reasonable* for us to follow, which we *should not* abandon." [Emphasis as in the original.]

One of the more interesting parts of Grice's analysis is his discussion of the uses to which people may put his maxims in the course of an ordinary conversation. Indeed, it was this part of his work that makes it a contribution to a science of communication. In science, the real tests of a new theory come when the scientist (1) checks the theory against further evidence, (2) attempts to base explanations on the theory, and (3) tries to use the theory to make predictions that can then be tested.

Grice made successful use of his maxims in analyzing a widespread conversational phenomenon he called *conversational implicature*.

9 Conversational Implicature

Conversational implicature occurs when a person says one thing and means something other than the literal meaning. For example, suppose Naomi says to Melissa, "I am cold" after Melissa has just entered the room and left the door wide open. Literally, Naomi has simply informed Melissa of her body temperature. But what she surely means is "Please close the door." Naomi's words do not actually say this; rather it is *implicated* by her words. Grice used the word "implicate" rather than "imply" for such cases since Naomi's words certainly do not *imply* the "close the door" meaning in any logical sense. Assuming Melissa understands Naomi's remark as a request to close the door, she does so because of cultural knowledge, not logic.

Conversational implicatures are ubiquitous in our everyday use of language. They can be intended by the speaker, or can be made by the listener. Traditional methods of analyzing language say virtually nothing about the way conversational implicature works. Grice used his maxims to analyze the phenomenon.

Although Grice makes no claim that people have any conscious awareness of his maxims, his discussion of conversational implicature establishes a strong case that the maxims capture part of the abstract structure of conversation. They enable the linguist to provide satisfactory, after-the-event explanations of a variety of conversational gambits.

According to Grice, a participant in a conversation, say Bill in conversation with Doris, may fail to fulfill a maxim in various ways, including the following.

(1) Bill may quietly and unostentatiously violate a maxim. In some cases, Bill will thereby mislead Doris.

(2) Bill may opt out from the operation both of the maxim and the Cooperative Principle, making it plain that he is unwilling to cooperate in the way the maxim requires. For example, he might say, "I cannot say more."

(3) Bill may be faced with a clash. For example, he may find it impossible to satisfy both the quantity maxim "Be as informative as required" and the quality maxim "Have adequate evidence for what you say."

(4) Bill may flout or blatantly fail to fulfill a maxim. Assuming that Bill could satisfy the maxim without violating another maxim, that he is not opting out, and that his failure to satisfy the maxim is so blatant that it is clear he is not trying to mislead, then Doris has to find a way to reconcile what Bill actually says with the assumption that he is observing the Cooperative Principle.

Case (4) is the one that Grice suggests most typically gives rise to a conversational implicature. For example, suppose Professor Alice Smith is writing a testimonial for her linguistics student Mark Jones, who is seeking an academic appointment at MIT. She writes a letter in which she praises Jones's well groomed appearance, his punctuality, his handwriting, and his prowess at tennis, but does not say anything about his ability as a student of linguistics. Clearly, Professor Smith is flouting the maxim "Be relevant." The implicature is that Professor Smith has nothing good to say about Jones's ability in linguistics, but is reluctant to put her opinion in writing.

Notice that Grice's analysis of conversation does not involve any use of mathematical notation. Nevertheless, it is inspired by axiomatic mathematics, and the methodology is very definitely mathematical.

10 Sociolinguistics

Another example of a bottom-up attempt to develop a "mathematical" analysis of language is provided by a seminal article published in 1972 called *On the Analyzability of Stories by Children*

[Sacks 1972]. In this article the sociologist Harvey Sacks tried to understand the way people use and understand ordinary language in an everyday setting, in particular the way a speaker and a listener make use of their knowledge of social structure to communicate. According to Sacks, the particular choice of words used by a speaker in, say, a description is critically influenced by her knowledge of social structure, and the listener utilizes his knowledge of social structure in order to interpret, in the manner the speaker intended, the juxtaposition of these words.

The principal data Sacks examined consists of the first two sentences uttered by a small child asked to tell a story:

The baby cried. The mommy picked it up.

As Sacks observes, when heard by a typical, competent speaker of English, the utterance is almost certainly heard as referring to a very small human (although the word *baby* has other meanings in everyday speech) and to that baby's mommy (even though there is no genitive in the second sentence, and it is certainly consistent for the mommy to be some other child's mother). Moreover it is the baby that the mother picks up (although the *it* in the second sentence could refer to some object other than the baby). Why do we almost certainly, and without seeming to give the matter any thought, choose this particular interpretation?

To continue, we are also likely to regard the second sentence as describing an action (the mommy picking up the baby) that follows, and is caused by, the action described by the first sentence (the baby crying). We do this even though there is no general rule to the effect the sentence order corresponds to temporal order or causality of events (even though it often does so).

Moreover, we may form this interpretation without knowing what baby or what mommy is being talked of.

Furthermore, we recognize these two sentences as constituting a "possible description" (Sacks' terminology) of an ordered sequence of events. Indeed it seems to be in large part because we make such recognition that we understand the two sentences the way we do.

As Sacks noted, what leads us effortlessly, instantaneously, and almost invariably, to the interpretation we give to this simple discourse, is the speaker and listener's shared knowledge of, and experience with, the social structure that pertains to (the subject matter of) this particular utterance. Specifically, it is our knowledge of the way mothers behave toward their babies in our culture that leads us to hear the two sentences the way we do. It is this underlying social structure that Sacks is after.

Sacks was the first to admit that the chosen example is extremely simple. But, he claimed, far from rendering his study trivial, this very simplicity makes his observations all the more striking. He observed: "the fine power of a culture . . . does not, so to speak, merely fill brains in roughly the same way, it fills them so that they are alike in fine detail."

It is that observation that the influence of culture on human behavior makes us *alike in fine detail* that makes the domain amenable to a mathematical (in the evolving sense I am talking about) analysis.

To begin his analysis, Sacks first of all introduced what he called *categories*. For example, the following are categories (of persons): *male, female, baby, mommy*. He then defined a (*membership*) *categorization device* to be a non-empty collection of categories that "go together" in some natural way, together with rules of application. For example, the categorization device *gender*. This device consists of the two categories *male* and *female*, together with the rule for applying these categories to (say) human populations. Other examples are the *family* categorization device, which consists of categories such as *baby*, *mommy*, *daddy*, etc. and the *stage-of-life* device, which consists of categories such as *baby*, *child*, *adult*, etc.

As examples of the rules of application that are part of a categorization device, Sacks gave the following. First, *the economy rule*:

(ER) A single category from any device can be referentially adequate.

For instance, the economy rule allows use of the phrase *the baby* to be referentially adequate. Sacks' second rule, *the consistency rule*, says:

(CR) If some population of persons is being categorized, and if a category from some device has been used to categorize one member of the population, then that category, or other categories of the same device, may be used to categorize further members of the population.

For instance, if the device *family* has been used to refer to some baby by means of the category *baby*, then further persons may be referred to by other categories in the same device, such as *mommy* and *daddy*.

Associated with the consistency rule, Sacks formulated the following *hearer's maxim*:

(HM1) If two or more categories are used to categorize two or more members of some population, and those categories can be heard as categories from the same device, then hear them that way.

For instance it is in this way that in the two sentences under consideration, "baby" and "mommy" are heard as from the *family* device. But notice that this does not preclude our simultaneously hearing "baby" as from the *stage-of-life* device—indeed, as Sacks himself argued, this is probably what does occur.

This hearer's maxim does not fully capture what goes on in the example under consideration. For it is not just that "baby" and "mommy" are heard as belonging to the same category *family*; rather they are heard as referring to individuals in *the very same* family. Sacks explained this by observing that the device *family* is one that is what he called *duplicatively organized*. In his own words:

When such a device is used on a population, what is done is to take its categories, treat the set of categories as defining a unit, and place members of the population into cases of the unit. If a population is so treated and is then counted, one counts not numbers of daddies, numbers of mommies, and numbers of babies but numbers of families— numbers of "whole families," numbers of "families without fathers,". etc. A population so treated is partitioned into cases of the unit, cases for which what properly holds is that the various persons partitioned into any case are "coincumbents" of that case.

The following hearer's maxim is associated with duplicatively organized devices:

(HM2) If some population has been categorized by means of a duplicatively organized device, and a member is presented with a categorized population which can be heard as coincumbents of a case of that device's unit, then hear it that way.

According to Sacks, it is this maxim that results in our hearing "the baby" and "the mommy" in our example as referring to individuals in the very same family.

Sacks' next point was that the phrase *the baby* is in fact heard not just in terms of the *family* device but simultaneously as from the *stage-of-life* device. The reason is, he claimed, that *cry* is, in his terminology, a *category-bound* activity, being bound to the category *baby* in the *stage-of-life* device.

Sacks codified what it is that leads us to hear "baby" as from the *stage-of-life* device in addition to the *family* device, by means of a further hearer's maxim:

(HM3) If a category-bound activity is asserted to have been done by a member of some category where, if that category is ambiguous (i.e., is a member of at least two devices), but where, at least for one of those devices, the asserted activity is category-bound to the given category, then hear at least the category from the device to which it is bound.

The final part of Sacks' analysis that I shall consider here concerns the way that an observer describes a particular scene. For instance, if you were to observe a very small human crying, you would most likely describe what you saw as "A baby is crying," or some minor variant thereof. You are far less likely to say "A person is crying" or, even if you could identify the gender of the baby as female, "A girl is crying." Again, if you subsequently saw a woman pick up that baby, and if that woman looked about the age to be the baby's mother, you would probably describe what you saw as "Its mother picked it up." You are less likely to say "A woman picked it up."

Sacks explained the first of these observations, the use of the phrase "the baby," by means of the following viewer's maxim:

(VM1) If a member sees a category-bound activity being done by a member of a category to which the activity is bound, then see it that way.

Since the activity of crying is category-bound to the category *baby* in the *stage-of-life* device, this is the natural way to see and to describe the activity, whenever such a way of seeing and describing is possible.

Turning to the remaining set of observations, there are social norms that govern, or can be seen to govern, the actions of members of the society, and one such norm is that a mother will comfort her crying baby. This is a very powerful social norm, and society generally demonstrates strong disapproval for a mother who fails to conform to it. Sacks' point was that in addition to governing behavior—where by "governing" we may mean nothing more than that the norm serves to describe a normal way of behaving—norms fulfill a further role; namely, viewers use norms to provide some of the orderliness of the activities they observe. In this case, we may capture such a use of a norm by means of a second viewer's maxim:

(VM2) If one sees a pair of actions which can be related via a norm that provides for the second given the first, where the doers can be seen as members of the categories the norm provides as proper for that pair of actions, then (a) see that the doers are such members, and (b) see the second as done in conformity with the norm.

By means of VM2(a), the viewer sees the person who picks up the baby as the baby's mother, provided it is possible to see it thus, and moreover, by VM2(b), takes it that this action is performed in accordance with the norm that says that mothers comfort their crying babies.

11 Why this Will be Viewed as Mathematics

Neither Grice's analysis nor Sacks' can be classified as mathematics. They are, however, clearly inspired by mathematics, and exhibit a definite mathematical flavor. They do, I believe, represent first attempts to develop, in a bottom-up fashion, analyses that may (and I think will) eventually lead to analyses that will in time (perhaps when they meet top-down approaches) be viewed as mathematics. Their approaches work because, for all the nondeterminism inherent in people, their behavior nevertheless exhibits regular, repeated, and for the most part predictable patterns. And that makes them amenable to a mathematical analysis, *providing only that you broaden your conception of mathematics to include the study of such patterns.* Today, we are not yet at the point of making such a leap. But as a result of the increasing production of such analyses, I believe that a hundred years from now, such reasoning will indeed be thought of as mathematics, just as probability theory is today, despite the fact that the ancient Greeks believed—and wrote—that a study of chance events was not within the realm of a mathematics.

The new mathematics I am speaking of is not yet here. At best, what we see today are blends of (contemporary) mathematics with other methods of reasoning and analysis, such as Bayesian reasoning or Black-Scholes analysis. More commonly, there is not even a meeting of the two approaches, let alone a blending. The top down approach involves the development of mathematical *models* (such as Chomksy's formal grammar) that reflect the domain of interest in a fairly crude fashion. The mathematical models typically ignore much of the complexity of the domains they are intended to model. Meanwhile, the bottom-up approach, which tries to reflect the full complexity of the domain (for example, Grice's maxims), can perhaps best be described as the adoption of a mathematical *approach* to an analysis, as opposed to an application of mathematics.

So why do I think that what we are seeing are the initial steps toward what will in due course be viewed as a *bona fide* part of mathematics? First, there is historical precedence. Mathematicians have always tried to extend their discipline to apply to new areas. The ancient Greeks tried to develop a "mathematical analysis" of language and reasoning (the Stoic and Aristotelian schools of Logic). Gottfried Leibniz in the 17th century and George Boole in the 19th further tried to develop a mathematics of language and reasoning, leading eventually to a rich branch of mathematics known as mathematical logic, developed in the first half of the 20th century. Then there are the examples of Bernoulli, Bayes, and Black–Scholes–Merton I considered earlier.

Mathematical logic eventually came to be regarded as a fully-fledged branch of mathematics. The work of Bernoulli and Bayes in probability theory is also generally viewed as mathematics. Black-Scholes theory is usually referred to as "financial mathematics," a terminology that arguably reflects acceptance of it as a branch of mathematics. Certainly many college and university mathematics departments offer course called "Financial Mathematics" and count completion of such a course towards a mathematics degree.

On the other hand, statistics and computer science, for all their heavy dependence on mathematics, are generally classified as outside of mathematics. This despite the fact that, in a great many colleges and universities, both are taught (only) in the Mathematics Department.

Moreover, statisticians, in particular, are often seen in large numbers at major mathematics conferences.

All this seems to suggest a certain degree of indecision on the issue.

However, the case I am trying to make in this essay is based not so much on historical precedent, but rather on a highly pragmatic view of the discipline as a human activity carried out by a human community. I titled my article "What will count as mathematics in 2100?" As I see it, the question then is, who gets to say what is and what is not to be called "mathematics"?

One answer is to leave it to the profession to *define* the answer—a solution that mathematicians (of all people) might be expected to find attractive. This was done for several years by the International Mathematical Union, in connection with the publication of the (now abandoned) annual *World Directory of Mathematicians*. The general criterion of admissibility to the Directory was two articles reviewed in *Mathematical Reviews*, *Referativnyi Zhurnal*, or *Zentralblatt fuer Mathematik* over the preceding five years, or the publication of five papers reviewed in these journals at any time. That certainly provides a precise, profession-certified definition of what counts as a "mathematician," and by extension yields a definition of what counts as mathematics. The problem is, the definition excludes the vast majority of people who earn their living as "professors of mathematics" at colleges and universities all over the world, and an even great number of people in industry, government, and elsewhere in society, who spend large parts of every day "doing math."

Another way to provide an answer is to abstract one from mathematical practice. Mathematics is the science of patterns. The mathematician extracts—or abstracts—patterns either from the world or from a mental discipline (including mathematics itself) and studies them. What makes such a study mathematics, as opposed to some other discipline, is the high degree of abstraction of both the patterns and the way they are studied. Mathematical patterns are the structural skeletons that lie beneath the world. Such is their abstraction that the only tool available for their study is pure human reasoning. Except in an occasional, peripheral way, observation and experiment are of little use. When applied to the deterministic, physical world, mathematics leads to results that can be given with absolute certainty. When we try to use the mathematical method to study nonmathematical objects—in particular, people—however, the results become less certain. But, I would suggest, that is the nature of the *domain*; it is not caused by the *method* we use to analyze it. It is only a matter of time, I believe, before we become so familiar with mathematically-inspired analyses such as the examples I have presented here that the mathematical community—or at least a substantial portion thereof—begins to accept them as a regular part of the discipline. I believe we put ourselves on that path the moment we took on board the definition of mathematics as the science of patterns.

Contemporary mathematics may have declared its goal to be the formulation of precise definitions and axioms and the subsequent deduction of theorems, but that is a fairly recent phenomenon, and likely just a passing fad. It is also, I suggest, a foolish one that serves no one particularly well. In the last hundred years or so, mathematics has parted company with (and even distanced itself from) theoretical physics, statistics, and computer science, and even split internally into "pure" and "applied" mathematics, only to find that some of the most exciting and productive new developments within core mathematics (i.e., those parts that have not been cast out) have come from those other disciplines.

Of course, there is a difference between "mathematics pursued for its own sake" and mathematics carried out in the course of studying some other domain. Of course there is a difference between formal, proof-oriented mathematics based on axioms, and applications of mathematics such as the work of Chomksy or of Black and Scholes. And of course there will always be some individuals who prefer to focus entirely on the pursuit of axiom-theorem type mathematics. The question is, will that relatively small group be able to define what the word "mathematics" means?

I believe the answer is no. In my view, what counts as mathematics will be determined not (solely) by one particular organization within mathematics (such as the IMU). Rather, I believe it will be determined for the most part on sociological grounds, by society as a whole. What classifies as "mathematics" will be determined by what gets done in university "mathematics departments" and by what society expects from people it categorizes as "mathematicians." For it is society as a whole that, one way or another, provides both the environment in which mathematics is done and the funds for its pursuit. Since the main importance of mathematics for humanity as a whole is the role it plays in human understanding, when the main objects of that understanding are *people*, as well as deterministic physical systems, as will be the case in the coming century, mathematics will change.

I should finish by observing that I am by no means the first to come to such a conclusion. For instance, the late Gian-Carlo Rota of MIT wrote (see [Rota/Schwartz/Kac 1985]):

> *Sometime, in a future that is knocking at our door, we shall have to retrain ourselves or our children to properly tell the truth. The exercise will be particularly painful in mathematics. The enrapturing discoveries of our field systematically conceal, like footprints erased in the sand, the analogical train of thought that is the authentic life of mathematics. Shocking as it may be to a conservative logician, the day will come when currently vague concepts such as motivation and purpose will be made formal and accepted as constituents of a revamped logic, where they will at last be allotted the equal status they deserve, side-by-side with axioms and theorems.*

If you take Rota's phrase "made formal and accepted as constituents of a revamped logic" as suggesting formalized mathematics in the sense currently understood in mathematics—as some have done—then Rota's suggestion seems doomed to fail, as I am sure Rota himself would have agreed. But if you interpret it in the light of a changing conception of what counts as mathematics, as I do and as I suspect Rota intended, then it predicts a rosy future for mathematics.

References

[Black/Scholes 1973] Fischer Black & Myron Scholes, "The pricing of options and corporate liabilities," *Journal of Political Economy*, Vol. 81 (1973), pp. 637–59.

[Chomsky 1957] Noam Chomsky, *Syntactic Structures*, The Hague/Paris: Mouton, 1957.

[Grice 1975] H. Paul Grice, "Logic and Conversation," in Peter Cole and Jerry L. Morgan (Eds.) *Syntax and Semantics, Vol. 3, Speech Acts*, New York: Academic Press, 1975.

[Montague 1974] Richard Montague, "The Proper Treatment of Quantification in Ordinary English," in Richmond Thomason (Ed.), *Formal Philosophy: Selected Papers of Richard Montague,* New Haven: Yale University Press, 1974.

[Rota/Schwartz/Kac 1985] Gian Carlo Rota, Jacob Schwartz & Mark Kac, eds., *Discrete Thoughts*, Basel: Birkhäuser, 1985.

[Sacks 1972] Harvey Sacks, "On the analyzability of stories by children," in John Gumperz & Del Hymes, eds., *Directions in Sociolinguistics: the Ethnography of Communication*, New York: Rinehart & Winston, 1972.

15

Mathematics Applied: The Case of Addition

Mark Steiner
Professor of Philosophy
The Hebrew University of Jerusalem

From the Editors

In this chapter, Mark Steiner first describes what makes the applicability of mathematics a thorny philosophical problem—and explains that it is really two problems, one that the philosophers have been discussing and a quite different one that physicists have been asking. He then concentrates on one aspect of the problem, which at first seems quite unproblematic, the question of addition. It leads to a better understanding of what Wigner might have meant when he asked, many years ago, about the "Unreasonable Effectiveness of Mathematics in the Natural Sciences."

Mark Steiner is a Professor of Philosophy in the Faculty of Humanities at the Hebrew University of Jerusalem (socrates.huji.ac.il/Prof_Mark_Steiner.htm). His interests include philosophy of mathematics, philosophy of science, Wittgenstein, and Hume. He has worked on philosophical issues in contemporary mathematics education.

In the first paragraph of this chapter, he has written "to the extent one can talk about a 'consensus' of the philosophical community, I'm not in it." This is because he has consistently worked, independently of the trends in the philosophy of mathematics, on issues involving how mathematics is actually practiced. His work has several times been far ahead of that of his contemporaries. His book, Mathematical Knowledge, *published in 1975, was one of the first works to say that none of the traditional foundational schools—logicism, formalism, intuitionism— describe how mathematicians come to have mathematical knowledge. Mathematical knowledge cannot be identified with formal proofs, as mathematicians very rarely give formal proofs. Also, some informal arguments have actually produced mathematical certainty, or near certainty, as in some of the work of Euler. Steiner appears to have been the first philosopher to recognize this. His essay on explanatory versus non-explanatory proofs from thirty years ago has now begun to motivate studies on explanation in the mathematical field. His second book,* The Application of Mathematics as a Philosophical Problem, *published in 1998, makes a very important distinction (also explained in this chapter) between what philosophers mean when they write of this problem and what physicists mean. Steiner was the first to point out that the two groups have been talking past each other for years. Steiner's may be the only philosophical book dedicated solely to the applications of mathematics in the twentieth century.*

Other articles of his which are likely to be of interest to the readers of this volume include "Mathematical explanation", Philosophical Studies *(1978), "Wittgenstein: Mathematics, Regularities, Rules" in* Benacerraf and His Critics *(1996), "Teaching elementary arithmetic through applications" in* A Companion to the Philosophy of Education, *edited by R. Curren (2003), "Mathematical intuition and physical intuition in Wittgenstein's later philosophy,"* Synthese *(2000), "Frege, the Natural Numbers, and Natural Kinds" in* Between Logic and Intuition: Essays in Honor of Charles Parsons, *(2000), and "Penrose and Platonism" in* The Growth of Mathematical Knowledge *(2000), and his forthcoming "Getting more out of mathematics than what we put there."*

When we speak of applying mathematics, we have one of two roles in mind: the logical role, and the empirical role. Philosophers tend to focus on the former role, scientists the latter. As a result, there is often a "communications gap" between the two communities, which this little essay will try to bridge. I make no claim, however, to "represent" the philosophical community, because to the extent one can talk about a "consensus" of the philosophical community, I'm not in it.

To illustrate the various kinds of roles that mathematics plays in application, I will focus on an example, that of addition (including its generalizations, such as complex addition, tensor sum, etc.). I will show how addition functions both logically and empirically. Some empirical applications are easily accounted for, and I will give an example. But there are others in which the physical basis for the empirical applications seems lacking, and I will give an example of this too.

One of the more striking kinds of application of mathematics is the use of mathematics to discover the application itself. For example, if two mathematical structures are isomorphic, physicists tend to assume (without further evidence) that the physical structures they describe are interchangeable. Here I will give an example from particle physics, using a generalization of the arithmetic identity $4 + 2 = 3 \times 2$ to the arithmetic of group representations. The kind of reasoning exhibited in these kinds of applications which create themselves reminds me of Pythagoreanism, understood broadly as the view that mathematics is the reality underlying all nature. I will set forward (merely for discussion) a number of Pythagorean principles of increasing strength to conclude the essay on a speculative note.

The logical role of mathematics is just that: mathematics resembles logic, in that it *facilitates* reasoning. It can be applied to anything, just as can logic—both are, in the philosophical jargon, "topic neutral." It resembles logic to such an extent, that some philosophers and logicians have actually asserted that mathematics is simply a more advanced branch of logic, i.e., that it *is* reasoning. Using elementary arithmetic, the theory of numbers, the most applicable theory of mathematics in the logical sense, we can balance—or unbalance—our checkbooks. My grandchildren live in a building with floors that include -3, -2, and -1. In order to know how many flights they need to ascend from -2 to 3, they need to invoke the arithmetic of the integers, positive and negative, at an extremely early age. Addition can be used to reason from the premise that I have 5 female students and 4 male students to the conclusion that I have a total of 9 students.

In fact, the last example can actually be formalized directly in the "predicate calculus with iden-tity," as a purely logical argument, with no direct reference to numbers at all. Not every form of arithmetic reasoning can be formalized so simply: even such a simple proposition as "Two apples and two more apples does not make seven apples" cannot be formalized in the predicate calculus with identity.[1] If we allow elementary set theory to be called "logic," however, we can regard arithmetic as a theory of properties of sets (e.g., their cardinality). Arithmetic so understood resembles logic in its abstractness and topic neutrality—i.e., arithmetic is not about any kind of set in particular: one can use arithmetic to reason about thoughts, as well as sticks and stones.

Arithmetic as logic is not likely to pose puzzles for physicists, or other scholars not working in the philosophy of mathematics. Nevertheless, most of the literature on the philosophy of mathematics in the last twenty years seems to focus on the logical side of arithmetic. Questions such as "How do we know there are sets? How do we know there are numbers? Even if there are numbers, how could they possibly be relevant to the empirical world?" have exercised some of the most intelligent people in philosophy today. A good survey of this literature (up to 1997) will be found in [Burgess/Rosen 1997], as well as the chapters by Balaguer, Chihara, and Shapiro in the present volume, but I will focus this essay on the empirical aspects of mathematical application.

Consider, for example, the addition of weights. What is involved in arguing that, if we combine on the scale a weight of 3 kg and another one of 4 kg, the total weight is 7 kg? This, certainly, is not a purely logical argument, but follows from the additivity of mass, and thus weight, which is an empirical hypothesis, used to great effect by Newton. The additivity of weight is so far from being "logic," that it is not even quite true: according to Einstein's theory, when we put two equal weights together, the system of the two loses gravitational energy, and thus mass, according to Einstein's famous equation, $E = mc^2$. Of course, this difference is undetectable in normal applications.

We have thus an example of the empirical role of mathematics, which is to supply the means for empirical descriptions of natural phenomena. When we add two groups of bodies, we can use addition to calculate their total number, meaning the cardinal number of the two sets (logical role), but also to calculate their weight (empirical role).[2]

Let us now ask the following question: is there anything mysterious about the applicability of addition?

We can see that the answer has to be divided into two parts. As for the logical role of addition (in counting), there is an obvious sense in which $7 + 5 = 12$ is a necessary truth (i.e., it could not have been false, there is no possible world in which it is false). And if necessary truths can be mysterious, they are mysterious in ways which do not interest thinkers outside philosophy itself. Let us, then, explore the empirical role of addition – in weighing, assuming for simplicity's sake that weighing is completely additive. Is that mysterious?

---

[1] The first order transcription of an addition identity is a conditional sentence: "If we have . . . then we have. . . . " Thus, if we negate the logical version of "Two apples and two more apples is seven apples", we get: "There are two apples and there are two more apples, but there are not seven apples," a sentence that may well be false (if there are no apples, say).

[2] Nevertheless, even in using addition to calculate the cardinality of a set (an abstract object), there is still an empirical aspect, since there are often empirical implications to the answer we get. (For a discussion of the empirical applications of elementary arithmetic, see [Steiner 2005].) Also, even in using addition to calculate weights, there is still the purely logical matter of adding natural numbers—it is an empirical fact, that weight is given by addition, but the result of that addition is given by "logic."

Not really. It seems quite reasonable, and seemed to Newton, that the sum of two masses is the mass of the sum, i.e., that there is no interaction between the masses that spoils the additivity. This argument is obviously not deductively conclusive (otherwise we would have a refutation of Einstein's Special Theory of Relativity), but it is good enough to make additivity quite plausible. We can generalize this kind of argument to extract from it the formal properties that make any magnitude additive, and there is literature on this subject in what is called the theory of measurement.[3]

Now let us look at masses in motion, accelerated by forces. We have, first, Galileo's famous thought experiment showing that light masses fall like heavy ones. Galileo imagines two equal masses connected by a long light string. The masses fall together. We now imagine the string shorter and shorter, till the masses fuse. Galileo reasons by continuity that the fused mass will fall just as each of the equal parts.

In the Newtonian conceptual scheme, Galileo's conclusion reads: to get the total force on a body (at a given time and place), we calculate each force acting on the body, and sum.

This principle is also at work in Galileo's thought experiment on the trajectory of the cannon ball. Suppose the angle of the cannon barrel is 45 degrees. We have gravity acting throughout; and, following the initial cannon blast, we have no other forces, i.e. zero force. So we have inertial motion in the direction of the barrel. Galileo resolves the inertial motion into horizontal and vertical components, and then splits the problem into two – one for vertical inertial motion plus gravity, and one for horizontal motion without gravity. He then takes the vector sum of both solutions. In this example, aside from the principle that at any given time and place we sum forces acting on a body to get the total force, we have special assumptions used by Galileo, namely that (again, in Newtonian terminology) the force of gravity is constant in space and time – an assumption approximately true for our military example. The more general assumption is the one calling for the summing of all forces acting on a body at a given point in space and time, to get the net force then and there. We can call this general assumption the **sum rule**.

The principle underlying the sum rule is, of course, that two forces operating together, operate independently of one another. This is an empirical principle, which is not always true, but, where it is true, it expresses itself mathematically in the sum rule.

Just as we saw that the additivity of various magnitudes corresponds to the non-numerical Suppes-Field axioms for scalar fields, so in general we can say that the numerical sum rule for force fields corresponds to a general, metaphysical, qualitative directive, which is vague, yet supremely useful: in studying a complex cause, try to analyze the cause into a finite number of component causes, each of which operates independently of the others. Let us call this principle the **Principle of Analysis**, the empirical, yet qualitative basis of the sum rule.

---

[3] Hartry Field ([Field 1980], Chapter 7) shows how to formulate nonnumeric axioms for scalar fields (ordered and nonordered) that determine uniquely (up to a scale factor) the numerical functions which we call additive magnitudes, such as temperature. These axioms can be regarded as the explanation for the utility of addition in a wide variety of theories. Field's treatment is an adaptation of the classic work, [Krantz et al. 1971], which presents a qualitative (non-numerical) system of axioms, based on "congruence" and "betweenness." This underlies – and I would say explains – our practice of *numerical* measurement of lengths, which is of course additive (before Einstein). Field's book, which I expect will be mentioned in other contributions to the present volume, was written to support the "nominalist" philosophy of mathematics, i.e. to show that there is no good reason to believe in the existence of mathematical entities like numbers. I think the book is much better regarded in a different light—as an exploration of the conditions under which certain mathematical concepts can be applied in physics.

If we formulate the the the Principle of Analysis as an existential statement—"There is" a way to break a system into noninterfering components—it is very hard to refute; nevertheless, it is an empirical, disconfirmable, statement. It is, furthermore, a plausible principle, even if not a priori. It is certainly a statement that physicists would hope is true; so it has the benefit of wishful thinking. And it certainty shows that using a sum rule is reasonable, not mysterious.

The waters get muddier when we look at the sum rule in quantum mechanics. A sum rule is certainly operating; the problem is, there seems to be no Principle of Analysis underpinning it.

In quantum mechanics, we have, indeed, a very simple sum rule: the sum of two solutions of the equation is again a solution. This follows from the linearity of the Schrödinger equation. The problem here is that we are dealing (as it turns out) with equations that have complex solutions, such that we can't even "separate out" the real from the imaginary parts. (No analysis here.) There is immediately an issue of what we mean physically by the solution in the first place, to say nothing of what the physical "meaning" of complex addition is.

The interpretation of the mathematical formalism generally adopted by working physicists is as follows. For a single particle in a potential field, a solution $\psi(x)$ of the Schrödinger equation[4], where x is position along the x axis, is said to mean: the probability (density[5]) of finding the electron at x is $|\psi(x)|^2$. (For this to work, the solutions of the equation have to be "normalized" to prevent probabilities greater than 1. I will pass over this problem.)

What now is the meaning of the sum of two solutions, $\psi(x) + \phi(x)$? The standard answer is: suppose that $\psi(x)$ is a solution of the equation under condition A, and suppose $\phi(x)$ is a solution of the equation under condition B, then $|\psi(x) + \phi(x)|^2$ is the probability (density) the electron is at x when either A or B holds.

For example, let a phosphorescent screen be placed behind a wall with two slits which can be opened or closed. Now suppose A is the condition that the electron shot from a gun can get through the wall only at slit A, and $\psi(x)$ is the solution corresponding to the possibility that the electron has gotten through slit A and landed at place x on the screen; and B is the condition that the electron can get through the wall only at slit B, and $\phi(x)$ is the solution corresponding to the possibility that the electron has in fact gotten through the wall at slit B and landed at x on the screen; then the solution $\psi(x) + \phi(x)$ corresponds to the condition that the electron can use either A or B (both are open) and the probability that the electron has arrived at x on the screen when both slits are open is therefore $|\psi(x) + \phi(x)|^2$.

Note that this means that a place on the screen which could be arrived at with slit A open, and with B open, might not be possible with both slits open, since the sum of the two (complex) functions might be zero at x. And this presents a serious problem with applying the Principle of Analysis in finding a nonmathematical explanation of why we use this formalism, under this interpretation. Our usual physical intuition would argue for the opposite: the more slits, the better.

To be sure, there are physicists who have attempted to use a "wave" concept to provide the missing analysis. In fact, if we allow water waves to travel through two slits, we can actually

---

[4] The nonrelativistic equation of quantum mechanics usually used in this connection was given by Schrödinger in the middle of the 1920's. I mention this because I actually know at least one mathematician who was not aware (till I told him) that this equation has application to physics, although he had studied for years its properties professionally. I know this sounds unbelievable, but you will have to take this on faith, since I won't reveal the name of the mathematician.

[5] There are difficulties defining this concept here, but I believe they can be gotten around by mathematicians, and ignoring them won't weaken my case; on the contrary, discussing the difficulties would only strengthen it.

produce interference phenomena which can cancel the wave at given points, where the wave height would not have been zero with one slit open. Yet nobody has succeeded in reducing quantum mechanics to the mechanics of waves; even in the case of two electrons in Euclidean space, where the solutions in general are functions in six variables, there is no way to make sense of "waves" in physical space. I think, then, it would be fair to say that working physicists have abandoned the wave concept. As Feynman says, the only thing we ever *observe* are particles, never waves. Even light "waves" are really photons. We are left with a sum rule which remains true as a mathematical constraint on theories, without any empirical backing for the Principle which could be stated without using the very mathematical formalism in question. To adapt a figure from Lewis Carroll, we have the grin, without the cat.

Before we go on, a remark. I have been referring to complex addition as though it were the same function as in real analysis. Actually, it might be said, we have a generalization of addition, not addition itself. I am artificially increasing the "wonder" involved in the Principle of Superposition. Yet the matter is not so simple. Complex addition is the analytic continuation of real addition into the entire complex plane, and is thus the only way to extend real addition while keeping its axioms intact. We could thus just as well say that addition on the reals is what you get by arbitrarily restricting "plus" to the real line. I find it rather mysterious, in fact, why functions on the reals which were known for hundreds of years before Cardan, turned out to be continuable. Why did the extension of these functions to the complex plane turn out to be largely "forced," in the sense of my colleague Meir Buzaglo? (See [Buzaglo 2001].)

Up to now, we have discussed the concept of (various kinds of) addition: addition as a kind of logic, addition as a quantification of a physical idea which we can understand qualitatively, by some very general, if empirical, principle; finally, addition as mathematical concept, whose application to quantum mechanics has no obvious physical substratum.

Our final example adds a new element: the use of addition to replace the entire notion of a physical substratum. (Even if you don't buy this, it's an interesting example nevertheless.)

Suppose we have two noninteracting particles, particle 1 and particle 2, whose location probabilities at x_1 and x_2 are defined, respectively, by complex valued functions $\psi(x_1)$ and $\phi(x_2)$. Then the function $\psi(x_1)\phi(x_2)$ gives the location probability (density) of the pair of noninteracting particles. We will call this the **product rule**. This rule is nothing special, since the probability of the conjunction of two independent events is the product of the probabilities of these events, and of course we have, for complex numbers u and v, $|uv|^2 = |u|^2|v|^2$. I just need it for what follows.

There is a generalization of the sum and product rule which is also important for what follows. We think of the functions as vectors in complex linear spaces (finite or infinite dimensional—in what follows, finite) and their values as the coordinates. If we have vectors that represents states of particles, then we substitute "tensor sum" and "tensor product" for the sum and product of two functions.

Let us now take a brief look at particle physics. In 1932, Heisenberg made a bold hypothesis that revolutionized nuclear physics: that the proton and the neutron, despite their different charges, and despite the slight difference in their mass, are two states of the same particle, called today a nucleon. The essential difference between neutron and proton was only in the different orientation of their "isotopic spin," an abstract quantity supposedly conserved in nuclear reactions. (The term "orientation" is here a metaphor, to be explained, in terms more familiar to mathematics, such as "double covering group.")

Subsequently, other families of particles were found, particles which respond to the nuclear (strong) force; today these particles are called *hadrons*. For example, we have the three *pions*: positive, neutral, and negative. These have comparable mass, but 15% of that of the nucleons. Again, these are regarded as three "orientations" of one particle. And we have the four *delta* particles: double positive, positive, neutral, negative. These are unstable and decay into a nucleon-pion pair rather quickly, so their presence must be inferred, but 10^{-23} seconds is still enough time to create a new political party where I live.

Let's imagine that we could set up an experiment where a positive delta decays into a pion and a nucleon.[6] By charge conservation, the two possibilities are a positive pion and a neutron; or a neutral pion and a proton. These two possibilities are not ruled out, either, by energy-momentum conservation. We would like, however, to know the probability of each of these possibilities. In other words it would be nice to express the result of the decay as a linear combination of the two possibilities (each possibility thought of as a basis vector of a linear space); squaring the complex coordinates of the two possibilities would give these probabilities, though at the present time we have no way of saying that one possibility is more likely than the other.

To go further with this we have to take a deeper look at the meaning of conservation laws. It is well known that, in physics, symmetries are associated with conservation laws. But in general, the conservation laws were established before the symmetries were discovered. Even where there are exceptions—for example, the conservation of the "Lenz vector" in Keplerian motion (which prevents precession of the orbits), on account of an "internal" O(3) symmetry of this motion[7]—the historical direction could have been reversed, since the property in question (the Lenz vector) could easily have been formulated before the mathematical symmetry of the Hamiltonian had been discovered (by Lenz).

In our case of hadrons, even after it was "discovered," the magnitude known as "isotopic spin," or isospin today, had no meaning other than "the property conserved under SU(2) symmetry," SU(2) being a group known, no doubt, to readers of this volume. Each type of hadron corresponds to a irreducible representation of this group: thus, nucleons correspond to the 2-dimensional representation; pions, to the 3-dimensional; deltas, to the 4-dimensional. In other words, if we ignore their mass and everything else about them except their isospin, we can regard the proton and the neutron as basis vectors of a 2-dimensional complex vector space, on which the group G acts by means of 2×2 complex matrices which in this case are themselves members of SU(2). In the case of the pions and the deltas, the action of SU(2) is represented by 3×3 and 4×4 complex matrices. The isospins of all the hadrons can thus be calculated using group theoretical methods discovered by Eli Cartan well before the advent of particle physics, and clearly described in [Sternberg 1994]; and in this classification the isospin of the proton is $+1/2$, that of the neutron is $-1/2$; that of the pions are $1, 0, -1$; that of the deltas are $3/2, 1/2, -1/2,$ $-3/2$. Even if we don't know what isospin "is" we can speak of its conservation in hadronic reactions.

Returning to our positive delta decay example, we easily see that isospin conservation also allows the two results: positive pion and a neutron (π^+, n) and a neutral pion and a proton (π^0, p).

---

[6] In actual experiments, we "create" the delta by a collision between a pion and an nucleon, but I want to keep things mathematically simple.

[7] For details see [Guillemin/Sternberg 1990].

The amazing difference is that isospin conservation, or rather SU(2) symmetry, will enable us to show that the latter possibility is twice as likely as the former!

Let us consider a system consisting of a pion and a nucleon; we use the "product rule." The pions "are" the basis vectors of a 3-dimensional irreducible representation of SU(2), and the nucleons, of a 2-dimensional irreducible of the same group. Taking the tensor product here means that we take the 6-dimensional space whose basis vectors are the Cartesian product of the two aforementioned bases. This, according to group theory, gives a 6-dimensional reducible representation of SU(2). We have the "tensor product" of the representations.

We can also derive another 6-dimensional reducible representation of SU(2) starting with the 4-dimensional representation which we identify with the set of deltas. By simply "padding" the basis with the two basis vectors p and n from the nucleonic 2-dimensional representation, we get a 6-dimensional vector space upon which we have a reducible representation of SU(2). We get the "tensor sum" of the two representations: a rather fancy way of saying $4 + 2 = 3 \times 2$.

We now pull the rabbit out of the hat. These two reducible representations are SU(2)-isomorphic (alternative terminology: SU(2) morphic) to one another. This means that there is an isomorphism between the two vector spaces that preserves the action of SU(2): any member of SU(2) maps corresponding vectors onto corresponding vectors.

We now make the following assumption: ignoring all other properties of matter, any two configurations of isospin which are SU(2)-morphic, can be regarded as essentially the same. Here is how we use this assumption.

Take the basis vector representing positive delta in the sum representation, and find its correspondent in the product representation. Then express this latter vector as a linear combination of basis vectors in the product representation. After a group theoretical calculation (which Sternberg carries out in great detail in 4.8), we find that

$$\Delta^+ = \sqrt{1/3}(\pi^+, n) + \sqrt{2/3}(\pi^0, p)$$

which leads to the conclusion that the neutral pion and proton is twice as probable as the positive pion and neutron. This is the type of "prediction" that group theory allows in particle physics. The conjunction, then, of tensor sums and products, with theory of group representations, yields a powerful engine of discovery.

The peculiar role of "sum" in this example is quite remarkable. Though we are interested in the decay of a positive delta, we cannot do the calculation unless we sum the delta representation with something that has nothing to do with this experiment – the two-dimensional nucleonic representation.

What is behind the leading assumption that there must be a vector in the product representation that *physically* represents the positive delta? The only thing I can think of is:

Pythagorean Principle A (PPA): At the deepest level of description, physical systems which are mathematically equivalent are physically equivalent—and thus one can be transformed into the other.

This principle does not follow from the sum or the product rules of quantum mechanics. It is left deliberately vague, since it is meant to be a principle of scientific inquiry. It doesn't even work all the time. An example of this is the fortuitous analogy between the mathematics of electronic "spin," and the mathematics of nucleonic "isospin." Both of these magnitudes correspond to

SU(2) symmetry, yet there does not seem to be any explanation for this analogy, *even though* it was used by Heisenberg to discover isospin!

On the macroscopic level, of course, PPA is violated all the time. On that level, the opposite thing happens: quite different systems are found to be describable by the same differential equations. To put it another way, once a mathematical structure is discovered, it is almost sure to be discovered somewhere else. The increasing applicability of mathematics to molecular biology is a result of this.

Another version of the Pythagorean Principle can be stated thus:

PPB: The ultimate classification of reality is by mathematical symmetries and structures.

From time immemorial, thinkers have tried to classify reality according to different schemes. One can't begin to formulate the laws of nature without such a scheme. One recalls the four elements of the Greeks: earth, air, fire, and water. A better one was the periodic table of the elements. Even better is the system of atomic numbers and atomic weights.

Yet in early antiquity there were those who looked to mathematics as the deepest method of classifying reality. The Pythagoreans went so far as to attempt to reduce all existence to numbers. There were attempts to use the characteristic concepts of number theory in biology, to predict the gestation period of animals by the use of such concepts as prime number or perfect number. The only successful Pythagorean application of this type was, of course, to music – in which the continuum of musical tones was related to the discrete domain of numbers.

In the last 100 years, however, it would seem that Pythagorean reasoning has returned with a bang. Classifying the elementary particle world using the classifications of Lie groups has been enormously successful, replacing the "empirical" type of classification hitherto seen. Instead of saying that atom consists of protons, neutrons, and electrons, we say the 6-dimensional reducible of SU(2) "consists" of the four and the two irreducibles by tensor sum.

Ending this essay with an even more speculative leap, we formulate:

PPC: The deepest language with which to describe reality, even qualitatively, is mathematical language.

I am not arguing that PPC is actually true, but only that it deserves consideration. At least one prominent physicist has actually adopted PPC: ". . . [A]t the deepest level, all we find are symmetries and responses to symmetries. Matter itself dissolves, and the universe itself is revealed as one large reducible representation of the symmetry group of nature."[8] It would be rather ironic if the two warring Greek schools, Democritean materialism and Pythagorean Platonism, should turn out to have been ultimately saying the same thing—everything is matter, but all matter is mathematics.[9]

---

[8] [Weinberg 1989], p. 80.

[9] One of the editors of this volume asked whether this means that Newton, standing under the apple tree, was hit by a tensor product. By this reasoning, however, we could "prove" that anger is not a state of the brain, since we never saw a furious neuron. The truth is, that on the level of "appearances," i.e., how things look to us, there is a tree, an apple, collisions, and, for that matter, Newton. On the deepest level of description, none of these things, including "hits," exist. Again, this is only speculation, since I don't know whether PPC is true.

Acknowledgements I would like to thank Shmuel Elitzur for saving me from serious errors in the accounts of physical theories in this paper. Thanks also to the editors of this volume for very helpful criticisms of earlier drafts. This work was written with the aid of a grant from the Israel Science Foundation.

References

[Burgess/Rosen 1997] Burgess, J., & Rosen, G., *A Subject With No Object: Strategies for Nominalistic Interpretations of Mathematics*, Oxford: Oxford University Press, 1997.

[Buzaglo 2001] Buzaglo, Meir, *The Logic of Concept Expansion*, Cambridge, UK: Cambridge University Press, 2001.

[Field 1980] Field, Hartry, *Science without numbers: a defense of nominalism*, Princeton: Princeton University Press, 1980.

[Guillemin/Sternberg 1990] Guillemin, Victor, and Shlomo Sternberg, *Variations on a theme by Kepler (American Mathematical Society colloquium publications, vol. 42)*, Providence, Rhode Island: American Mathematical Society, 1990.

[Steiner 2005] Steiner, Mark, "Mathematics—application and applicability," in *OUP handbook for the philosophy of mathematics*, S. Shapiro, ed., Oxford University Press, 2005.

[Sternberg 1994] Sternberg, Shlomo, *Group theory and physics*, Cambridge University Press, 1994.

[Krantz et al. 1971] Krantz, David M., Luce, Duncan, Suppes, Patrick, Tversky, Amos, *Foundations of measurement*, Academic Press, 1971.

[Weinberg 1989] Weinberg, Steven, "Towards the final laws of physics," in *Elementary particles and the laws of physics: the 1986 Dirac Memorial Lectures*, Cambridge: Cambridge University Press, 1989.

16

Probability—A Philosophical Overview

Alan Hájek
Professor of Philosophy
Australian National University

From the Editors

The work in the philosophy of probability seems much closer to the mathematical content than most work by philosophers of mathematics. This could be because probability is a relatively recent addition to the set of mathematical subjects, or maybe because of its origins in topics such as gambling. Perhaps because everyone has studied at least some mathematics in school, but not everyone has studied much probability, there are fewer philosophers working in the philosophy of probability. Alan Hájek is a philosopher with a deep interest in the philosophy of probability. His chapter is an introduction to many of the issues currently being discussed in the philosophy of probability. It should be accessible to anyone who has taken the standard undergraduate probability and statistics course. Neither of the editors of this volume have done any work in the field, but the questions here seem very natural to us.

Alan Hájek is a Professor of Philosophy in the Research School of Social Sciences at the Australian National University (philrsss.anu.edu.au/people-defaults/alanh/index.php3). His research interests include the philosophical foundations of probability and decision theory, epistemology, the philosophy of science, metaphysics, and the philosophy of religion. His paper "What Conditional Probability Could Not Be" won the 2004 American Philosophical Association Article Prize for the best article published in the previous two years by a younger scholar. The Philosopher's Annual *selected his "Waging War on Pascal's Wager" as one of the ten best articles in philosophy in 2003. Other articles of interest are "Perplexing Expectations" (with Harris Nover),* Mind *115, (2006) and "The Cable Guy Paradox",* Analysis, *65: 2, (2005). He has a book in preparation,* Arrows and Haloes: Probabilities, Conditionals, Desires, Beliefs *to be published by Oxford University Press.*

1 Personal and Pedagogical Prologue

Once upon a time I was an undergraduate majoring in mathematics and statistics. I attended many lectures on probability theory, and my lecturers taught me many nice theorems involving probability: '*P* of this equals *P* of that', and so on. One day I approached one of them after a lecture and asked him: "What is this '*P*' that you keep on writing on the blackboard? *What is probability?*" He looked at me like I needed medication, and he told me to go to the philosophy department. In the interests of pedagogy, in retrospect I think that *he* could have benefited from some discussions with philosophers. For when I now teach those same theorems to my students, I hope that I can imbue them with deeper meaning and motivation when I point out what is at stake philosophically.

Anyway, I *did* go to the philosophy department. (Admittedly, my route there was long and circuitous.) There I found a number of philosophers asking the very same question: what is probability? All these years later, it's still one of the main questions that I am working on. I still don't feel that I have a completely satisfactory answer, although I like to think that I've made some progress on it. For starters, I know many things that probability is *not*, namely various highly influential analyses of it that cannot be right—we will look at them shortly. As to promising directions regarding what probability *is*, I will offer my best bets at the end, concluding with some further personal and pedagogical thoughts.

2 Introduction

Bishop Butler's dictum [Butler 1736] that "Probability is the very guide of life" is as true today as it was when he wrote it in 1736. It is hardly necessary to point out the importance of probability in statistics, physics, biology, chemistry, computer science, medicine, law, meteorology, psychology, economics, and so on. Probability is crucial to any discipline that deals with indeterministic processes, any discipline in which evidence has a non-deductive bearing on hypotheses, indeed any discipline in which our ability to predict outcomes is imperfect—which is to say virtually any serious empirical discipline. Probability is also seemingly ubiquitous outside the academy. Probabilistic judgments of the efficacy and side-effects of a pharmaceutical drug determine whether or not it is approved for release to the public. The fate of a defendant on trial for murder hinges on the jurors' opinions about the probabilistic weight of evidence. Geologists calculate the probability that an earthquake of a certain intensity will hit a given city, and engineers accordingly build skyscrapers with specified probabilities of withstanding such earthquakes. Probability undergirds even measurement itself, since the error bounds that accompany measurements are essentially probabilistic confidence intervals. We find probability wherever we find uncertainty—that is, almost everywhere in our lives.

It is surprising, then, that probability is a comparative latecomer on the intellectual scene. To be sure, inchoate ideas about chance date back to antiquity—Epicurus, and later Lucretius, believed that atoms occasionally underwent indeterministic swerves. In the middle ages, Averroes had a notion of 'equipotency' that might be regarded as a precursor to probabilistic notions. But probability theory was not conceived until the 17th century, when the study of gambling games motivated the first serious mathematical study of chance by Pascal and Fermat in the mid-17th century, culminating in the *Port-Royal Logic*. Over the next three centuries, the theory was developed by such authors as Huygens, Bernoulli, Bayes, Laplace, Condorcet, de Moivre, Venn,

Johnson, and Keynes. Arguably, the crowning achievement was Kolmogorov's axiomatization in 1933, which put probability on rigorous mathematical footing.

When I asked my professor "What is probability?", there are two ways to understand that question, and thus two kinds of answer that could be given (apart from bemused advice to seek attention from a doctor, or at least a doctor of philosophy). First, the question may mean: *what are the formal features of probability?* That is a mathematical question, to which Kolmogorov's axiomatization is the widely (though not universally) agreed upon answer. I review this answer in the next section as it was given to me at great length in my undergraduate statistics courses. Second, the question may mean: *what sorts of things are probabilities*—what, that is, is the subject matter of probability theory? This is a philosophical question, and while the mathematical theory of probability certainly bears on it, the answer must come from elsewhere—in my case, from the philosophy department.

3 The Formal Theory of Probability

3.1 Unconditional Probability

Kolmogorov begins his classic book ([Kolmogorov 1933]) with what he calls the "elementary theory of probability": the part of the theory that applies when there are only finitely many events in question. Let Ω be a set (the 'universal set'). A *field* on Ω is a set of subsets of Ω that has Ω as a member, and that is closed under complementation (with respect to Ω) and finite union. Let Ω be given, and let $\mathcal{F}$ be a field on Ω. Kolmogorov's axioms constrain the possible assignments of numbers, thought of as *probabilities*, to the members of $\mathcal{F}$. Let P be a function from $\mathcal{F}$ to the real numbers obeying:

1. (Non-negativity) $P(A) \geq 0$ for all $A \in \mathcal{F}$.
2. (Normalization) $P(\Omega) = 1$.
3. (Finite additivity) $P(A \cup B) = P(A) + P(B)$ for all $A, B \in \mathcal{F}$ such that $A \cap B = \emptyset$.

Such a triple $(\Omega, \mathcal{F}, P)$ is called a *probability space*.

Here the arguments of the probability function are sets, often called *events*. (Note that this is a technical sense of the term that may not neatly align with ordinary usage—for example, it is not clear that 'events' in the latter sense have the required closure properties.) Kolmogorov's probability theory is thus dependent on set theory.

We could instead attach real-valued probabilities to members of a collection $\mathcal{S}$ of *sentences* of a language, closed under finite truth-functional combinations, with the following counterpart axiomatization:

I. $P(A) \geq 0$ for $A \in \mathcal{S}$.
II. If T is a tautology, then $P(T) = 1$.
III. $P(A \vee B) = P(A) + P(B)$ for all $A \in \mathcal{S}$ and $B \in \mathcal{S}$ such that A and B are logically incompatible.

Note how these axioms take the notions of 'tautology,' 'logical incompatibility' and 'implication' as already understood. To this extent we may regard probability theory, so formulated, as dependent on deductive logic.

Now let Ω be infinite. A non-empty collection $\mathcal{F}$ of subsets of Ω is called a *sigma algebra* (or *sigma field*, or *Borel field*) on Ω iff $\mathcal{F}$ is closed under complementation and *countable* union, i.e.

$$A_1, A_2, \ldots \in \mathcal{F} \Rightarrow \bigcup_{n=1}^{\infty} A_n \in \mathcal{F}.$$

Kolmogorov introduces a further 'infinitary' axiom.

4. (Continuity) If $E_1, E_2, \ldots$ is a sequence of sets such that $E_i \supseteq E_{i+1} \forall i$ and $\bigcap_{n=1}^{\infty} E_n = \varnothing$ then $P(E_n) \to 0$ (where $E_n \in \mathcal{F}$ for all n).

That is, if $E_1, E_2, \ldots$ is a sequence of non-increasing sets (according to the set-inclusion relation), with empty infinite intersection, then $\lim_{n \to \infty} P(E_n) = P(\bigcap_{n=1}^{\infty} E_n)$. Now, define a *probability measure $P(-)$ on $\mathcal{F}$* as a function from $\mathcal{F}$ to $[0, 1]$ satisfying axioms 1–3, as before, and also the new axiom 4.

Equivalently, we can replace the conjunction of axioms 3 and 4 with a single axiom:

3'. (Countable additivity) If $\{A_i\}$ is a countable collection of (pairwise) disjoint sets, each $A_i \in \mathcal{F}$, then

$$P\left(\bigcup_{n=1}^{\infty} A_n\right) = \sum_{n=1}^{\infty} P(A_n).$$

Thanks to the assumption that $\mathcal{F}$ is a sigma algebra, we are guaranteed that the probability on the left hand side is defined.

De Finetti ([de Finetti 1972] and [de Finetti 1974]) marshals a battery of arguments against countable additivity, most of them variations on these:

The infinite lottery: Suppose a positive integer is selected at random—we might think of this as an infinite lottery with each positive integer appearing on exactly one ticket. We would like to reflect this in a uniform distribution over the positive integers (indeed, proponents of the principle of indifference would seem to be committed to it), but if we assume countable additivity this is not possible. For if we assign probability 0 in turn to each number being picked, then the sum of all these probabilities is again 0; yet the union of all of these events has probability 1 (since it is guaranteed that some number will be picked), and $1 \neq 0$. On the other hand, if we assign some probability $\varepsilon > 0$ to each number being picked, then the sum of these probabilities diverges to ∞, and $1 \neq \infty$. If we drop countable additivity, however, then we may assign 0 to each event and 1 to their union without contradiction.

Biased assignments to denumerable sets: Countable additivity allows one to assign uniform probability $1/n$ to each member of an n-celled partition (for example, 1/6 to each result of tossing a die). However, it requires one to assign an extremely biased distribution to a denumerable partition of events. Indeed, for any $\varepsilon > 0$, however small, there will be a finite number of events that have a combined probability of at least $1 - \varepsilon$, and thus the lion's share of all the probability.

See [Seidenfeld 2001] for further discussion of countable additivity.

It is often thought that the only part of the axiomatization that is not merely conventional stipulation is the third axiom, in either its finite or countable form. (For example, it is tempting to say that it is *purely* conventional to set $P(\Omega) = 1$, rather than $P(\Omega) = 100$, say.) That is too quick. For each of the following involves a substantial mathematical assumption:

(i) Probabilities are defined by *functions* (rather than by one-many or many-many mappings).

(ii) These are functions of *one variable* (unlike primitive conditional probability functions, which are functions of two variables): there is just a single argument for a probability function.

(iii) Such a function is defined on a *field* (rather than a set with different closure conditions).

(iv) Probabilities are to be represented *numerically* (rather than qualitatively, or comparatively).

(v) Their numerical values are *real* numbers (rather than those of some other number system).

(vi) These values are *bounded* (unlike other quantities that are treated measure-theoretically, such as lengths).

(vii) Probability functions attain *maximal* and *minimal* values (thus prohibiting open or half-open ranges, such as $(0, 1)$ or $(0, 1]$).

For a discussion of rival theories that relax or replace (ii), (iii), (iv) and (vi), see [Fine 1973]. Complex-valued probabilities are proposed by Feynman and Cox (see [Mückenheim et al. 1986]); infinitesimal probabilities (of non-standard analysis) by Skyrms ([1980]) and Lewis ([1980]) among others; unbounded probabilities by Renyi ([1970]).

3.2 *Conditional Probability*

Kolmogorov also defines the *conditional probability* of A given B by the ratio formula:

$$(1) \quad P(A \mid B) = \frac{P(A \cap B)}{P(B)} \quad (P(B) > 0).$$

Thus, we may say that the probability that the toss of a fair die results in a 6 is 1/6, but the probability that it results in a 6 *given* that it results in an even number, is $^{1/6}/_{1/2} = 1/3$. In straightforward applications in which the requisite unconditional probabilities are well-defined, and the denominator $P(B)$ is greater than 0, this formula seems to be impeccable.

But not all applications are straightforward, and in some these conditions are not met. Consider first the proviso that $P(B) > 0$. As probability textbooks repeatedly drum into their readers, probability zero events need not be impossible, and indeed they can be of real significance. It is curious, then, that some of the same textbooks glide over (1)'s proviso without missing a beat. In fact, interesting cases of conditional probabilities with probability-zero conditions are manifold. Consider an example due to Borel: a point is chosen at random from the surface of the earth (thought of as a perfect sphere); what is the probability that it lies in the Western hemisphere, given that it lies on the equator? 1/2, surely. Yet the probability of the condition is 0, since a uniform probability measure over a sphere must award probabilities to regions in proportion to their area, and the equator has area 0. The ratio formula thus cannot deliver the intuitively correct answer. Obviously there are uncountably many problem cases of this form for the sphere. (For more discussion see [Hájek 2003b].)

Probability theory and statistics are shot through with cases of non-trivial zero-probability events. Witness the probabilities of continuous random variables taking particular values (such as a normally distributed random variable taking the value 0). Witness the various 'almost sure' results—the strong law of large numbers, the law of the iterated logarithm, the martingale convergence theorem, and so on. They assert that certain convergences take place, not with certainty, but with probability 1. A fair coin may land tails forever. But despite this event's having probability 0, various probabilities conditional on it are intuitively well defined—for example, the probability that the coin lands tails forever, *given* that it lands tails forever, is surely 1. More generally, it is surely a triviality that the probability of any possible outcome, *given itself*, is 1. This is about as fundamental an intuition about conditional probability as there could be. The fact that the ratio formula cannot respect this intuition is a major strike against it.

The difficulties that probability zero conditions pose for the ratio formula for conditional probability are well known (which is not to say that they are unimportant). Indeed, Kolmogorov himself was well aware of them, and he offered a more sophisticated account of conditional probability as a random variable conditional on a sigma algebra, appealing to the Radon-Nikodym theorem to guarantee the existence of such a random variable. But my complaints about the ratio analysis are hardly aimed at a straw man, since (1) is by far the most commonly used analysis of conditional probability (especially in philosophical applications of probability). Moreover, the move to Kolmogorov's more sophisticated theory of conditional probability does not lay to rest the problem of zero-probability conditions. In particular, even Kolmogorov's more sophisticated account of conditional probability does not respect the fundamental intuition above concerning conditional probability, as evidenced by the existence of so-called *improper* conditional probability random variables. Seidenfeld et al. ([2001]) show just how extreme and how widespread violations of the intuition can be.

Hájek ([2003b]) goes on to consider further problems for the ratio formula: cases in which the unconditional probabilities that figure in the ratio are *imprecise* or are *undefined*, and yet the corresponding conditional probabilities are defined. Consider first the problem of imprecise unconditional probabilities: $P(A|B)$ does not have a sharp value when $P(B)$ is imprecise. Example: presumably, your probability that it rains tomorrow is not a sharp value, such as 0.3. After all, that value is *infinitely* sharp, precise to infinitely many decimal places: 0.30000.... Rather, your probability is spread out over a range of values—say, the interval [0.25, 0.35]. Philosophers commonly represent such imprecision with a set of probability functions—in this case, the set of all such functions that assign some value in the interval [0.25, 0.35] to it raining tomorrow, and that agree with your opinions in all other respects. (See [van Fraassen 1990], which develops proposals by Levi [1980] and Jeffrey [1983], among others.) Nevertheless, you assign various perfectly sharp conditional probabilities, *given* that it rains. For example, the probability that a particular fair coin toss lands heads, given that it rains tomorrow, is a sharp 1/2.

The problem of undefined conditional probabilities is even more widespread: $P(A \mid B)$ is undefined when either or both of $P(A \cap B)$ and $P(B)$ are undefined. Suppose that P(Sam has just watered the garden) is undefined. Still, the probability that the garden is dry, given that Sam has just watered the garden, is 0. Hájek ([2003b]) discusses a plethora of such cases, ranging from quantum mechanics to decision theory, from non-measurable sets to probabilistic causation.

The ratio formula goes silent where our intuitions cry out what the answers should be. Moreover, these prove to be problematic also for the more sophisticated account; so therein lies no solution either.

I believe that the right response is to turn the tables on Kolmogorov's analysis: rather than regarding unconditional probabilities as fundamental, and later defining conditional probabilities in terms of them, we should regard conditional probability as the fundamental notion. And there are various ways to define primitive conditional probabilities as total functions from $\mathcal{F} \times \mathcal{F}$ to $[0, 1]$ (for a given sigma field $\mathcal{F}$). Popper ([1959a]) presents a general account of such conditional probabilities. Among other things, any such account codifies the intuition that the conditional probability of anything, given *itself,* is 1. See [Hájek 2003b] for further discussion.

3.3 Independence

If $P(X \mid Y) = P(X)$, then X and Y are said to be independent. Intuitively, the occurrence of one of the events is completely uninformative about the occurrence of the other. Thus, successive spins of a roulette wheel are typically regarded as independent. When $P(X) > 0$ and $P(Y) > 0$, the definition of independence is equivalent to

$$P(Y \mid X) = P(Y)$$

and to

$$P(X \cap Y) = P(X)P(Y).$$

The latter formulation can be used even when $P(X)$ or $P(Y)$ is 0.

The locution 'X is independent of Y' is somewhat careless, encouraging one to forget that independence is a 3-place relation that events or sentences bear *to a probability function*. Furthermore, this technical sense of 'independence' should not be identified unreflectively with *causal* independence, or any other pre-theoretical sense of the word, even though such identifications are often made in practice.

Independence plays a central role in probability theory—indeed, it is that theory's distinctive add-on to the more general measure theory on which it is based. Many of those theorems that my statistics professors taught me, and which I now teach, hinge on it—witness again the various laws of large numbers, for instance. It should come as no surprise that my misgivings about the ratio analysis of conditional probability carry over to the present definitions of independence, which presupposes that analysis.

4 Interpretations of Probability

This section turns to the so-called *interpretations* of probability: attempts to answer the central *philosophical* question: what sorts of things are probabilities? The term 'interpretation' is misleading here. Various quantities that intuitively have nothing to do with 'probability' obey Kolmogorov's axioms—for example, length, volume, and mass, each suitably normalized—and are thus 'interpretations' of it, but not in the intended sense. Nevertheless, we will silence our scruples and follow common usage in our quick survey of the 'interpretations' of probability. (See [Hájek 2003a] for a far more detailed survey.)

4.1 Classical Interpretation

The *classical* interpretation, historically the first, can be found in the works of Pascal, Huygens, Bernoulli, and Leibniz, and it was famously presented by Laplace ([1814]). Cardano, Galileo, and Fermat also anticipated this interpretation. Suppose that our evidence does not discriminate among the members of some set of possibilities—either because that evidence provides equal support for each of them, or because it has no bearing on them at all. Then the probability of an event is simply the fraction of the total number of possibilities in which the event occurs—this is sometimes called the *principle of indifference*. We may think of this as the *rational subjective probability* appropriate for someone in the evidential situation described. This interpretation was inspired by, and typically applied to, games of chance that by their very design create such circumstances—for example, the classical probability of a fair die landing with an even number showing up is 3/6. Probability puzzles typically take this means of calculating probabilities for granted.

Unless more is said, the interpretation yields contradictory results: you have a one-in-a-million chance of winning the lottery, but either you win or you don't; so each of these possibilities has probability $\frac{1}{2}$! We might look for a "privileged" partition of the possibilities, but we will not always find one. For example, in this case, the million-celled partition corresponding to each of the possible lottery outcomes seems more natural than the win/don't win partition, if only because the former is more fine-grained. But Bertrand's paradoxes ([Bertrand 1889]) show that a particular problem may have competing, equally natural, partitions. They all turn on alternative parametrizations of a given problem that are non-linearly related to each other. The following example (adapted from [van Fraassen 1989]) nicely illustrates how Bertrand-style paradoxes work. A factory produces cubes with side-length between 0 and 1 foot; what is the probability that a randomly chosen cube has side-length between 0 and 1/2 a foot? The tempting answer is 1/2, as we imagine a process of production that is uniformly distributed over side-length. But the question could have been given an equivalent restatement: a factory produces cubes with face-area between 0 and 1 square-feet; what is the probability that a randomly chosen cube has face-area between 0 and 1/4 square-feet? Now the tempting answer is 1/4, as we imagine a process of production that is uniformly distributed over face-area. And it could have been restated equivalently again: a factory produces cubes with volume between 0 and 1 cubic feet; what is the probability that a randomly chosen cube has volume between 0 and 1/8 cubic-feet? Now the tempting answer is 1/8, as we imagine a process of production that is uniformly distributed over volume. What, then, is *the* probability of the event in question?

4.2 Logical Interpretation

The *logical* interpretation of probability, developed most extensively by Carnap ([1950]), sees probability as an extension of logic. Traditionally, logic aims to distinguish valid from invalid arguments by virtue of the syntactic form of the premises and conclusion. (E.g., any argument that has the form

p
If p then q
Therefore, q

is valid in virtue of this form.) But the distinction between valid and invalid arguments is not fine enough: many invalid arguments are compelling, in the sense that the premises strongly support the conclusion—we will see an example of such an argument shortly. Carnap described this relation of "support" or "confirmation" as the logical probability that an argument's conclusion is true, given that its premises are true. He had faith that logic, more broadly conceived, could also give it a syntactic analysis. So according to this program, probability is a measure of the degree to which a sentence supports another sentence, where this could be determined by the syntactic forms of the sentences themselves.

The program did not succeed. A central problem is that changing the language in which items of evidence and hypotheses are expressed will typically change the confirmation relations between them—for example, adding further predicates or names to a given language will typically revise how probabilities are shared around individual sentences. Moreover, Goodman ([1983]) showed that inductive logic must be sensitive to the meanings of words, for syntactically parallel inferences can differ wildly in their inductive strength. For example,

> *All observed snow is white.*
> *Therefore, all snow is white.*

is an inductively strong argument: its premise gives strong support to its conclusion. However,

> *All observed snow is observed.*
> *Therefore, all snow is observed.*

is inductively weak, its premise providing minimal support for its conclusion. It is quite unclear how a notion of logical probability can respect these intuitions.

4.3 Frequency Interpretations

Frequency interpretations date back to Venn ([1876]). Gamblers, actuaries and scientists have long understood that relative frequencies bear an intimate relationship to probabilities. Frequency interpretations posit the most intimate relationship of all: identity. In a sound bite, *probabilities are relative frequencies* according to this view. Thus, the probability of '6' on a die that lands '6' 3 times out of 10 tosses is, according to the frequentist, 3/10. In general:

the probability of an outcome A in a reference class B is the proportion of occurrences of A within B.

Frequentism is still the dominant interpretation among scientists who seek to capture an objective notion of probability, heedless of anyone's beliefs. It is also the philosophical position that lies in the background of the classical Fisher/Neyman-Pearson approach that is used in most statistics textbooks. Frequentism does, however, face some major objections. For example, a coin that is tossed exactly once yields a relative frequency of heads of either 0 or 1, whatever its true bias—an instance of the infamous 'problem of the single case'. A coin that is tossed twice can only yield relative frequencies of 0, 1/2, and 1. And in general, a finite number n of tosses can only yield relative frequencies that are multiples of $1/n$. Yet it seems that probabilities can often fall between these values. Quantum mechanics, for example, posits irrational-valued probabilities such as $1/\sqrt{2}$.

Some frequentists (notably Reichenbach [1949] and von Mises [1957]) address this problem by considering infinite reference classes of hypothetical occurrences. Probabilities are then defined as limiting relative frequencies in suitable infinite sequences of trials. Von Mises offers a sophisticated formulation based on the notion of a *collective:* an (hypothetical, or virtual) infinite sequence of 'attributes' (possible outcomes) of a specified experiment that is performed infinitely often. He goes on to lay down two requirements for such an infinite sequence ω to be a collective. Call a *place-selection* an effectively specifiable method of selecting indices of members of ω, such that the selection or not of the index i depends at most on the first $i - 1$ attributes. The axioms are:

Axiom of Convergence: *the limiting relative frequency of any attribute exists.*

Axiom of Randomness: *the limiting relative frequency of each attribute in a collective ω is the same in any infinite subsequence of ω which is determined by a place selection.*

The probability of an attribute A, relative to a collective ω, is then defined as the limiting relative frequency of A in ω.

Collectives are abstract mathematical objects that are not empirically instantiated, but that are nonetheless posited by von Mises to explain the stabilities of relative frequencies in the behaviour of actual sequences of outcomes of a repeatable random experiment. Church ([1940]) renders precise the notion of a place selection as a recursive function.

If there are in fact only finitely many trials of the relevant type, then this kind of frequentism requires the actual sequence of outcomes to be extended to a hypothetical or 'virtual' sequence. This creates new difficulties. For instance, there is apparently nothing that determines how the coin in my pocket would have landed if it had been tossed indefinitely—it *could* yield any hypothetical limiting relative frequency that you like. Moreover, a well-known problem for any version of frequentism is the *reference class problem*: relative frequencies must be relativized to a *reference class*. Suppose that you are interested in the probability that Collingwood will win its next match. Which reference class should you consult? The class of all matches in Collingwood's history? Presumably not. The class of all recent Collingwood matches? That's also unsatisfactory: it is somewhat arbitrary what counts as 'recent', and some recent matches are more informative than others regarding Collingwood's prospects. The only match that resembles Collingwood's next match in every respect is that match itself. But then we are saddled again with the problem of the single case, and we have no guidance to its probability in advance. The reference class problem can also be a very practical problem—insurance companies face it on a daily basis. After all, the premiums that they set for a given individual are based on frequencies of claims of people of that type; but the individual is a member of many classes of people, whose relevant frequencies may differ wildly.

4.4 Propensity Interpretations

Propensity interpretations, like frequency interpretations, regard probability as an objective feature of the world. Probability is thought of as a physical propensity, or disposition, or tendency of a system to produce given outcomes. This view, which originated with Popper ([1959b]), was motivated by the desire to make sense of single-case probability attributions on which frequentism apparently foundered, particularly those found in quantum mechanics. Propensity theories fall into two broad categories. According to *single-case* propensity theories, propensities measure a system's tendencies to produce given outcomes; according to *long-run* propensity theories, propensities are tendencies to produce long-run outcome frequencies over repeated trials. See [Gillies 2000] for a useful survey.

Single-case propensity attributions face the charge of being untestable. Long-run propensity attributions may be considered to be verified if the long-run statistics agree sufficiently well with those expected, and falsified otherwise; however, then the view risks collapsing into frequentism, with its attendant problems. A prevalent objection to any propensity interpretation is that it is uninformative to be told that probabilities are 'propensities.' For example, what exactly is the property in virtue of which this coin has a 'propensity' of 1/2 of landing heads (when suitably tossed)? Indeed, some authors regard it as mysterious whether propensities even obey the axioms of probability in the first place. (See [Hájek 2003a].)

4.5 Subjectivist Interpretations

Subjectivist interpretations—sometimes called 'Bayesian'—pioneered by Ramsey ([1926]) and de Finetti ([1937]), see probabilities as *degrees of belief,* or *credences* of appropriate agents. These agents cannot be actual people since, as psychologists have repeatedly shown, people typically violate probability theory in various ways, often spectacularly so. Instead, we imagine the agents to be ideally rational.

But what are credences? De Finetti identifies an agent's subjective probabilities with his or her betting behavior. For example,

your probability for the coin landing heads is $\frac{1}{2}$

if and only if

you are prepared to buy or sell for 50 cents a ticket that pays \$1 if the coin lands heads, nothing otherwise.

All of your other degrees of belief are analyzed similarly.

The analysis has met with many objections. Taken literally, it assumes that opinions would not exist without money, and moreover that you must value money linearly; but if it is just a metaphor, then we are owed an account of the literal truth. Even if we allow other prizes that you value linearly, problems remain. For your behavior in general, and your betting behavior in particular, is the result of your beliefs and desires working in tandem; any such proposal fails to resolve these respective components. Even an ideally rational agent may wish to misrepresent her true opinion; or she may particularly enjoy or abhor gambling; or, like a Zen master, she may lack a desire for worldly goods altogether. In each case, her betting behavior is a highly misleading guide to her true probabilities.

A more sophisticated approach, championed by Ramsey, seeks to fix an agent's utilities and probabilities simultaneously by appeal to her preferences. Suppose that you have a preference ranking over various possible states of affairs and gambles among them, meeting certain conditions required by rationality (for example, if you prefer A to B, and B to C, then you prefer A to C). Then we can prove a 'representation' theorem: these preferences can be represented as resulting from an underlying probability distribution and utility function. This approach avoids some of the objections to the betting interpretation, but not all of them. Ramsey's method essentially appeals to preferences over gambles, raising again the concern that the wrong quantities are being measured. And notice that the representation theorem does not show that rational agents' opinions *must* be represented as probabilities; it merely shows that they *can* be, leaving open that they can also be represented in *other,* radically different ways.

Radical subjectivists such as de Finetti recognize no constraints on initial, or 'prior,' subjective probabilities beyond their conformity to Kolmogorov's axioms. But they typically advocate a learning rule for updating probabilities in the light of new evidence. Suppose that you initially have a probability function $P_{initial}$, and that you become certain of E (and of nothing more). What should be your new probability function P_{new}? The favoured updating rule among Bayesians is conditionalization; P_{new} is related to $P_{initial}$ as follows:

$$(2)\ P_{new}(X) = P_{initial}(X \mid E)\ (\text{provided } P_{initial}(E) > 0).$$

Radical subjectivism has been charged with being too permissive. It apparently licenses credences that we would ordinarily regard as crazy. For example, you can assign, without its censure, initial probability 0.999 to your navel ruling the universe—provided that you remain coherent (and update by (2)). Radical subjectivism also seems to allow inferences that are normally considered fallacious, such as 'the gambler's fallacy' (believing, for instance, that after a surprisingly long run of heads, a fair coin is more likely to land tails). Rationality, the objection goes, is not so ecumenical.

A standard defense (e.g., [Savage 1954], [Howson/Urbach 1993]) appeals to famous 'convergence-to-truth', and 'merger-of-opinion' results. Roughly, they say that in the long run, the effect of choosing one prior probability function rather than another is washed out: successive conditionalizations on the evidence will, with probability one, make a given agent eventually converge to the truth, and thus initially discrepant agents eventually come to agreement. Unfortunately, these theorems tell us nothing about how quickly the convergence occurs. In particular, they do not explain the unanimity that we in fact often reach, and often rather rapidly. We will apparently reach the truth 'in the long run;' but as Keynes quipped, "in the long run, we shall all be dead."

5 Conclusion

In this limited space I have tried to convey how tendentious the mathematical and philosophical foundations of probability remain, despite some 350 years of research in the area. The interested reader will find more discussion of some of the liveliest current debates, trends, and prospects for the future in [Hájek/Hall 2002] and [Fitelson et al. 2005].

Feller ([1957], p. 19) writes: "All possible definitions of probability fall short of the actual practice." Certainly, a lot is asked of the concept of probability. In a suitably self-referential post-modern moment, I will complete this survey with some of my own bets on the uncertain future of the field.

I wager that we will continue to appeal to some *quasi-logical* notion of probability—for the evidential relations between various sentences or propositions are hardly exhausted by 'entailment' and 'refutation,' the stock-in-trade of deductive logic. Confirmation theory, pioneered by Hempel ([1945]) and Carnap ([1950]), is making a big comeback in philosophy. Arguably, the leading approach is probabilistic—sometimes called *Bayesian confirmation theory.* Its central idea is simple: confirmation relations are identified with dependence relations. Thus, we may say that, relative to probability function P:

- E confirms H iff $P(H \mid E) > P(H)$.

Note that this notion of confirmation is *incremental* in the sense that *E increases* the amount of evidence for *H*, without necessarily leaving *H* highly supported. Thus, a coin's landing heads on the first toss confirms its landing heads 100 times in a row. Similarly,

- *E* disconfirms *H* iff $P(H \mid E) < P(H)$.
- *E* is evidentially irrelevant to *H* iff $P(H \mid E) = P(H)$.

See [Hájek/Joyce forthcoming] for a survey of confirmation theory.

I see a healthy future for *objective probability*, or *chance*, underpinning the indeterministic aspects of the mind-independent world, such as we apparently find in radioactive decay. I find especially promising approaches that ground chance in physical *symmetries*—see e.g. [Strevens 1998]. Think of how fundamental symmetries are to probabilistic reasoning. These approaches seem to capture what's right about the principle of indifference, without inheriting what's wrong about frequentism.

And we will need the notion of *degrees of belief* or *credences* as long as there is uncertainty—which is to say, as long as there is human thought. But radical subjectivism is, to my mind, *too* radical—remember the navel ruling the universe! It needs to be constrained by something objective. For example, Lewis's ([1980]) *Principal Principle* says roughly that rational credences strive to align with chances, so that if a rational agent knows the chance of a given outcome, her degree of belief will agree with it. More generally, where '*P*' is the subjective probability function of a rational agent, and '*ch*' is the chance function,

$$P(A \mid ch(A) = x) = x, \text{ for all } A \text{ and for all } x \text{ such that } P(ch(A) = x) > 0.^{1}$$

For example, my degree of belief that this coin toss lands heads, given that its chance of landing heads is 1/4, is 1/4.

Perhaps one would do better to think of these quasi-logical, objective, and subjective notions as distinct *concepts* of probability, admittedly with some important interrelations—we have already seen one such interrelation in the Principal Principle (and see [Hájek 2003a] for more). Each of the leading interpretations, then, attempts to illuminate one of these concepts, while leaving the others in the dark. In that sense, the interpretations might be regarded as complementary, although to be sure each will need some further refinement.

Clearly, much work remains to be done on the foundations of probability. Equally clearly, we have come a long way since the *Port-Royal Logic* and Bishop Butler.

6 Personal and Pedagogical Epilogue

I began with some brief personal and pedagogical reflections, and so will I end. You have just taken a crash course in the philosophical foundations of probability, a high-speed version of a course that filled 10 weeks at Caltech when I used to teach there. My students were typically budding scientists and engineers, and I tried to bring the material to life for them by emphasizing how ubiquitous probability is, and how often high-stakes decisions are made on the basis of probability judgments. I used to begin with this example:

---

[1] There are subtleties that I cannot go into here, including the notion of admissibility, the relativity of chances to times, and Lewis' revised version of the Principle.

On January 28, 1986 at 11:38 A.M., the space shuttle Challenger was launched in Florida. Seventy-three seconds later it exploded, setting back the American manned space program by years. Managers made the decision to launch, against the advice of engineers, on the basis of a superficial and flawed analysis of the probability that the two solid rocket motors would fail at low temperatures, leading to a serious underestimate of that probability ([Dalal et al. 1989]). Lacking a clear conception of probability—and with it, a well understood, universally accepted methodology for determining probabilities—otherwise careful engineers and managers resorted to ad hoc calculations and dubious rules of thumb that resulted in tragedy. In particular, I believe that none of the parties concerned truly understood the notion of the single-case probability of disaster that was appropriate for Challenger. And yet its launch that day, in exactly the conditions that prevailed, was by its very nature unrepeatable.

Nor have the scientists (even at Caltech!) succeeded in understanding probability. I used to set my students the following question on the final exam for my course:

> In Feynman's *Lectures on Physics*, Volume 1, we find the following "definition" of probability:
>
>> By the "probability" of a particular outcome of an observation we mean our estimate for the most likely fraction of a number of repeated observations that will yield that particular outcome.
>
> There are many problems with this definition. Briefly indicate several of them.

Now that you, dear reader, have seen some of the problems with frequentism, you should be able to make a good start on this question. Here are some further hints:

- the definition is circular;
- it is easy to come up with cases in which there is more than one "most likely fraction;"
- irrational probabilities, such as $1/\sqrt{2}$ are excluded—yet our best physical theory, quantum mechanics, freely assigns such probabilities!

Finally, to bring home the subjective interpretation of probability in a way that I hope the students will *never* forget, I used to give them a multiple-choice test with a twist. Rather than choosing a correct answer, they had to assign *credences* to each possible answer.[2] The test began with the following explanation:

> Each of the following questions has exactly one correct answer among the choices a–d. I would like you to assign subjective probabilities to each of the choices, representing in each case your own probability that *that* choice is correct. For example, suppose you are nearly certain that b. is the correct answer to a given question, and the other choices look about equally implausible to you. Then you might represent your opinion as follows:
>
> a. 0.01
> b. 0.97
> c. 0.01
> d. 0.01

---

[2] David Dowe of Monash University devised a similar 'probabilistic football betting' system, and I am grateful to him for suggesting the scoring rule.

For each question, you will receive a score determined by the probability you gave to the correct answer. Let that probability be p. Your score for that question will be:

$$\text{Score} = 1 + \frac{1}{2} \log_2 p$$

Thus, if you give probability 1 to the correct answer to a question, and 0 to the rest, you get a perfect score of 1 for that question; if you give 0 to the correct answer, you get a score of $-\infty$ for that question. (Totals less than 0 will be rounded up to 0.) Make sure your probabilities for a given question are nonnegative, and add up to 1—otherwise you get 0 for that question automatically!

You, dear reader, might like to try your hand at the first question on my test, reprinted below. Good luck!

Q1. Let Ω be a non-empty set. Which of the following provides a correct characterization of a set F of subsets of Ω being a **sigma algebra on** Ω?

    a. $\Omega \in$ F; if $A \in$ F, then $\neg A \in$ F; and if $A_1, A_2, \ldots$ is a sequence of pairwise disjoint sets, each one $\in$ F, then their countable union $\cup A_n \in$ F.

    b. F is non-empty, closed under complementation (with respect to Ω) and under countable intersection.

    c. $\emptyset \in$ F; F is closed under complementation (with respect to Ω) and under finite union.

    d. F is the power set of Ω.

Postscript: almost every year at least one student would get a score of $-\infty$.

Acknowledgements Many thanks to Ned Hall, who years ago slaved with me on a research proposal that was an early ancestor of some parts of this paper, and also for letting me borrow from parts of our ([Hájek/Hall 2002]). Thanks also to Bonnie Gold and the editorial committee for this project for perceptive comments on earlier drafts, and to Ralph Miles for his meticulous proof-reading.

Bibliography

[Bertrand 1889] Bertrand, J., *Calcul des Probabilités*, 1st edition, Gauthier-Villars, 1889.

[Butler 1736] Butler, Joseph, *Analogy of Religion*, 1736; reprinted by Frederick Ungar Publishing Company, 1961.

[Carnap 1950] Carnap, Rudolf, *Logical Foundations of Probability*, 2nd ed., Chicago: University of Chicago Press; 2nd. Edition, 1950.

[Church 1940] Church, A., "On the Concept of a Random Sequence", *Bulletin of the American Mathematical Society* 46 (1940), pp. 130–135.

[Dalal et al. 1989] Dalal, S. R., E. B. Fowlkes, and B. Hoadley, "Risk Analysis of the Space Shuttle: Pre-Challenger Prediction of Failure", *Journal of the American Statistical Association* 84 (1989), pp. 945–957.

[de Finetti 1937] de Finetti, Bruno, "La Prévision: Ses Lois Logiques, Ses Sources Subjectives", *Annales de l'Institut Henri Poincaré* 7 (1937), pp. 1–68; translated as "Foresight. Its Logical Laws, Its Subjective Sources", in H. E. Kyburg, Jr. and H. E. Smokler, eds. *Studies in Subjective Probability*, New York: Robert E. Krieger Publishing Co., 1980.

[de Finetti 1972] ——, *Probability, Induction and Statistics*, New York: John Wiley, 1972.

[de Finetti 1974] ——, *Theory of Probability*, New York: John Wiley, 1974. Reprinted 1990.

[Feller 1957] Feller, William, *An Introduction to Probability Theory and Its Applications*, Vol. 1, 2nd ed., John Wiley & Sons, Inc., 1957.

[Fine 1973] Fine, Terrence, *Theories of Probability*, New York: Academic Press, 1973.

[Fitelson et al. 2005] Fitelson, Branden, Alan Hájek, and Ned Hall, "Probability", in *The Philosophy of Science: An Encyclopedia*, Sahotra Sarkar and Jessica Pfeiffer, eds., Routledge, 2005.

[Gillies 2000] Gillies, Donald, "Varieties of Propensity", *British Journal for the Philosophy of Science* 51 (2000), pp. 807–835.

[Goodman 1983] Goodman, Nelson, *Fact, Fiction, and Forecast* (4th ed.), Cambridge, MA: Harvard University Press, 1983.

[Hájek/Hall 2002] Hájek, Alan and Ned Hall, "Induction and Probability", in *The Blackwell Guide to the Philosophy of Science,* eds. Peter Machamer and Michael Silberstein, Blackwell, 2002, pp. 149–172.

[Hájek 2003a] Hájek, Alan, "Probability, Interpretations of", *The Stanford Encyclopedia of Philosophy,* Edward N. Zalta, ed., plato.stanford.edu/entries/probability-interpret/, 2003.

[Hájek 2003b] ——, "What Conditional Probability Could Not Be", *Synthese* 137, No. 3 (December 2003), pp. 273–323.

[Hájek/Joyce forthcoming] Hájek, Alan and James M. Joyce, "Confirmation", in the *Routledge Companion to the Philosophy of Science,* Stathis Psillos and Martin Curd, eds., forthcoming.

[Hempel 1945] Hempel, Carl, "Studies in the Logic of Confirmation", *Mind* 54 (1945), pp. 1–26, 97–121.

[Howson/Urbach 1993] Howson, Colin and Peter Urbach, *Scientific Reasoning: The Bayesian Approach,* 2nd ed., Chicago: Open Court, 1993.

[Jeffrey 1983] Jeffrey, Richard, "Bayesianism with a Human Face", pp. 133–156 in *Testing Scientific Theories,* John Earman, ed., *Minnesota Studies in the Philosophy of Science,* vol. **X** (1983).

[Kolmogorov 1933] Kolmogorov, Andrei. N., *Grundbegriffe der Wahrscheinlichkeitrechnung, Ergebnisse Der Mathematik,* 1933; translated as *Foundations of Probability,* New York: Chelsea Publishing Company, 1950.

[Laplace 1814] Laplace, Pierre Simon, *A Philosophical Essay on Probabilities;* English edition, New York: Dover Publications Inc., 1951.

[Levi 1980] Levi, Isaac, *The Enterprise of Knowledge*, Cambridge, MA: The MIT Press, 1980.

[Lewis 1980] Lewis, David, "A Subjectivist's Guide to Objective Chance", pp. 263–293 in R. C. Jeffrey (ed.) *Studies in Inductive Logic and Probability*, Vol II., Berkeley and Los Angeles: University of California Press, 1980; reprinted in *Philosophical Papers Volume II,* Oxford: Oxford University Press.

[Mückenheim et al. 1986] Mückenheim, W., Ludwig, G., Dewdney, C., Holland, P., Kyprianidis A., Vigier, J., Petroni, N., Bartlett, M., and Jaynes. E., "A Review of Extended Probability", *Physics Reports* 133 (1986), pp. 337–401.

[Popper 1959a] Popper, Karl, "The Propensity Interpretation of Probability", *British Journal of the Philosophy of Science* 10 (1959), pp. 25–42.

[Popper 1959b] ——, *The Logic of Scientific Discovery*, London: Hutchinson & Co., 1959.

[Ramsey 1926] Ramsey, Frank P., "Truth and Probability", pp. 156–198 in *Foundations of Mathematics and other Essays,* R. B. Braithwaite, ed., Routledge & P. Kegan, 1931; reprinted in *Studies in Subjective Probability*, eds. H. E. Kyburg, Jr. and H. E. Smokler, 2nd ed., New York: R. E. Krieger Publishing Co., 1980, pp. 23–52; reprinted in *Philosophical Papers*, ed. D. H. Mellor. Cambridge: University Press, Cambridge, 1990.

[Reichenbach 1949] Reichenbach, Hans, *The Theory of Probability*, Berkeley and Los Angeles: University of California Press, 1949.

[Renyi 1970] Renyi, Alfred, *Foundations of Probability*, San Francisco: Holden-Day, Inc., 1970.

[Savage 1954] Savage, L. J., *The Foundations of Statistics*, New York: John Wiley, 1954.

[Seidenfeld 2001] Seidenfeld, Teddy, "Remarks on the Theory of Conditional Probability: Some Issues of Finite Versus Countable Additivity", pp. 167–178 in *Probability Theory,* V. F. Hendricks et al., eds., Kluwer Academic, 2001; also available at: philsci-archive.pitt.edu/archive/00000092/

[Seidenfeld et al. 2001] Seidenfeld, Teddy, Mark J. Schervish, and Joseph B. Kadane, "Improper Regular Conditional Distributions", *The Annals of Probability* 29, No. 4 (2001), pp. 1612–1624.

[Skyrms 1980] Skyrms, Brian, *Causal Necessity,* New Haven: Yale University Press, 1980.

[Strevens 1998] Strevens, Michael, "Inferring Probabilities From Symmetries", *Noûs* 32 (1998), pp. 231–46.

[van Fraassen 1989] van Fraassen, Bas, *Laws and Symmetry,* Oxford: Clarendon Press, 1989.

[van Fraassen 1990] ——, "Figures in a Probability Landscape", pp. 345–356 in J.M. Dunn and A. Gupta, eds., *Truth or Consequences*, Dordrecht: Kluwer, 1990.

[Venn 1876] Venn, John, *The Logic of Chance*, 2nd ed. Macmillan and Co, 1876; reprinted, New York, 1962.

[von Mises 1957] von Mises, Richard, *Probability, Statistics and Truth*, revised English edition, New York: Macmillan, 1957.

Glossary of Common Philosophical Terms

If you start reading books, or articles in journals, written by philosophers of mathematics, you will find many of the terms listed below thrown about with abandon. In this volume, we usually asked authors either to replace them by what they mean, or at least to decrease dramatically the density of these terms in a given paragraph. But you will be expected to be very familiar with these words if you start reading articles written by philosophers for other philosophers. Thus, the glossary is provided here only partly to help with the chapters in this volume; it is also here to help those who would like to read further.

In addition to the words discussed here, quite a few standard philosophical terms are defined in the chapters by Balaguer, Chihara, and Shapiro.

Warning: this is a rough-and-dirty glossary. For formal definitions, see any introduction to philosophy (or Wikipedia, or the Stanford Encyclopedia of Philosophy (plato.stanford.edu/)).

Abstract objects are the opposite of concrete objects, which are the objects of everyday life (tables, books, etc.). Abstract objects are usually taken to be "causally inert" (see that entry), and do not exist in space-time. Mathematical objects are often taken to be the quintessential examples of abstract objects.

Acausal is another word for "Causally inert;" see that entry.

Causally inert/causally isolated/ (and their opposite, **causally efficacious**): an object is causally inert if it does not interact with anything in the world in a cause-and-effect way. That is, nothing in the world changes that object, and the object cannot cause any change in some real-world object. Mathematical objects are usually taken to be causally inert (even though, for example, graphical properties of the bridges of Königsberg appear to cause us not to be able to complete an Euler tour of them).

Desideratum: something that is desired, something you want to be true, or that you want to find.

Empirical: subject to experimental or observational verification; using the methods of the sciences.

Entails: what mathematicians usually call "implies." If A entails B, B follows from A. Entailments of a theory are propositions that are implied by that theory.

Epistemology (epistemological, epistemic): epistemology is the study of the basis for assertions that we know something: how we acquire knowledge, what must we do to be able to say we know something. An epistemological investigation is an investigation into how we can justifiably say we know something, and an epistemic support would be a support of a claim to knowledge.

Existential import: a statement has existential import if it implies that something exists; see "ontological commitment."

Fa: this is how philosophers apparently designate that *a* is an *F*, or that property *F* is true of *a*, or that *a* has property *F*. In particular, usually *F* is an elementary predicate not further logically analyzable.

Instantiated: a concept is instantiated if some object (usually concrete) is an example, or instance, of it.

Metaphysical/metaphysics: metaphysics is study of the nature of objects, of the basic structure of reality. It includes ontology and epistemology.

Modal (logic, notions, statements): modal statements are statements that concern what is necessary or possible. For example, "it is *necessary* that 17 is a prime number, but it is *not necessary* that there are 9 planets around our sun—in fact, they recently decided that there are just 8. Yet it is *possible* that another will one day be discovered and we will have 9 again." One way to interpret statements such as "it is necessary that 17 is a prime number" is to say that "in all possible worlds, 17 is a prime number." See Shapiro's and Chihara's chapters.

Mereology/mereological sum: mereology is the study of the relationship between parts of a whole, each other, and the whole. The mereological sum of two objects is the whole that consists of just those two objects.

Nominalism is the view that there are no abstract objects; in particular, there are no mathematical objects. See, particularly, Balaguer's chapter, where he describes various versions of nominalism, and Chihara's chapter (as he is a nominalist).

Obtain: "there does obtain the following facts . . . " is philosophical jargon for "the following are facts . . . "

Occam's razor, due to William of Ockham in the 14th century, is that one should not multiply entities unnecessarily. It is also called the principle of parsimony. If your house burns down, you could explain it by saying that a fire genie got hold of your house and set it afire, or you could observe that you let a burning match drop on your carpet and did not notice it until it was too late. The former is multiplying entities (in this case, a fire genie) unnecessarily. (Occam presumably used his razor to slash away the vast undergrowth of unnecessary entities.) In Latin, it is "*entia non sunt multiplicanda praeter necessitatem*": literal translation, "*entities should not be multiplied beyond necessity.*"

Ontology (**ontological, ontological commitment**): ontology is the study of what things exist, what things there are. If a theory has an ontological commitment to a certain object, that theory implies/assumes that the particular object exist. (This is also called "existential import.")

Physicalism is the thesis that everything is physical. Physicalists do not deny that the world might contain many items that at first glance do not *seem* physical—items of a psychological, moral, or social nature. But they insist that, when more carefully analyzed, it will be clear that such items are wholly physical. As a philosophy of mathematics, it is usually a variety of nominalism. However, there was an attempt by Penelope Maddy to find mathematical objects in the physical world (a set of 12 elements in a carton of eggs, etc.), and she is one of the main proponents of physicalism in mathematics.

Platitudes (platitudinous): obviously or trivially true facts.

Platonism (mathematical platonism): the term originates from several of Plato's dialogues, in which a realm of Forms is described (examples being Beauty, Justice, and Goodness), which are eternal, unchanging, and acausal (causally inert). Mathematical objects were also viewed as Forms. Mathematical platonism is the belief that (1) there are mathematical objects; (2) these objects are non-physical, non-mental, abstract objects that have always existed and are independent of people; and (3) properties and theorems about mathematical objects are true independently of whether people are aware of them. Thus, mathematical facts and objects are discovered, rather than invented. See the chapters by Balaguer, Chihara, and Shapiro in this volume for substantial discussions of mathematical platonism. Another term for platonism is "realism."

Posit: to posit an object is to affirm its existence, generally in order to discuss issues relevant to it.

Prima facie: apparent, self-evident; "it would seem obvious that."

Realism is another word used to describe views sometimes called "Platonism." Using this term allows one to separate different parts of platonism, as Shapiro does in his chapter (realism in ontology, realism in epistemology) and investigate variations on the traditional platonic view.

Reference (the problem of reference): how can we pick out and refer to specific mathematical objects if they are not part of our physical world, and we can not see, hear, touch, or otherwise interact with, them? (See Chihara's chapter for a discussion of this problem.)

Reify: to make real, or treat as if it were a real object. On some views, going from "there are two objects on the table" to talking about the number two is reifying the number two.

Semantic(s): this word is used differently by philosophers than its most common use in English. When most people say, "the difference is semantic," the word "just" is understood prior to "semantic," and it means that it is just a difference of how we say it, not a real difference. However, in philosophy, the "semantic problem," how we know what someone means, is a substantial one. A semantic theory is a (presumably) empirical theory about what certain expressions mean (or refer to) in ordinary discourse.

Social constructivism is the view that mathematical objects and mathematical truths are products of social mathematical activity, rather than existing outside of space and time.

Spatiotemporal: involved in space and/or time. Platonism (see that entry) includes a belief that mathematical objects are not spatiotemporal.

Token: a concrete object that stands in for, or represents, something else. For example, a "sentence token" is what non-philosophers would call a sentence. When you see on this page just now "There are only three moons of Jupiter," this is a sentence token standing for the (false) sentence that says that there are only three moons of Jupiter. Similarly, the 3 here is a token for the number three. Gertrude Stein's "Rose is a rose is a rose is a rose" has ten word-tokens (four tokens for "rose," and three each for "is" and "a"), though just three word-types ("rose," "is," and "a").

True-in-the-story-of versus **true *simpliciter***: "Sherlock Holmes, the detective, smokes a pipe" is true in the detective stories of Conan Doyle, but is not true *simpliciter* because there is no person with that name and thus he cannot smoke a pipe. "True in a model" is a mathematical version of "true-in-the-story-of."

About the Editors

Bonnie Gold was born in New York City and attended high school at the Bronx High School of Science, received her A.B. degree from the University of Rochester with highest honors in mathematics, her M.A. in mathematics from Princeton University, and her PhD in mathematics from Cornell University. She taught for twenty years at Wabash College in Indiana, and since 1998 has been in the Mathematics Department at Monmouth University in New Jersey. While at Wabash College, she received the McLain-Turner-Arnold Award for Excellence in Teaching, as well as a Lilly Open Faculty Fellowship to study and begin research in the philosophy of mathematics. In 2006, she received the Distinguished Teaching Award from the New Jersey section of the Mathematical Association of America. At Monmouth University, she was co-director of the 21st Century Science Teachers' Skills Project. She is a member of the American Mathematical Society, the Mathematical Association of America (MAA), the Canadian Society for the History and Philosophy of Mathematics, is a founding member of the Special Interest Group of the MAA for the Philosophy of Mathematics (POMSIGMAA) (and was its first Chair and is its current Public Information Officer), and is currently Vice-Chair for Speakers of the New Jersey section of the MAA. She was co-editor (with Sandra Keith and William Marion) of *Assessment Practices in Undergraduate Mathematics*. She is the editor of MAA Online's Innovative Teaching Exchange, and is the director of NJ-NExT, a state version of a national project for new college faculty in mathematics. She has written one article in the philosophy of mathematics, "What Is the Philosophy of Mathematics, and What Should It Be?" *Mathematical Intelligencer*, 1994.

Roger A. Simons was born in Detroit, Michigan. When he was 3 years old, his family moved to Los Angeles, where he received all his early education. He graduated from University High School and received an A.B. degree with honors from UCLA as a mathematics major. He received an M.A. and PhD from the University of California, Berkeley in mathematics and also an Sc.M. from Brown University in computer science. During his PhD program, he worked two summers at Aerospace Corporation on NASA's Gemini and Apollo Projects. He was on the faculty of the University of Wisconsin, Green Bay for ten years and is just retiring after 27 years as a professor of mathematics and computer science at Rhode Island College. He has been a visiting faculty member in mathematics at the University of Hawaii 6 times, an adjunct professor of computer science at the University of Rhode Island, and has consulted to three different companies, including Chemical Bank, helping with their in-house software development. He has refereed for the *ACM Symposium on Theory of Computing, Algorithmica, Communications of the ACM, IEEE Computer Society Transactions on Knowledge and Data Engineering, Journal of Computer System Sciences, Theoretical Computer Science*, the Consortium for Computing in Small Colleges, and the Association for Symbolic Logic. He has also evaluated grant proposals for the NSF CCLI program in Computer Science. He appeared in *Who's Who Among America's Teachers 2005*, a recognition resulting only by the nomination of an honor student as the one teacher in her academic career who made a difference in her life. He is a member of the Mathematical Association of America, American Mathematical Society, Association for Symbolic Logic, Association for Computing Machinery, Sigma Xi, and Pi Mu Epsilon. He served one term as president of the Rhode Island College Chapter of Sigma Xi and has been its treasurer for 15 years. His PhD studies in logic and foundations aroused his curiosity about several issues not explained or not addressed by logic or set theory. Thirty years later some reading on the philosophy of mathematics brought back his former thoughts and raised his interests enough to become active in this field. He is a founding member of the Special Interest Group of the MAA for the Philosophy of Mathematics (POMSIGMAA) and was its second Chair.